DATE DUE

~~NV 7 '99~~			

Computer Systems for Automation and Control

Prentice Hall International
Series in Systems and Control Engineering

M. J. Grimble, Series Editor

BANKS, S. P., *Control Systems Engineering: modelling and simulation control theory and microprocessor implementation*
BANKS, S. P., *Mathematical Theories of Nonlinear Systems*
BENNETT, S., *Real-time Computer Control: an Introduction*
BITMEAD, R. R., GEVERS, M. and WERTZ, V., *Adaptive Optimal Control*
BUTLER, H., *Model Reference Adaptive Control*
CEGRELL, T., *Power Systems Control*
COOK, P. A., *Nonlinear Dynamical Systems*
ISERMANN, R., LACHMANN, K. H. and MATKO, D., *Adaptive Control Systems*
KUCERA, V., *Analysis and Design of Discrete Linear Control Systems*
LUNZE, J., *Feedback Control of Large-Scale Systems*
LUNZE, J., *Robust Multivariable Feedback Control*
McLEAN, D., *Automatic Flight Control Systems*
OLSSON, G. and PIANI, G. *Computer Systems for Automation and Control*
PARKS, P. C. and HAHN, V. *Stability Theory*
PATTON, R., CLARK, R. N. and FRANK, P. M. (editors), *Fault Diagnosis in Dynamic Systems*
PETKOV, P. H., CHRISTOV, N. D. and KONSTANTINOV, M. M., *Computational Methods for Linear Control Systems*
SÖDERSTROM, T. and STOICA, P., *System Identification*
SOETERBOEK, A. R. M. *Predictive Control: A unified approach*
STOORVOGEL, A., *The H_∞ Control Problem*
WATANABE, K., *Adaptive Estimation and Control*
WILLIAMSON, D., *Digital Control and Instrumentation*

Computer Systems for Automation and Control

Gustaf Olsson

Professor in Industrial Automation, Lund Institute of Technology

Gianguido Piani

Consultant in computer control applications

Prentice Hall

New York   London   Toronto   Sydney   Tokyo   Singapore

First published 1992 by
Prentice Hall International (UK) Ltd
Campus 400, Maylands Avenue,
Hemel Hempstead, Herts. HP2 7EZ
A division of
Simon & Schuster International Group

© Prentice Hall International (UK) Ltd, 1992

try

Printed and bound in Great Britain at
Dotesios Limited, Trowbridge

Library of Congress Cataloging-in-Publication Data

Olsson, Gustaf.
   Computer systems for automation and control / Gustaf Olsson,
Gianguido Piani
      p.   cm. — (Prentice Hall International Series in Systems and
Control Engineering)
   Includes bibliographical references and index.
   ISBN 0-13-457581-4
   1. Automatic control — Data processing.   2. Process control — Data
processing.   I. Piani, Gianguido.   II. Title.   III. Series.
TJ223.M53047   1992
629.8′9 — dc20   91-36248
                                                                              CIP

British Library Cataloguing in Publication Data

A catalogue record for this book is available
from the British Library.

ISBN 0-13-457581-4

1   2   3   4   5   96   95   94   93   92

From Gustaf to Kirsti and
to our children Tova, Max, Mia, John, Sara, Anja

From Gianguido to the generation of Claudia, Ilaria and Elisa

Contents

List of Abbreviations

ABS	automatic braking system
a.c.	alternating current
A/D, ADC	analog to digital (converter)
AI	analog input; artificial intelligence
AM	amplitude modulation
ANSI	American National Standards Institute
AR	autoregressive (digital filter)
ARMA	autoregressive moving average (digital filter)
ASCII	American Standard Code for Information Interchange
ASN.1	Abstract Syntax Notation no. 1
BBM	break-before-make (contact)
BER	Basic Encoding Rules
Biϕ-L	biphase level
CCITT	Comité Consultif International de Télegraphie et Téléphonie
CIM	computer-integrated manufacturing
CMOS	complementary metal-oxide semiconductor
CPU	central processing unit
CRC	cyclic redundancy check
CSM	central service module (in Multibus II)
CSMA/CD	Carrier-sensing Multiple Access/Collision Detection
D/A, DAC	digital to analog (converter)
d.c.	direct current
DCE	data communication equipment
DDC	direct digital control
DDDC	distributed direct digital control
DEC	Digital Equipment Corporation
DI	digital input
DIN	Deutsche Industrie Normen (German industrial standard)
DIP	dual in-line package
DMA	direct memory access
DO	dissolved oxygen (concentration)
DoD	(US) Department of Defense

DSP	digital signal processor
DTE	data terminating equipment
EBCDIC	Extended Binary-Coded Decimal Interchange Code
EIA	Electrical Industries Association
EISA	Extended Industry Standard Architecture
EMC	electromagnetic compatibility
FCS	Frame Check Sequence
FDM	frequency division multiplexing
FEP	front-end processor
FET	field-effect transistors
FIP	Factory Instrumentation Protocol, also Flux Information Processus (French)
FM	frequency modulation
FMS	flexible manufacturing systems
FPLA	field-programmable logic arrays
FSK	frequency shift keying, same as FM
FTAM	File Transfer Access and Management
FUP	FUnction Plan
GM	General Motors
Grafcet	GRAphe de Commande Etape-Transition
HDLC	high-level data link control
HS	high selector (for control)
IC	integrated circuit
IEC	International Electrotechnical Commission
IEEE	Institute of Electrical and Electronics Engineers
I/O	input/output
ISA	Industry Standard Architecture (meaning the IBM AT bus)
ISA	Instrument Society of America
ISDN	Integrated Systems Digital Network
ISO	International Standardization Organization
K	degrees Kelvin
LAN	local area network
LAPB	Link Access Procedure-Balanced
LAPD	Link Access Procedure, D-channel
LED	light-emitting diode
LLC	logical link control
LS	low selector (for control)
LSI	large scale integration
LS-TTL	low-power Schottky transistor–transistor logic
MA	moving average (digital filter)
MBB	make-before-break (contact)
ms^{-1}	meters per second
MAC	medium access control
MAP	Manufacturing Automation Protocol

M&C	monitoring and control
MCA	Micro Channel architecture
mg l^{-1}	milligrams per litre
MHS	Message and Handling System
MIT	Massachusetts Institute of Technology
MMS	Manufacturing Message Specification
MOSFET	metal oxide field effect transistor
MPC	message passing coprocessor (in Multibus II)
MSI	medium scale integration
MTBF	mean time between failures
n.c.	normally closed (switch)
NC	numerically controlled
n.o.	normally open (switch)
NRZ	non-return to zero
ODE	ordinary differential equation
OEM	original equipment manufacturers
op-amp	operational amplifier
ORP	oxygen−reduction potential
OSI	Open Systems Interconnection
PAL	programmable array logic
PI	proportional-integral (controller)
PID	proportional-integral-derivative (controller)
PIPI	two PI controllers in series, or a PI controller in series with a low pass filter
PLC	programmable logical controller
PLD	programmable logic devices
PLS	pulse instruction
PM	phase modulation
PROFIBUS	Process Fieldbus
PROM	programmable read-only memory
PSK	phase shift keying, same as PM
PTT	post, telegraph and telephone
PWM	pulse width modulation
QAM	quadrature amplitude modulation
RAM	random access memory
RC	resistance−capacitance (circuit)
RL	resistance−inductance (circuit)
ROM	read-only memory
RTD	resistance temperature detector
RX	Reception, Receiver
RZ	return to zero
SCADA	supervisory control and data acquisition
SCR	silicon-controlled rectifiers
SDLC	synchronous data link control

S/N	signal to noise ratio
SPDT	single-pole double-throw (contact)
SPST	single-pole single-throw (switch)
SR	set−reset (flip-flop)
TAS	Test__And__Set
TDM	time division multiplexing
TOP	Technical and Office Protocol
TTL	transistor−transistor logic
TX	Transmission, Transmitter
VMD	virtual manufacturing device
VME	VERSA module Eurocard
VMS	VME serial (bus)
VMX	VME extension (bus)
VSB	VME subsystem (bus)
VT	Virtual Terminal
VXI	VME extended instrument (bus)

Preface

Computer Systems for Automation and Control deals with the principal aspects of computer applications in system control, with particular emphasis on industrial automation. The description is not limited to the hardware and software components of process control computers but covers additional topics such as systems modelling, signal processing, digital control and data communication. Accent is put on the integration of the different components, as well as of the computer system, with the external environment, including interface with the human user.

The book is intended for readers with widely different backgrounds. Some who have process-related experience and who know quite a lot about traditional control methods, may feel insecure about the use of digital technology. Other readers may be computer experts, entirely at home with software but less so with process control. A third group of readers could be project managers who need a broad picture of the subject without getting too involved with specific details. Therefore, the topics have been selected and presented in such a way as to integrate readers' knowledge with computer-oriented or application-oriented information, depending on their particular needs.

We present automation as an integrated concept and deal with the major issues of each related field of knowledge. Each chapter is self-contained to allow an easy selection of the desired topics and presents the key theoretical issues together with practical details. In the practical examples, emphasis is given to standard components and methods accepted by both industry and official bodies, as we believe that technical solutions supported by several manufacturers stand the best chance of success in the future. Moreover, standards support modularity — one of the most important factors in the systems approach.

Process control can be realized in many different ways and with different technologies, from specifically-built integrated circuits to system boards and to large turnkey systems. This book focuses on design from board level upwards, because in practice system integration at different levels is the most common way to work. We do **not** claim to teach how to design perfect systems, but rather how to look for acceptable and reasonable solutions using readily available components.

Research on real-time computing has been carried out at the Department of Automatic Control of the Lund Institute of Technology since the end of the 1960s. Based on this experience, from the early 1980s a new course on computer applications in control could be offered. The purpose of the course is to combine theoretical and practical issues and present a broad view of computer and control engineering. Thanks to the traditionally close ties between industry and universities in Sweden, the course has profited largely from industrial field experience. Prime movers in the development of the course were Hilding Elmqvist, Leif Andersson, Sven-Erik Mattsson, Johan Wieslander and Gustaf Olsson.

The course has been appreciated, not only by the students but also by professional engineers, one of whom was Gianguido Piani. This was his first contact with the Institute in Lund, and after a few years spent in different European countries, Piani returned — this time as guest lecturer for the course. Together, Gustaf Olsson and Gianguido Piani have worked to produce this book, and only the reader can tell whether we have succeeded in combining theory and research data with real-life issues.

The authors hope that automation techniques will increasingly help in the practical solution of many of today's most pressing problems such as resource conservation, environmental control and improving the living standards for populations of the world's poorest countries.

In a few instances we have used 'he' to also include 'she' following traditional grammar and it does not imply anything about the authors' views. In fact, we regret that most people in this area are men and would welcome a greater interest in engineering by women.

Acknowledgments

A number of people have contributed to the preparation of this book. They include Gunnar Lindstedt, Ulf Jeppsson, Sven-Göran Bergh, Anders Peltomaa and Dr Jaroslav Valis at the Department of Industrial Automation, Lund Institute of Technology. Leif Andersson, with his broad experience in almost every aspect of real-time computing, has provided very valuable comments. The first draft of the manuscript, partly 'Swenglish', partly 'Italiese', was read by Professor Otto J.M. Smith of the University of California in Berkeley, Professor John F. Andrews at Rice University in Houston, Dr Robert 'Bob' Hill, City of Houston, Walt Schuk, Cincinnati, and by Professor Don Wiberg at the University of California in Los Angeles. We want to thank them for their comments.

Further feedback has been provided by Dag Brück, Ingvar Ekdahl, Richard Franz, John Lawrence, Mike Neumann, Tomas Schönthal and Bengt Simonsson. Thanks to his long experience in industrial automation projects, Dr Krister Mårtensson helped considerably to improve the quality of the final manuscript.

Several of the proposed ideas were iterated in endless discussions with many friends, among them Helena Rosenlind, Stefano Castelli and Cornelia Verbeek. Many companies have also supported our work by providing information and material and by giving us the opportunity to discuss current trends in technology with their engineers. Finally, we wish to thank many others who in different ways helped us in this initiative, in particular, Lena Somogyi at the Department, who gave us a lot of practical support.

The team at Prentice Hall has been most helpful and cooperative, especially considering that contact had to take place across different countries.

Lund, Sweden and Trento, Italy, February 1992
Gustaf Olsson and Gianguido Piani

1

Introduction

The Buddha, the Godhead, resides quite as comfortably in the circuits of a digital computer or the gears of a cycle transmission as he does at the top of a mountain or in the petals of a flower. To think otherwise is to demean the Buddha, which is to demean oneself.

Robert M. Pirsig, *Zen and the Art of Motorcycle Maintenance*

1.1 COMPUTERS IN PROCESS CONTROL

Today the use of digital computers for automatic control is of fundamental importance for the technological infrastructure of the modern society. Industrial production, transportation, communication and — though still minor — environmental protection depend largely on computer-supported control systems. Without some form of internal control, very few technical realizations could operate at all — and this holds for the train as well as the nuclear reactor. Digital computers play a fundamental role; in many situations there is no viable alternative to computer-supported process control.

To describe the particular role played by the computer in process control, it is necessary first to define what we mean by **process**. A **physical process** is a combination of operations performed in order to act on, and change, something in the physical world. Movement, chemical reactions and heat transfer are all processes in this sense. Examples of processes are any industrial or chemical production, room conditioning (i.e. control of the physical variables temperature and humidity), and transportation, which consists in the controlled change of speed and position in a vehicle. Information processing alone does not bring changes in the physical world and thus cannot be considered to be a physical process in this sense.[1]

The term **process** can be used to describe transformation of matter and energy as well as a program in execution. In general there should be no problem in understanding from the context what process is referred to. Where there might be doubt, we will use the term **physical process** to indicate transformation of matter/energy and **program module** or **task** for the other meaning of the term.

Every physical process is characterized by its input and output elements in terms of:

* materials
* energy
* information

Examples of these are given in Table 1.1

Materials and **energy** are obvious basic components of a physical process. **Information** is also a fundamental quantity in the same way as materials and energy. As such, information is an indispensable part of any process. Many other factors that cannot be controlled or manipulated also influence a process. These are **disturbances** that lead the process away from the desired operating point (Figure 1.1). Disturbances are not physical quantities in themselves but appear as sporadic variations in material, energy and information flows.

An industrial process outputs some product from the raw materials and energy input. The input information to the process is the set of variables that controls and works with the process instructions; and the output information of the process is the set of measured variables and parameters that describe the evolution of the process. A large amount of information also lies in the 'know-how' of the process and the final product. Information is therefore not restricted to monitoring and control data but includes software and company know-how, down to the memos circulating between offices and delivery orders. This information is as important as any other element for balancing and optimizing any production process.

We will consider the input and output elements in a broad sense. For instance, in the case of transportation it might not be immediately obvious what the output is. In fact, transportation involves a change in geographical position (i.e. a physical state) so that **work** is produced — and work is a form of energy.

Information is important in the control of physical processes because it enables better use to be made of the other two factors, matter and energy. Considering the paramount

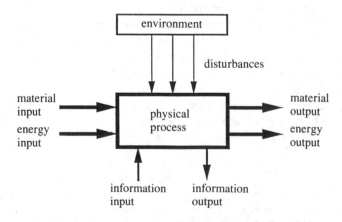

Figure 1.1   The model of a generic physical process.

Table 1.1 Example of process input/output

	Material input	Material output	Energy input	Energy output	Information input	Information output
Chemical reactor	reagent feed stream	one or more new product(s)	added heat or cooling	heat generation from the reaction	control of feed flow rates and added heat	measurements: temperature, flow rates, pressure, concentration(s)
Room conditioning			added heat or cooling	heat radiation	heating/cooling fluid flow rates, heat control	measurements: temperature
Aircraft control			fuel to power the engines	the aircraft movement	control of engine speed and position of the aerodynamic surfaces	measurements: angle of attack, roll, pitch, speed, altitude

Figure 1.2   Operation of a digital computer.

problems resulting from industrial activities, with related resource depletion, waste production and environmental pollution, anything that improves efficiency and reduces waste is of great interest. In fact, information processing that leads to more efficient processes is already in high demand.

Digital computers are essentially devices to process information (Figure 1.2), and may operate on particular information related to physical processes (Figure 1.3). In most current applications, computers are used to check that the operating parameters of a process remain within given limits and control the process in such a way that its output stays within the desired limits even in the presence of external disturbances.

Industrial process control differs in many respects from conventional data processing. For common data processing applications such as accounting and text editing, input and output are information that can be recorded and transmitted on any medium able to carry it. The computer can set its own timing to produce the output; more powerful computing resources will lead to a faster result, but not a different one from that for a slower device.

The situation is different for process computers. Here the computer does not control the execution flow but instead must follow the pace at which things happen in the outside world; the computer must react within a short time to external events. Control systems must continuously process the data on input, very often without the possibility of changing the amount and the rate at which they are received. At the same time they might have to perform other functions; for instance, information exchange with human operators, data

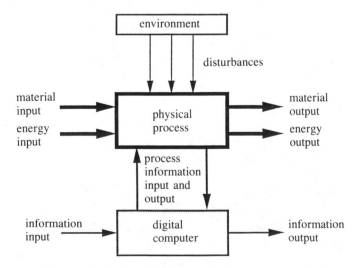

Figure 1.3   Use of the computer in process control.

presentation, and reaction to particular signals. This aspect is so important that this kind of operation has its own name: **real-time**.

1.2 A BIT OF HISTORY

The first practical example of a process computer application dates back to 1959; it concerned some functions at the Texaco petrochemical plant in Port Arthur, Texas (USA). This pioneering work was a cooperative effort by the companies Thomson Ramo Woolridge and Texaco. The RW300, a tube-based computer, controlled flows, temperatures, pressures and compositions in the refinery. The computer calculated the desired control signals based on its input data and changed the setpoints of analog regulators or indicated to the operators the controls that had to be implemented manually. The machine had a very small capacity compared with the computers of today: its typical addition time was 1 ms and multiplication took around 20 ms. Reliability was not a strong point either: the mean time between failures (MTBF) was a matter of hours, or at most days.

The first applications of process computers had all the odds against them. In order for a computer investment of the order of $US1 million to be economically motivated, the plant to be controlled had to represent an investment of an order of magnitude larger than the computer. This meant that the process necessarily had to be very complex, as in the case of large chemical plants. Structuring of the computer control problem was — and remains — a key issue.

Economy was not the only problem. Hardware reliability was low because it used electronic tubes. Software was written in Assembler language and not a single bit of the scarce memory available was left unused. Still these pioneering efforts were supported by many computer manufacturers who saw a great potential market for these applications.

Computer control is an area where a constructive interaction between theory and practice has taken place (this does not happen often). As we will see later, computer control has special features of its own. Conventional mathematical methods based on continuous time analysis cannot be used directly for the design of control loops to be implemented in computers. Computer control showed early on the necessity for a special theory of sampled control, providing the foundation for its development. The coincidence of the US space programme of the 1960s, and especially the Apollo project, acted as catalyst for the theoretical and practical work.

An important step was taken in 1962, when ICI (Imperial Chemical Industries) in England introduced the concept of direct digital control (DDC). Basically, the idea was to replace all the conventional analog control loops with one central computer. A large and costly controller panel consisting of hundreds of analog controllers and recorders could be replaced by the computer terminal. The computer simulated numerically the analog controllers in a way that was not much different from today's solutions.

The basic idea of DDC is still applied in many of today's industrial computer systems. A reason for its success is that the structuring of the problem had been defined already by the analog techniques. Unlike many pioneering implementations the objectives of computer control were well understood and clearly defined. The obvious drawback with

DDC was that system reliability depended on the centralized computer. However, for an additional investment a second computer could be installed to take over, should the first fail. The computer of 1962, a Ferranti-Argus, was considerably better than the 1959 generation. Addition and multiplication times had improved by a factor of 10, while reliability had also increased an order of magnitude.

The advent of the transistor in the 1960s led to notable progress for computer applications. The introduction of transistor-based minicomputers in the mid 1960s brought computer power to a price that was significantly cheaper than for mainframes, even though a typical minicomputer was still priced at higher than \$US100,000. Computer control could also be considered for smaller and less complex applications. Minicomputers could be installed close to the process and therefore became popular in small automated production and testing units. The three factors: better computer hardware, less complex processes and more developed control theory, combined to accelerate the success of computer control.

Every decade seems to bring a new computer technology with considerable advantages in performance and price over the preceding one. Powerful card-mounted microprocessors have been available since the mid 1970s and today even the most simple process control units are economically realized with digital techniques. Current personal computers are more powerful than the minicomputers of the 1960s and 1970s (central processing units with 16 or 32-bit word length, at least 10^6 bytes primary memory, 10^7-10^8 bytes secondary storage), at costs of at least an order of magnitude lower.

Computer applications largely popular in industrial control are the open buses. Here the accent is put not on specific components, but on the interface (the **bus**) between operational modules. The hardware modules in a bus system are selected to fit the intended application without leaving unused resources.

On the other hand, computing power is not everything. The on-board computer of the first lunar spaceship, Apollo 11 in 1969, had 64 kbyte of primary memory. Today, nobody would give a second look at such a machine, but the lesson to learn by considering what that computer helped accomplish, is that attention has to be given to the goals rather than just to the size of computer hardware.

In a digital control system, it is comparatively easy to try new control strategies since a digital computer may completely change the way it works when its programs are modified, without having to redesign and recable the whole system. Thus digital control systems offer not just a new way to implement established principles for control, but an entirely new technique that has greater flexibility and opens new possibilities. Process knowledge, system dynamics and control theory are necessary ingredients for a computer control project to be successful, but they represent only half of the picture. The structuring of the entire solution in terms of hardware units, software modules and communication remains a major challenge.

After 30 years of computer usage, we now have enough experience to ask ourselves about the real advantages brought by computers in industry and administration. In the 1950s and 1960s, before computers became so widespread, one income-producing person could support a whole family including paying housing mortgages. In the typical family of today, at least two people must work full time to achieve the same goal. The intense focus on computer optimization distracts attention from where the real problems might be.

1.3 SYSTEMS

Processes and their control systems consist of different parts interacting in complex ways. The components are connected together so that matter, energy and information can be exchanged in order to obtain a certain result. It is not easy to give a general definition of a system, but the following could be important aspects:

- The idea of a system is a mental aid to interpret and understand the function of any complex body where different components interact with each other.
- There are many ways to divide a system in order to describe it: any of its components can be divided into smaller parts, or the divisions may take a different form. It is important to always select the right level of detail.
- It is usually not necessary to know the internal workings of an element to foresee the global functions of the system it belongs to. It is sufficient to know its input/output relations (this is known as the **black box** principle).
- The goal of a system is to reach a result quantitatively or qualitatively higher than the sum of the results of the single components taken alone. The system adds 'something more', to justify its function (the principle of **synergy**). This 'something more' is not the result of a single system component, but rather of the way the single parts interact.

Systems may be of many different types, e.g. electrical, chemical, mechanical or biological. A personal computer is a system built with basic functional components (central processing unit, memory and peripheral devices). Taken alone, these units do not do anything. Put together with the operating software, we have a computer that can accomplish many things. The human body is an extremely complex system built with organs that serve different functions. The body is able to fulfill its basic goal, to sustain its own life, by each organ functioning together with the others.

Different mathematical methods can be used to describe systems and their elements. Although strongly emphasized in most textbooks, such methods can be applied only where all the elements of a system as well as its external environment can be described with quantitative relations. We will see later in Chapter 3 that mathematical descriptions are not the only ways to describe systems. When they can be used, they obviously have the advantage of the correctness of formalism.

An important aspect of systems knowledge is **system dynamics**, the prediction of future system behaviour and selection of appropriate control actions in order to influence that behaviour. System dynamics is tricky because of the many interactions among the different system components that have to be taken into consideration. So the evolution of a system often seems to lead in the 'wrong' direction, or at least against 'intuitive' expectations. Every car driver knows instinctively what system dynamics is. The driver can plan to increase or reduce the pressure on the accelerator at the sight of a hill, so that the car speed remains constant. The car behaviour is normally known and predictable, but might turn out to be unpredictable if not altogether uncontrollable if the road is wet or icy. Similar problems are encountered every day in industrial automation.

In large applications, one of the major problems is how to *structure* the system. Many people must cooperate, the computer systems should be renewed gradually and new facilities added. In the approach to complex systems it is important to have both *deep* and *broad* views. Problems cannot usually be solved at one level only, instead they have to be put in the right perspective and focused on at the right level. This does not mean knowing all the details of any given situation, but rather that one should be able to investigate specific details when necessary. A solution found at the wrong level is often not a solution at all; it could even make things worse. It does not help to check a program for bugs for a missed printout, if the printer is switched off. Every engineer has their own anecdotes in this respect.

The boundaries between what used to be the private realms of electrical engineers, programmers, application experts and users are today becoming more and more fuzzy. A special approach is needed to understand and build complex systems consisting of many different parts that interact. We present how to use such an approach in the analysis and design of automated systems. System thinking has general validity independent of the size of the application, but it obviously becomes more and more important for larger projects.

1.4 SOME TYPICAL APPLICATIONS OF COMPUTER CONTROL SYSTEMS

Applications of computer control range from consumer products to high technology industry. Today a standard car applies computer control to both ignition and air/fuel ratio control in the engine as well as temperature control of the passenger compartment; not even the radio tuning is left alone to the driver.

Control systems may seem to differ widely: the control systems for a chemical process and for railroad junctions at a large train station may seem to have little in common with a robot for painting cars or the space shuttle board computers. However, in all these systems it is possible to identify some similar basic functions: data acquisition, execution of clock or interrupt-driven functions, feedback control and communication to a human operator or to other computers.

The general structure of a process computer interacting with a physical process consists of the following parts (Figure 1.4):

- central data processing unit
- process communication channels
- A/D and D/A converters
- sensors and actuators
- physical process

The physical process is observed with **sensors**, devices that convert physical variables, such as temperature, pressure or position, into electrical variables that can be measured directly, such as resistance, current or potential difference. Conversely, a process can be influenced through **actuators** that function in an opposite manner to the sensors. Actuators transform electrical signals in physical actions, mainly movement (displacement and rotation) or heat production. Examples of actuators are servo motors, hydraulic valves and pneumatic positioners.

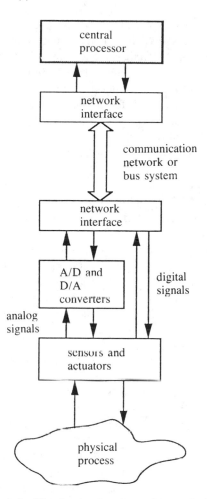

Figure 1.4   The basic structure of a control system.

A digital control system works only on information in numerical form, so that collected electric variables have to be converted via analog to digital (A/D) converters. The inverse operation is less problematic because the computer can control directly actuators, like motors and valves, through electric signals.

Information from different source points distributed in space is brought to the central unit in communication channels. The central control unit interprets all incoming data from the physical process, sends control signals, takes decisions on the base of program instructions, exchanges data with the human operators and accepts their commands.

Not unexpectedly, manufacturing industry shows many applications of digital computers. **Numerically controlled (NC)** machines produce high-precision mechanical parts by following well determined sequences of operations. The NC units are built for multiple operations; their operations depend on the stored software, which can be changed at comparatively reduced time and costs. The flexibility and sophistication of industrial robots owes a major part to computer control. When a machine alone cannot process a part, **flexible manufacturing systems (FMS)** can take over in a work cell or a section of a plant. In FMS the operations of each machine, the mutual interactions and the transport of components from one machine to the other are all controlled by computers.

The process industry, including chemical, metallurgical and the paper and pulp industry, offers many challenges. The different unit processes are often interconnected and large material flows are continuously transported between them. The processes mostly operate around the clock, so reliability is a key factor. The number of measurement variables is high, timescales range all the way from seconds to days and the physical size of the plants is large. The capital and material costs are often extremely high, and even small changes in operating conditions and quality have an impact on the economy and competitiveness of a company. The quality of a computer control system is crucial.

Electrical power systems range among the most complex systems ever built. They include primary power sources, such as hydroelectric plants, steam boilers and nuclear reactors, as well as large distribution networks. To describe the interconnected power systems of a nation of medium size, one needs thousands of equations for generators, turbines, power lines, loads, etc. Electrical power has to be produced in the same moment as it is consumed, so that the control of power transmission and generation must be extremely fast and accurate. Hence power companies need to keep track of when industries open and close for the day or even when popular television programmes start, in order to match the expected power load.

Transportation includes many computer control problems. The control of traffic lights is well known, but it is far from trivial to find the best control strategy. How many of us have cursed the red lights and been unhappy with the priorities which *always* seem to be given to a different traffic direction. To control all the traffic lights in a section of a city is a true challenge since the traffic intersections are more or less strongly coupled to each other.

In this book we also have another scenario in mind. Automation means not only fancy advanced applications like automated factories and robots. In most of the world these applications make little sense because of the lack of technological basis and of the necessary know-how. On the other hand, automation is needed in less and medium-developed countries in order to save scarce energy resources and improve the efficiency of environmentally dangerous, highly polluting industrial processes.

1.5 A GUIDE FOR THE READER

1.5.1 The Purpose of the Book

The main goal of this book is to present an integrated view of computer control of industrial processes. There are so many disciplines that have to be combined in order to realize functioning systems, that it is impossible to cover all aspects in detail in one book. Due to the breadth of the topic and its interdisciplinary character, each chapter can easily be expanded to a thick book (in most cases such thick books already exist). The problem is here to limit the scope of each chapter to help the reader combine different disciplines in order to get the complete picture of industrial computer control.

The book does not attempt to present new theories for computer hardware, software or control principles, rather it attempts to integrate known ideas into a common frame. Hardware and software concepts are combined to make implementation of computer control understandable, the theoretical concepts being presented together with their practical consequences. The reader of this book may not directly be a developer of real-time programming systems or new control methods, but instead will hopefully be a 'competent customer' or user, sufficiently knowledgable to combine the right pieces of hardware and software into a functioning computer control system as well as to understand its potential and limitations. Developers or specialists can hopefully use the book to widen their perspective and to bring a new focus to their own speciality.

1.5.2 Necessary and Desired Background

We have tried to present the material in such a way that the reader does not need specialized knowledge of any of the particular fields that are treated. However, we had to assume some *necessary* theoretical background, such as:

- elementary calculus
- elementary electric circuit theory
- elementary feedback control theory
- basic principles of computer structure and operations
- programming in some high level language, for instance, BASIC, FORTRAN or C

In order to get more from the book, a further understanding of differential equations and matrix calculus is *recommended*, though it is not an absolute requirement.

1.5.3 Book Outline

The chapters each present different parts of a complete computer control system. Figure 1.5 shows how these parts are coupled.

In Chapter 2 the concepts of real-time programming and computer process control are defined. The idea of concurrent processes is illustrated through a simple example. Different process control problems are then classified with respect to their character and to their complexity. The identified problem areas act as pointers to the later chapters of the book.

In Chapter 3 different system descriptions and concepts are given with the emphasis on continuous and time-discrete descriptions of dynamic systems. Different approaches for the description of uncertainties are illustrated.

Measurement, data acquisition and signal transmission is discussed in Chapter 4. Basic measurement and principles for sensors and actuators are described together with the issues related to electrical noise, disturbances and signal conditioning.

Chapter 5 deals with signal processing, mostly filtering to limit the noise content and

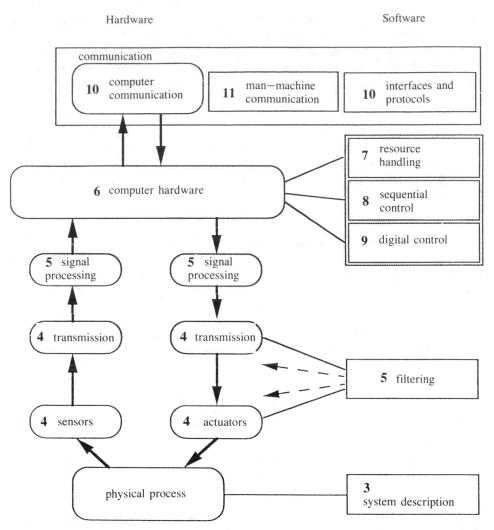

Figure 1.5   Structure of the book. Diagram of the components of a computer control system, the numbers give the chapters in which they are presented.

sampling techniques to convert analog signals to digital and vice versa. Digital methods for basic signal testing and filtering are described.

Bus-based systems and architectures are discussed in Chapter 6, with particular emphasis given to the VME and Multibus II standards.

Chapter 7 is an introduction to the specific methods of multiprogramming and real-time programming: resource protection, synchronization and data exchange between processes. It presents how an operating system supports the execution of programs under real-time requirements. The chapter does not describe a specific programming language.

Chapter 8 deals with sequencing control and its implementation in programmable logical controllers (PLCs) that operate on binary measurements and control signals. Despite, or thanks to, their simplicity, PLCs play a significant role in many industrial applications. Function charts as a way to structure the programming code are also shown.

Chapter 9 describes the computer implementation of regulators. In this chapter the structures of different control algorithms are described. The purpose is to give readers an idea of the required level of complexity for a digital controller and help them appreciate the potential of different control structures.

Data communication is an important topic in industrial automation. In Chapter 10 the Open System Interconnection (OSI) model is used as a framework to describe the different aspects and levels of the communication process. Some standards often found in industrial applications are described. The Manufacturing Automation Protocol (MAP) scheme and Fieldbus network solutions are treated in more detail because of their importance.

Man—machine communication plays a crucial role in the practical operations of control systems and is often an essential factor for the success of an application. Its principles, related to applications in industrial environments, are described in Chapter 11.

In Chapter 12 the topics of Chapters 3—11 are put together in a practical approach to control systems: which components are available, how do they perform and how are they to be interconnected? In industrial practice, digital control systems are not built starting anew each and every time. Several systems are available on the market that may be programmed with help of parameters instead of with ordinary programming code. Off-the-shelf software can be used for many applications. Some typical industrial applications are described in order to illustrate the different problems and approaches

Each chapter contains recommendations to guide the reader in further studies.

FURTHER READING

The quote at the beginning of this chapter comes from a book (Pirsig, 1974) that was a classic in the 1970s. Many of his comments about technology and how to deal with it are worth considering. The author warns: 'Don't use this book as a reference guide for the oriental doctrine Zen. It is not very factual on motorcycles, either.'

The masterwork of the 1980s about complex structures, information, communication and artificial intelligence is Hofstadter (1979). Countless connections between music, art, old languages, biology and computers develop naturally in the narration, and constantly lead the reader to look at things under new perspectives.

Numerous good articles have been published in *Scientific American* about process control and its technological implications, e.g. Spector (1984) and Ginzberg (1982). For a highly qualitative description of the possibilities offered by software, check Kay (1977, 1984).

Seborg, Edgar and Mellichamp (1989) and Stephanopoulos (1984) give a comprehensive but very readable introduction into process control applications in the chemical industry. On the more practical side of chemical process control, the book by Shinskey (1967) is still one of the best. For an introduction to electrical power systems the book by Elgerd (1982) is recommended.

2

Typical Process Computer Problems

Aim: To familiarize the reader with the basic issues and challenges of computer systems in automation and control.

This chapter is an introduction to the typical problems encountered with computer control systems. Each problem or part of a problem will show the necessity for solutions and act as introductions to the material given in later chapters. To exemplify the problems encountered in practice and their solutions, two case studies, an electrical drive system and the control of a biological wastewater treatment plant, are described in detail and are referred to throughout the book. These case studies were chosen because of their different timescales and because they embody many of the problems experienced in other control applications.

An analysis of the real-time problem from the point of view of computer programming is given in Section 2.2. Section 2.3 illustrates typical applications including the case studies and Section 2.4 outlines the major problem areas of computer control, bearing in mind how they are dealt with later in the book.

2.1 THE REAL-TIME CONTROL OF PROCESSES

Computers that are connected to physical processes present different problems from computers used for 'conventional' information processing. To begin with, computers interacting with an external process must operate at a speed compatible with the process under control (Figure 2.1). The very definition 'real-time' originates from the fact that no appreciable delays must be noticed in the reaction of the computer system to external events.

The other main peculiarity in computer process control is that the execution order of a program cannot be determined beforehand. External signals may interrupt and change the program sequence, so that it will be different for every new execution. In addition, the resources of real-time computer systems must be used effectively and the timing

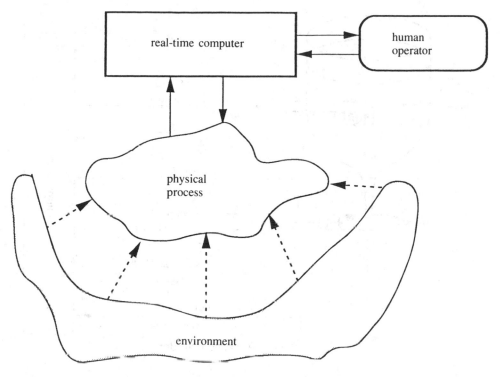

Figure 2.1   A real-time computer system.

constraints be respected. Because of this, special programming methods are required. The lack of a predictable execution order makes testing much more difficult in real-time systems than in conventional systems.

2.2 CAN REAL-TIME PROBLEMS BE SOLVED BY CONVENTIONAL PROGRAMMING?

The control of a plastic extruder will serve as an example for a typical process computer problem. The computer has to control the temperature and the sequence of operations at the same time. If programming is by conventional methods, very soon the problems of structuring the code become overwhelming. A different approach is thus called for.

2.2.1 Example: a Plastic Extruder

A plastic extruder is shown in Figure 2.2. A container holds fluid plastic material; the temperature of the container must be kept within a narrow range. The control computer

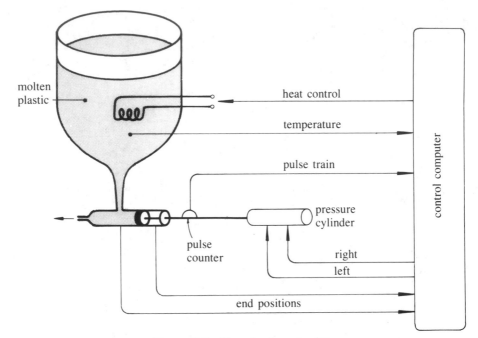

Figure 2.2   The plastic extruder.

reads regularly the actual temperature and calculates the necessary heat to keep it at the desired value. Heat is provided by an electrical element controlled by a relay, where the closing time corresponds to the produced heat.

The lower part of the extruder consists of a piston that pushes a certain amount of molten plastic through the mouthpiece. When the piston is in the right end position, the cylinder is filled with plastic. The piston is then moved to the left to eject the desired amount of plastic. The position of the piston is monitored by a pulse counter that gives a specified number of pulses per millimeter displacement, so that the number of pulses corresponds to the volume of ejected plastic material. The piston movement has to be interrupted as soon as the required number of pulses has been reached.

The computer system must control the temperature and the piston movement simultaneously. The temperature is checked by measuring a continuous signal from a sensor, while the piston position is detected from the pulse counter and with binary signals that indicate when one of the end positions has been reached. The computer has no internal clock and must therefore represent the time with an internal counter.

2.2.2 Control by Conventional Programming

The temperature control routine is shown in the flow diagram of Figure 2.3. The computer updates the time counter (represented by C) in a waiting loop until either the desired time

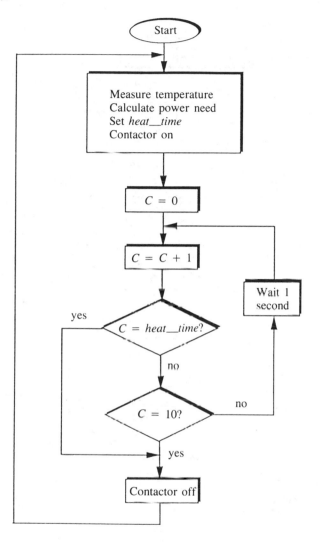

Figure 2.3   Flow chart for temperature control in the plastic extruder.

for heating (*heat__time*) or a maximum of 10 seconds have elapsed. Note that in this way the computer cannot do anything else while it waits for the temperature to be adjusted. Obviously, this is not the most efficient way to use the computer.

The control sequence for the piston movement is shown in the flow sheet of Figure 2.4. The computer starts a right movement and then keeps reading the end position sensor a number of times until confirmation that the end position has been reached. As in the case of temperature control, the computer cannot do anything else while it is waiting in the loop. The piston movement is then reversed to the left and the computer must again test the pulse signals, waiting for one pulse at a time (indicated by *n*). The sequence is

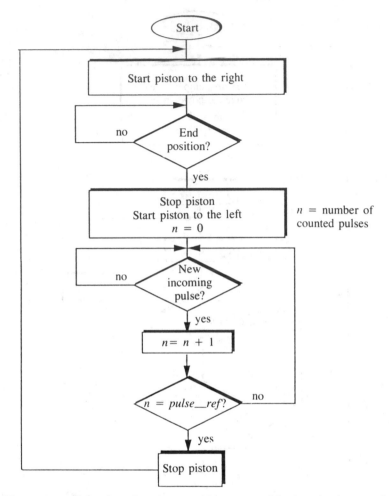

Figure 2.4   Flow chart for piston position control in the plastic extruder.

completed when the prescribed number of pulses (*pulse_ref*) has been reached. The sequence is then repeated.

Each of the two different tasks can be solved with a straightforward sequential code. However, to combine the two tasks is no longer trivial. Waiting loops cannot be allowed, since computer resources do not allow long time delays, and one activity can not wait for the other. In principle, it is possible to design a code without waiting loops, jumping between the two tasks all the time while checking what to do next. Such code soon becomes cumbersome and difficult to analyse. The forced serial disposition of instruction blocks which should be executed in parallel introduces strong ties between functions which should remain separated.

2.2.3 Control Using Interrupts

The major difficulty in writing a control program for the extruder process lies in how to tell the computer that it is time to change from one task to the other. The problem is solved in practice by writing two independent tasks to run on the same machine, one for the temperature regulation and the other for the control of the piston movement. A signalling method, called **interrupt**, is used to tell the processor when it is time to run one or the other task. With an interrupt feature, the waiting loops can be substituted with waits for interrupts.

A particular type of interrupt is related to time. A clock generates an interrupt signal when a specified time has been reached or a time interval has elapsed. The central processing unit of the computer is relieved from the task of checking the time.

With help of interrupts, the temperature control problem takes a more straightforward solution. A time interrupt is related to the variable *heat__time*. After the heater is switched on, the task waits for *heat__time* seconds before its execution needs to be resumed. The corresponding instruction is **wait__time** (*heat__time*), as illustrated in Figure 2.5.

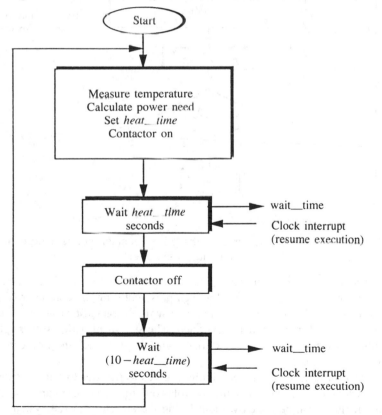

Figure 2.5   Use of time-driven interrupts for temperature control in the plastic extruder.

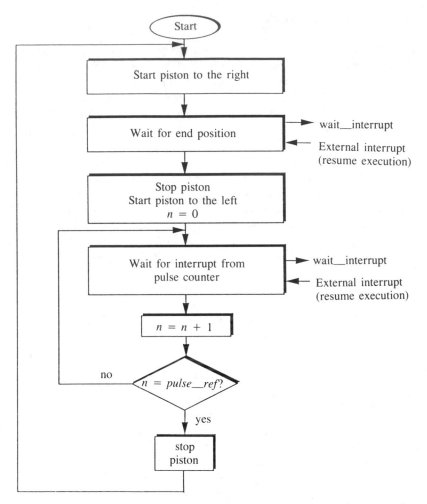

Figure 2.6   Use of interrupts for the piston position sequencing control
in the plastic extruder.

In a similar way, the piston sequencing control waits for two interrupts, one from the end position sensor and the other for each pulse entering the computer (Figure 2.6). This is expressed with a command of the type **wait_interrupt**(*x*) where *x* denotes the channel where the interrupt signal is expected. The statement **wait_interrupt** suspends the execution of the task. When an external signal is produced, the task can resume its execution.

The interrupt feature and the possibility it gives to transfer execution from one task to another is what is needed to solve the control problem in an elegant and straightforward way. New features can be added without losing the structure of the system, where the execution of each program module is independent of the others. The problem related to multiprocessing and the methods for real-time programming are treated in Chapter 7.

2.3 TYPES OF PROCESS COMPUTER APPLICATIONS

Process computer applications span over many fields. In this section we will give some examples to illustrate the major features.

2.3.1 Sequential and Binary Control

The simple chemical reactor in Figure 2.7 is an example of a sequential control application. Chemical components are mixed in the reactor. The flow of the feed components and of the effluent fluid are controlled through on/off valves A, B and C and the level in the tank is monitored by the pressure sensor P. The tank temperature is determined by letting hot or cold water flow in a mantle surrounding the tank; the water flows are controlled by valves D and E.

In our example, the following steps take place in the reactor:

1. Open valve A and fill Component 1.
2. When the pressure sensor P indicates that the desired level is reached, close valve A.
3. Start the mixer.
4. Repeat Steps 1 and 2 with Component 2 and valve B.
5. Open valve D to heat the tank to the desired temperature.
6. Start a timer for the duration of the chemical reaction.
7. When the time is over, the reaction is terminated. Stop the mixer. Open valve E to cool the tank.

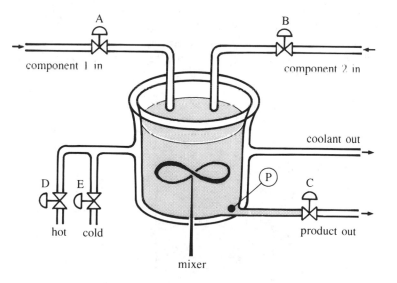

Figure 2.7   A simple chemical reactor with temperature control.

8. Check the tank temperature. When its value is below a defined limit, open valve
    C to empty the tank.

Many computer control applications are based on logical instructions like those
described here. The control computer input and output data are binary, which means that
sensors indicate two-state (or limit) values such as valve open or closed, an indicator
activated or not, or a button pushed or not. The decisions taken by the computer are of
on/off type, based on the given conditions.

When the control task consists only of logical decisions, the computations are
particularly simple. There are computers, called **programmable logical controllers (PLC)**,
that are specifically constructed to carry out this kind of task. These are treated in more
detail in Chapter 8.

2.3.2 A Simple Control Loop: the Temperature Regulator

Consider a tank filled with a fluid that has to be kept to a constant temperature (Figure
2.8). In this example, the signals are analog, i.e. the measurements are represented by
continuous instead of binary variables and the control actions are smooth and not of *on/off*
type.

The temperature is measured by a sensor giving output as a voltage proportional to
the temperature, at least within a certain range. Measurements are fed into the computer
regularly (e.g. once every second) and the actual value of the temperature is compared
with that of the desired (reference) temperature, also stored in the computer. The required
heating or cooling action to keep the temperature at the desired value is computed from
the difference between the desired and actual temperatures (Figure 2.9).

The final control elements can have different forms and accordingly control signals
will be different. The fluid can be heated by switching on a heater for a determined time
interval, while in another process the control could consist of a steam line and a cold water
line. In the first case, the control action is the time that the heater has to remain switched

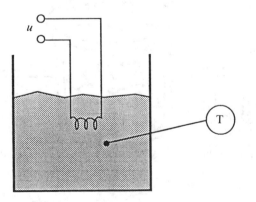

Figure 2.8   A simple temperature control system.

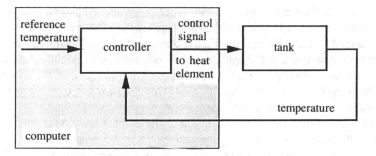

Figure 2.9   Example of a simple control loop — a temperature control system.

on; in the second case, control takes place by opening and closing the valves to let more or less steam or coolant fluid change the process temperature.

The temperature regulator shows some elementary features of a control loop. The temperature has to be measured periodically at a rate determined by the time constant of the process. If the heat capacity of the tank is large, the time constant is relatively long. Conversely, if the volume is small and the heater is powerful, the control system must be able to measure the temperature and influence the heating more often. The basic properties of process dynamics thus have to be observed in computer control. Control algorithms are discussed further in Chapter 9.

2.3.3 Reference Value Generation

In order to produce a certain compound in a chemical reactor, the required reaction temperature may have to follow a reference (setpoint) value that is continuously recalculated while the reaction is taking place. This calculation cannot be delayed: it has to be completed in time, so that the current temperature can be compared with the updated reference value. This is shown schematically in Figure 2.10.

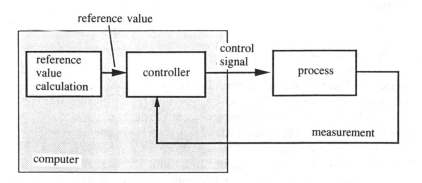

Figure 2.10   Reference value generation.

In servo systems the **setpoint** (the reference — or desired — value of the controlled variable) can be calculated or tabulated as a function of time. A **trajectory** in robot control typically describes the joint variables as functions of time. The trajectories have been computed from a **path** (a curve in space or a contour) and tabulated in the computer memory to be available as reference values for the controllers. In many cases the trajectories must be calculated on-line while the robot moves. This requires a lot of computing because of the complex geometry of a robot arm.

Once a new reference position is defined, the current position is measured and on the basis of the new values the computer can send new correction signals to the motors controlling the robot joints. The inverse operation may also be required, to find the gripper position from the angles of the joints. Both computations are resource-intensive and time-critical. A system, such as the robot described here, that is designed to follow a given command signal as fast or as accurately as possible is called a **servo mechanism** or **servo** for short.

2.3.4 Several Control Loops

Consider an office or apartment building where the temperature of each individual room has to be controlled. The actual temperature in each room depends on the influence of external factors (windows and doors that are opened, number of people in the rooms, etc.). One single computer can be used to perform the task, by devoting its attention to one room at a time. The computer executes the same control program many times over, each time on the basis of different input/output data.

There are many applications where there is a large number of control loops for temperature, levels, pressure, positions, etc. In most cases, all these individual tasks can be treated independently of each other and are consequently carried out by local controllers, each one with the typical features shown in Figure 2.9. In an alternative solution, a central control computer executes the same control routine based on different parameters and operational data for each loop. These control routines may also be executed at different time intervals. The computer must have sufficient capacity to process all the data in the required time.

2.3.5 Integrated Systems

In the process and manufacturing industry, the different types of control, from open-end sequencing tasks to digital and analog feedback, interact in every possible way. An industrial process is usually started up under sequential control (Section 2.3.1). Once the process has reached the desired operating state, a feedback control system (Sections 2.3.2−2.3.4) may take over for precise regulation. The electric motor drive system is one example of this. The operation of a chemical reactor also follows a similar pattern: the reactor temperature is brought to the operating range with sequential control and then feedback controller takes over to keep the temperature close to the desired value.

Another type of control structure interaction can be taken from the manufacturing industry. For example, on a production line where a robot moves parts between some numerically controlled machines, the position and velocity of each machine unit, including the robot, is controlled via several feedback loops of the types shown in Figures 2.9 and 2.10. Obviously, the machines do not work independently of each other, but have to be coordinated. A supervising binary control system can synchronize the machines and the robot. The machines send signals of the *on/off* type to the supervising system, to indicate 'operation completed', 'robot blocked', 'machine ready to receive new part', etc. The supervising system selects the most appropriate controls for efficient use of the machines and the robot, avoiding deadlocks and conflicting situations.

Sequential and feedback control are each treated in Chapters 8 and 9, respectively, and system integration is discussed in Chapter 12.

2.3.6 Time-critical Operations

Many processes require extremely fast control performance. To illustrate the problem, consider the speed control of a rolling mill. The different motors along the machinery have to be synchronized with extremely good accuracy, otherwise the steel band may break or bend too much. The idea is to allow for a small slack of the steel band. Considering the high speed ($10-100$ ms^{-1}) of the material, the correction of the speed of an individual motor has to be made within a few milliseconds. Of course, this sets an extreme demand on the execution speed of the control computer.

2.3.7 The Complexity of the Process Control Problem

A control computer is not only used for regulation and operation sequencing tasks of the kind shown here but must perform additional operations, for example, quickly recognize alarm situations and react to them appropriately. A control computer also collects operational data, calculates statistical parameters, shows current information to the process operator and accepts their commands. The most important computer tasks are illustrated in Figure 2.11

The very number of sensors, control loops and indicators are sufficient to make process control a complex task, but this is still only part of the truth. Additional complexity is added by the characteristics of the process itself, where the most important factors are the following:

- non-linearities
- varying environmental conditions
- changing process conditions
- long time delays
- internal process couplings

Almost all processes are inherently **non-linear**. In fact, linear relations are mostly

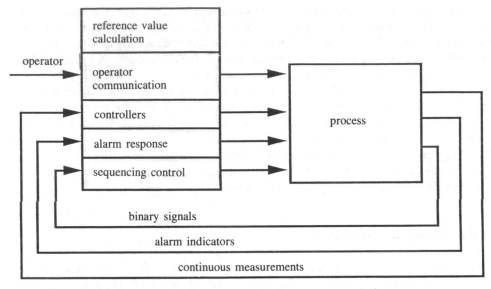

Figure 2.11   Computer tasks in process control

an artificial simplification of the real state of the things. For instance, in mechanical systems the relation between spring force and extension is often non-linear, i.e. if the spring extension is doubled the force will not be doubled. The reaction rates for most chemical processes depend on temperature in a non-linear way. Raising the temperature by 10 per cent does not mean that the reaction becomes 10 per cent faster. Yet linear models, because of their simplicity (at least compared with non-linear descriptions) are a very useful approximation of physical systems. Linear and non-linear models will be discussed in Chapter 3.

An important kind of non-linearity is **saturation** of magnetic materials in electrical machines. The magnetization of the armature is not even a single-valued function, but depends on the history of the motor, i.e. how the actual operating state was reached, so a **hysteresis** effect is often present. There is a difference in half-speed operation between that started from zero and that reduced from full speed. The design of a control system should take these factors into consideration.

Non-linearities play a role not only in the physical processes, but also in the interface to the computer, sensors and actuators. An on/off valve is a typical case, its states are fully closed and fully open. A computer might well indicate on the base of sophisticated mathematics that the optimal input flow to a process is 46 or 107 per cent of the full scale value, but the valve may only be able to operate at either 0 or 100 per cent.

Varying environmental conditions appear for instance in aircraft dynamics. An aircraft behaves differently at low altitudes than at high altitudes due to the different air pressures. The response to a wing flap movement is stronger at low altitudes where air is denser. Consequently an autopilot has to take altitude together with tens of other factors, into consideration to control the aircraft under varying conditions.

A power boiler is an example of a process with **varying dynamic behaviour**. Due to inherent non-linearities it responds quite differently at low and high power levels. This means that the controller parameter settings have to depend on the power level. Typically the parameters can be stored in a table as function of power, in what is known as **gain-scheduling control**.

Time delays are a big problem in control. They result in the controller taking decisions on the basis of old data, up to the point that it might send out the wrong commands. Time delays are present in all those processes where the value of a variable cannot be measured directly. For example, to control the concentration of a fluid, the concentration value is measured downstream along a pipe and fed back to a control valve. Due to the delay in the flow reaching the measuring point, the information is late, which may cause instability, i.e. difficulty in reaching and holding the desired value. Not only long pipes create time delays. Many types of sensor may need a long time to produce the measurement value, leading to delayed control actions which may be cause for instability.

The consequence of long measurement delays can be illustrated by a suggestive experiment, shown in some science museums. You speak into a microphone and listen to your own voice in a headphone. If the signal from the microphone is delayed for more than a few fractions of a second, you get confused very soon and cannot keep talking. This demonstates the instability caused by time delays. A similar effect is encountered when talking on the phone via satellite circuits. The time delay may lead to difficulties for the talkers to coordinate their conversation.

A controller in such a system has to remember old control actions, in other words, old information has to be stored and used in the calculations. There are controllers that can compensate for time delays (also called dead time), which is discussed further in Chapter 9. They contain in some form a model of the process under control and have to estimate the current value for a variable which cannot be measured directly without delay.

Internal couplings add a lot of complexity to a process, even when it would be basically simple. Consider the temperature control of the rooms in a building. If a window is opened in one room, the temperature will change not only locally, but to some extent also in the neighbouring rooms (Figure 2.12). Internal couplings can be illustrated in block diagram form as an input/output relationship from one input to several outputs (Figure 2.13).

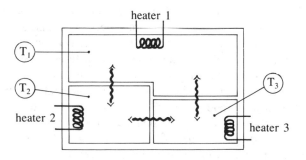

Figure 2.12   Schematic illustration of the internal couplings between rooms due to heat transfer.

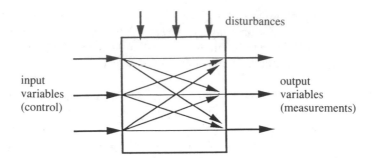

Figure 2.13   Internal couplings in a process.

A power generation and transmission system combines most of the difficulties that have been mentioned. The system is extremely complex in most terms: it has a large number of components, shows complex dynamics, its operations are time-critical, it is subject to changing environmental conditions, and the demands on responsiveness and reliability are extremely high. There is much more than one could imagine behind the availability of electric power at any outlet, with constant voltage and frequency, and without interruptions all year round! Effective control of large systems such as the electric power network can only be realized with the help of computers.

2.3.8 Case Studies

The following two systems, an electical drive system and a biological wastewater treatment plant, will be referred to in several later chapters. The systems are vastly different in timescale, but illustrate many typical problems in computer control applications.

Case study 1: Control of electrical drive systems

The problem of converting electrical power to mechanical power is extremely important. Mechanical power is needed everywhere, from industry to households. The conversion of electrical into mechanical energy takes place in motors that (as known from physics) base their operation on the electromagnetic force generated by an electric current. There are different types of motors depending on the driving electric power: d.c., a.c. monophase and a.c. multiphase. The working point of a motor depends not only on the the input current but also on its magnetization history, the load and the friction among other factors. We will not describe in detail the theory for a motor, but note that the working point for optimal energy transfer, i.e. with losses kept to a minimum, can be reached with an appropriate selection of amplitude, frequency and phase for the input power.

   The combination of motor, power electronics and the control unit is called an **electrical drive system**; it is schematically shown in Figure 2.14. Electrical

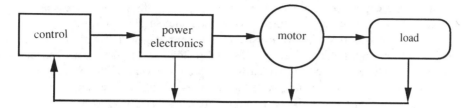

Figure 2.14 The main components of an electrical drive system.

drive systems are available for a very broad spectrum of power, speed and torque ratings. The construction of an electrical drive system is a challenge for both mechanical and electrical engineers. In fact, to achieve good results, close cooperation between people with specializations ranging from application to electrical machinery, mechanical systems, electronics, and control and computer engineering is necessary.

From here on, electrical drive systems will serve to illustrate those situations where there is a need for fast control reactions.

Case study 2: Biological wastewater treatment — the activated sludge process

In modern societies large amounts of water are consumed and the treatment of wastewater has become one of the major environmental issues. Wastewater can be treated by mechanical, chemical and biological methods in order to remove or reduce the amount of pollutants; in most modern treatment plants all of these

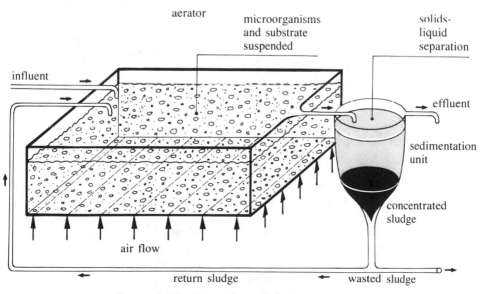

Figure 2.15 The activated sludge process.

methods are combined. Here we will briefly consider the control of biological
treatment which is used for both municipal and industrial wastewater.

In the **activated sludge process**, microorganisms suspended in the aerator
react with organic pollutants in the wastewater and with oxygen dissolved in the
water to produce more cell mass, carbon dioxide and water. The process
installation contains two main parts, the aerator and the sedimentation unit
(Figure 2.15). The aerator is the biological reactor that contains the
microorganisms and in the sedimentation unit the activated sludge (living and
inactive organisms and other inert mass) is separated from the rest of the liquid.
A portion of the concentrated sludge is recycled and inserted in the basin in
order to maintain enough mass of viable organisms in the system and a
reasonable food-to-organism ratio. Part of the sludge is removed from the system
for special treatment. The process effluent consists of the clarified overflow from
the sedimentation tank.

The timescale of a biological process is usually in the order of hours to days
and is therefore not at all critical for a process computer. This does not mean
that biological processes are easier to control, for they present their own
problems. The composition and concentration of the influent raw wastewater is
usually unknown. The concentrations of the pollutants are sometimes so small
that they can hardly be measured: the dirtiest water is still 99.95 per cent water!
Yet, even small concentrations of pollutants may be harmful for the environment
since they can accumulate in living organisms.

There are many types of microorganisms in a biological reactor; only a few
of them are actually known. Many of the microorganisms are in competition, if
one colony outgrows another the performance of the process might change.
Dissolved oxygen concentration, substrate, type of pollutants, pH and
temperature are only a few of the factors that influence the growth rate of
microorganisms.

To keep the aerator environment rich in oxygen, air must be pumped into
the basin; the air flow rate is of primary importance for the overall process
efficiency. If the dissoved oxygen level is below a certain minimum (in the
range of $1-2$ mg l^{-1}), the microorganisms will not be able to process ('eat up')
the pollutants at the normal rate and the process is slowed down.

The choice of a suitable dissolved-oxygen concentration will determine the
type of organisms that will dominate and consequently how the pollutants of the
wastewater will be removed. In a biological reactor, the relative composition of
the microorganisms depends on changes in the substrate or dissolved oxygen
concentrations and might lead to **significant process changes**. A toxic influent
may either inhibit the growth or kill some organisms. From a control point of
view this means that new types of control actions have to be selected depending
on the changing operating conditions. **Diagnostic methods** or **estimation
schemes** to identify the state of the process are needed.

Simple models to describe the dynamics of biological systems are described
in Chapter 3. The measurement problems are discussed in Chapter 4 and control
schemes are treated in Chapter 9.

Table 2.1   Some of the problems in computer process control. Factors in the system to be controlled that influence the type of solution

Original factor	Influenced parameter in the control solution	See Chapter
Timescale	System dynamics	3
	Modelling	3
	Frequency of measurements	5
	Frequency of control actions	5,9
	Hardware requirements	6
	Software requirements	7
Type of measurements	Measurement hardware, sensors	4
	Frequency of measurements	4,5
Measurement disturbances	Filtering, type of processing	3,5
Type of actuators, system controllability	Control hardware	4
System complexity	Couplings inputs/outputs	3,8,9
	Controller complexity	3,8,9
	Hardware requirements	6
	Software requirements, operating system, programming languages	7
	Communication requirements	10
Goals to achieve	Control strategy	8,9,12
Spatial localization of information	Data collection Communication Networks, protocols	10
Ease of utilization	Psychological factors, man—machine interface	11
System integration	Distribution of resources, reliability	12

2.4 GENERAL PROBLEM AREAS OF COMPUTER CONTROL SYSTEMS

The processes described in Section 2.3 show some aspects which have to be considered in computer control systems. The process for control is only part of the problem; another part is the control computer itself. The basic consideration when selecting a control system is that its capacity must match that required by the system for control. Many other parameters also have to be considered to estimate the required capacity for the process computer (Table 2.1).

2.4.1 Time-related Factors

The amount of data collected with measurements is closely related to the dynamics of the process, i.e. the choice of measurement frequency is of great importance. To determine the relevant sampling rate is far from being a trivial task. We will examine this problem in more detail in several chapters of the book.

A high sampling rate means a larger load on the computer. In many cases each sampling will also cost money as, for example, in concentration measurements that need chemical reagents. This means that the minimum number of measurements are desirable while keeping their frequency sufficiently high to be able to detect important changes in the variable under study. In other words, one has to compromise between the cost of measurement and the cost of not detecting important changes.

It is not only the measurement frequency but also the complexity of the calculations between the sampling instances that determine the computer load.

2.4.2 Measurements and Signal Processing

All measurement signals contain both information and disturbances. Measurements are not perfect due to calibration errors, inaccuracy of the sensors or electrical noise. The transmission of a signal from a sensor to a computer via an electrical conductor may be disturbed by electrical noise.

Signal filtering and information extraction are important tasks, as we know from everyday experience. If a lot of people sit around a table and talk, a microphone and recorder would reveal just a great noise of voices without any meaningful information. Yet a human ear can 'filter' out a specific voice from the rest of the noise and extract the relevant information. The same must be done by a filter with measurement information. A filter is basically a device or a calculation that works on a signal and extracts some information out of it according to predefined criteria. Obviously, the filter should be so designed to let the *desired* information pass through and block the *undesired*. Filters can be realized both with analog and digital techniques. We will examine both in Chapter 5.

Even if the sensor is accurate and the transmission good the measurement may not always represent the desired process variable correctly. For example, the measurement of a level may deteriorate due to ripple, or a concentration measurement may be misleading because of inhomogeneities in the fluid.

2.4.3 System Complexity

The complexity of the process can be characterized by many parameters. The number of sensors and actuators influences such factors as the number of input and output ports, the processing power, internal and external memory size. The bus system is a central factor for the computer system. The couplings between the process variables and the inputs and outputs determine the complexity of the controller software. Some of the control problems are considered in the Chapters 3, 8 and 9. The program structure, suitable programming languages and operating systems to deal with complex tasks are further described in Chapter 7.

2.4.4 Localization of Information

Complex control systems are based on several computers connected together, making computer—computer communication a central issue in modern process computer systems. In order to use the available capacity rationally, it is necessary to determine the amount of information that has to be exchanged between computers. Not all local computers have to know everything that is taking place in a plant. Communication reliability is very important and a system has to be worked out where the sender and the receiver know of each other and can acknowledge any message.

Communication also raises the question of standardization. Obviously cables and contacts have to fit; signal levels must mean the same thing in all computers; and even more intricate is to make computers that understand the same meaning in the messages. The communication issues soon become intricate and are dealt with in Chapter 10.

2.4.5 Ease of Utilization

With every process computer there is some communication with the human user. The operator has to be able to extract information from the computer and to input commands.

Graphical output grows increasingly sophisticated. Modern displays offer a fantastic potential to show advanced information to the user. Colours, shapes, pictures, curves and other diagrams can be used to inform the operator about the current state of the system. However, presentations like this need computing power and when operator communication competes with other tasks, the human operator may have to wait for the information. On the other hand, not all information can be kept waiting as alarms and other important messages have to be given priority. Therefore the software has to be selective and choose the appropriate information for the operator. This is considered further in Chapter 11.

2.4.6 System Integration

A key issue in all computer control is system reliability. This concept was already a central issue in the pioneering years, as mentioned in Section 1.2. The concept of using only a central computer in DDC was much criticized because of its low reliability. Even if the overall computer quality has increased dramatically since the 1960s, the problem of system reliability is still of central interest. The obvious solution to this problem is to decentralize the computing power by letting individual computers control local parts of a large process. The integration process is further explained in Chapter 12.

Reliability does not only depend on the hardware structure of the computer system.

Software reliability is a crucial matter in many large systems. In January 1990, the US telephone network could dispatch only about 50 per cent of its traffic under about 9 hours. The reason was an undetected error, or 'bug', in a very complicated computer program.

We will discuss some realistic aspects of how to influence system reliability, both by defining hardware structures that are relatively insensitive to component failures as well as by structuring the real-time code so that at least predictable errors do not arise. Real-time programming will always be much more complex to test than conventional sequential programming. The code has to be so well structured that faults can be detected. In software development today there is a clear tendency to make the languages require so much information that the compilers may detect serious coding errors early.

2.5 SUMMARY

Real-time control problems consist of several tasks, that have to be executed concurrently. Particularly:

- the execution order of real-time programs can not be determined in advance
- the execution order can be influenced by using interrupts, either from a clock or from external signals

A computer may be used for both sequential or feedback control. In many systems these tasks are integrated. The capacity of the computer system depends on many factors such as time-related parameters, number of sensors and actuators, internal process couplings, process dynamics and controller complexity. A major task of the process computer system is to keep the plant running by synchronizing many individual tasks.

Communication is a central issue, and encompasses both the coupling between the physical process and the computer, the internal computer connections, the coupling between different computers and the interface to the human user.

FURTHER READING

The example of the plastic extruder in Section 2.2 has been inspired by Hassel and Tuvstedt (1978). The failure in the US telephone system of January 1990 was widely reported in the press, see, for example, *Newsweek* (1990).

Journals

The list of specialized journals that address the 'no-man's land' between theory and practice in automation and control is not as long as it should be. *Control Engineering* and the *ISA* (Instrument

Society of America) *Journal* are dedicated to new applications of automatic control with attention to practical, on-field issues. Interesting articles on the subject are also found in *Chemical Engineering* and *IEEE Spectrum*.

Two good German journals are *at* (Automatisierungstechnik) and *atp* (Automatisierungstechnische Praxis), published by Oldenbourg Verlag in Munich (Germany).

3

Systems

Aim: To understand the fundamental issues in systems analysis based on
dynamic models and their purpose

A system description or a model contains condensed knowledge of a physical process. The reason we need a model is that, based on the measurement information the computer receives, a control strategy has to be devised. The model can tell how the system reacts on a specific control action. For simple control tasks a quantitative model may not be needed, for instance, filling a tank or starting a motor. Other control tasks can be much more complex and an elaborate quantitative model of the process is needed for the control (e.g. a robot following a certain path). A **dynamic system** behaves in such a way that the result of input manipulation can not be seen immediately, but takes some time.

Models and modelling principles for physical systems are described in Section 3.1. There are two main ways to find dynamic models, starting from basic physical principles or by using measurements. Some examples of basic modelling principles are shown in Section 3.2. Based on the examples, we shall formulate general descriptions of continuous dynamic systems in Section 3.3. The state/space approach and input/output descriptions are introduced. Time discretization of systems is of fundamental interest for computer control and some basic structures are discussed in Section 3.4.

By using model knowledge together with measurements it is possible to calculate internal process variables. The procedure of reconstruction or estimation is described in Section 3.5. It is closely related to observability, which shows whether the available set of sensors is adequate to give information about the system. The controllability issue addresses the problem if the control variables suffice to control the process in the desired direction.

Uncertainty in process models is always present, and in some cases it can be described in order to be dealt with more successfully. In Section 3.6 uncertainty is described both in statistical terms and with linguistic expressions as in fuzzy systems. Sequencing networks are extremely common in process control; the general principles for their analysis are described in Section 3.7.

We will refer to dynamic or sequencing models throughout the book. In particular, they are applied for measurement systems (Chapter 4), signal processing (Chapter 5), in sequencing control (Chapter 8) and for the derivation of control algorithms (Chapter 9). A word of warning is needed about the use (and abuse) of the term **system**. It is one of those general words that can mean everything or nothing. This chapter is mainly devoted to methods for analysis and description of physical systems, i.e. the processes we want to control. Chapter 12 is also dedicated to systems but in a broader sense: how to connect a physical process with its control circuitry in order to achieve the desired result. The 'system' of Chapter 12 is therefore more complicated. Also the approach is different. This chapter shows ways to describe and analyse systems while Chapter 12 shows how to build them.

3.1 MODELS FOR CONTROL

A model is of fundamental importance for control. Any control scheme is based on some understanding of how the physical process will react to an input signal. Therefore, the ability to model dynamic systems and to analyse them are two basic prerequisites for successful feedback control.

3.1.1 Types of Model

There are many different ways to describe systems. The choice of one way or another depends on the information at hand, the possibility of collecting further information and — most important of all — the purpose for which modelling is done. Contrary to science, where the purpose of modelling is to gain insight of a system, a model in control engineering is adequate if the control process based on it leads to the desired purpose (small variations around a given value, reproducibility of an input signal, etc.).

Example 3.1 Models for an ignition engine

A combustion engine is an extemely complex dynamic system. Depending on the user, the model may look quite different.

A *scientific* model which aims to get an insight into the intricate details of the combustion process has to consider phenomena such as the geometry of the cylinder, the mixing of air and fuel just as they meet in the cylinder compartment, the chemical composition of the fuel, the propagation in space and time of the combustion, and the resulting movement of the piston. The timescale is from the millisecond range upwards.

A model for the *design* of a *control* system for the air/fuel ratio will reflect a much more macroscopic view of the motor. Here the flow ratio of air to fuel has to be controlled close to the stoichiometric relation. The spatial distribution of the combustion is not considered, only the mass flows of air and fuel. The timescale is no longer in the millisecond range, but rather is 10—100 times longer.

The *driver* needs still another model of the motor. How car acceleration responds to the throttle pedal becomes more important and the details of the combustion phenomena or of the air/fuel mixing process may be neglected.

In control applications we are interested in dynamic systems. There are many different ways to model these, and the most important types of models are:

- **Continuous time description** of the system in terms of linear or non-linear differential equations, giving a quantitative description of mass, energy, force or momentum balances of a physical process. In many cases, non-linear equations can be linearized under reasonable assumptions.
- **Sampled time description** in terms of linear or non-linear difference equations. This means that information is available only at specified discrete time instants. Sampling is necessary when using process computers working sequentially in time. Choice of the sampling period is part of the modelling.
- **Discrete event** or **sequential systems**. Sequential control was mentioned in Section 2.4. Typically, the amplitudes of the inputs and outputs of the system are discrete and are often of the *on/off* type. Sequential systems can be described as queueing systems and modelled by so-called **Markov chains** or **Markov processes**.
- **Systems with uncertainty**. The system itself or the measurements are corrupted by undesired noise. In some cases this noise can be given a statistical interpretation. The statistical description may model either real random disturbances or modelling imperfections. In other cases, the uncertainty can be described by a more linguistic approach rather than numerical or mathematical terms. One way is to use a special algebra, called fuzzy algebra, to describe the uncertainties. Another way is to apply non-numerical descriptions of the type 'if-then-else' rules.

Clearly, there are different model approaches depending on how the model is to be used. Different controllers need different process models. Therefore we will look at systems both from a **time domain** (transient) approach and a **frequency domain** approach. Since a computer works in time-discrete mode it is important to formulate the physical process descriptions accordingly.

There is a common misunderstanding that only one model can ultimately describe the process. Instead, the model complexity and structure has to relate to the actual purpose of it.

3.1.2 Timescale in Dynamic Models

The timescale is probably the most important single factor in the characterization of a dynamic process, as we have already noted in Sections 2.3 and 2.4. Many industrial plants include a wide spectrum of response times, which makes it important to consider which ones are relevant for the actual purpose.

An example from the manufacturing industry may illustrate the point. The control tasks can be structured into different levels with different timescales. At the machine level there are events that take place within fractions of a second, such as the control of a robot arm or a machine tool. At the next level, the cell control level, one concentrates on the synchronization of machines, such as using a robot to serve two machine tools. Here the timescale is of the order of several seconds or minutes. On the cell level one assumes that the individual machine control tasks are taken care of at the machine level. In the cell timescale one may ask if a machine is supplied with material, if the robot is free to pick up a machined detail, etc. The synchronization control variables can be considered as command signals to the machine or robot control systems. In an even slower timescale we find the production planning stage, i.e. which articles to produce and when to produce them. The relevant timescale here may be days or weeks and the dynamics of a single machine is considered instantaneous (see also Section 12.2.3).

Another example is found in biological wastewater treatment. The timescales are wide apart. Compressed air is injected in the treatment plant to keep aerobic microorganisms alive, in a matter of minutes. Concentration changes due to influent flow disturbances take place in a few hours, while the organism metabolism is changed over days and weeks. If we need to study weekly changes in the metabolism, the hourly phenomena can be considered instantaneous. On the other hand, to control the air supply we need to measure on a minute-to-minute basis, while the metabolism in the short timescale is considered unchanged.

The relevant timescale of a dynamic model depends finally on the user, where an automatic controller is also one kind of 'user'. Then a plant operator may have to know the process variations and how to change its operating conditions in a minute-to-hour timescale. A process engineer or supervisor may be interested only in daily production rates or slow changes, and thus needs yet another time frame for the process description. The plant manager, finally, may want to look at completely different aspects of the plant operation, such as variations of plant productivity or seasonal demand changes.

3.1.3 Modelling Dynamic Systems

Many processes are well known and their fundamental behaviour has long since been studied, while other processes are poorly known and difficult to quantify. For example, the dynamics of an aircraft or of a nuclear reactor core have been studied extensively and accurate (though complex) models are available. Many industrial processes, however, may be difficult to quantify in mathematical models. For instance, a laboratory fermentation process with only a single type of microorganism fed by a well defined substrate may be described quite accurately. On the contrary, a biological wastewater treatment process contains a complex mixture of many types of organisms feeding on substrates that are difficult to characterize. Such a process can only partially be described by conventional quantitative models. Semantic descriptions of its behaviour offer further possibilities for characterization. Other examples of poorly known processes are steel processes, solids/liquid separation processes, digesters and rotating kilns.

Processes with time variable parameters present special problems. For example, in a biological system a new substrate may enter the process and new species of organisms will compete with the current organisms. This may change the whole dynamic behaviour of the system.

In many cases the modelling of complex systems is difficult, expensive and time consuming, especially when the important steps of experimental verification are included. In principle there are two main routes to develop a system model. In the **physical modelling approach** the model is derived from physical relations and balance equations. It will be demonstrated in Section 3.2 by some simple examples. The other way is to derive a dynamic model from measurements. The process is purposefully disturbed by different types of inputs, and the input and output time series are analysed in a procedure known as **parameter identification**. If the analysis is made on-line as the experiment progresses, the procedure is called **recursive estimation**.

Most often modelling practice is a combination of physical modelling and identification. With more insight into the fundamental properties of the process there is a greater potential to obtain an accurate dynamic description. Note, however, that even the most elaborate model based on physical insight has to be verified by experimentation.

Many systems have a spatial distribution. For example, the concentration of a liquid in a tank may not be homogeneous, but is a function of both space and time. The balances describing such systems have to be expressed by **partial differential equations**. In process control applications such systems are mostly approximated by finite differences in space so that the system can be described by ordinary differential equations.

3.1.4 Modelling Discrete Event Systems

The modelling of sequential or discrete event systems is quite different in character from dynamic systems modelling. A dynamic feedback control is often based on representative variables such as temperature, level or pressure measurements. The control is made so that the actual variable is kept close to some desired value with a certain accuracy.

Sequential control modelling often lacks a coherent theory. Binary control has to consider all possible non-normal states or alarm states that may appear. What happens if a pump breaks, if a measurement is not available, if the power fails, etc.? To obtain an exhaustive list of possible events is not trivial and can not be achieved with systematic theory.

The manufacturing machines served by a robot (Section 2.3.5) need a model for synchronization. This is a resource handling problem similar to those appearing in operating systems and where queuing theory can be applied. Note that this control problem is quite different in character from a simple feedback control problem. The synchronization has to be **exact** in the sense that a certain piece has to be delivered to the right machine at the right time and in proper order.

3.2 ELEMENTARY ASPECTS OF DYNAMIC SYSTEMS

The physical modelling approach of dynamic systems is based on elementary principles of force and torque balance as well as mass and energy balance equations. A few elementary examples of dynamic systems are presented to illustrate some of the general principles of such systems.

3.2.1 Mechanical Systems

The cornerstone for obtaining dynamic models for any mechanical system is **Newton's law**. The force **F** is the sum of all forces to each body of a system and is a vector that is described by both its amplitude and direction. Application of this law typically involves defining convenient coordinates to describe the body's motion, including position, velocity and acceleration. The acceleration **a** is also a vector, having a direction parallel to **F**. The mass of the body is m, and the vector **z** is the position. The force balance is:

$$\mathbf{F} = m\mathbf{a} = m\frac{\mathrm{d}^2\mathbf{z}}{\mathrm{d}t^2} \tag{3.1}$$

Newton himself actually stated the more general form that relates to velocity **v**:

$$\mathbf{F} = \frac{\mathrm{d}}{\mathrm{d}t}(m\mathbf{v}) \tag{3.2}$$

The force equation can alternatively be written as a system of first order differential equations (called the **state/space form**). Assuming the direction of the force is given, the position z and velocity v are written as scalars $\mathrm{d}z/\mathrm{d}t = v$ and $\mathrm{d}v/\mathrm{d}t = F/m$.

Example 3.2 Mechanical system

Many mechanical systems can be described as in Figure 3.1. A mass m is connected to a fixed wall by a spring and a damper. The spring acts in proportion to its relative displacement, while the damper yields a force proportional to its velocity.

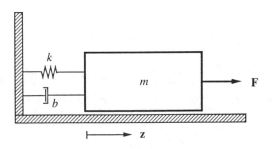

Figure 3.1   Newton's law for translation.

Newton's law then states:

$$m\frac{d^2z}{dt^2} = -b\frac{dz}{dt} - kz + F \tag{3.3}$$

After simple rearranging we obtain:

$$\frac{d^2z}{dt^2} + \frac{b}{m}\frac{dz}{dt} + \frac{k}{m}z = \frac{F}{m} \tag{3.4}$$

Many systems in servo mechanisms can be characterized by Newton's law in Example 3.2. The qualitative solution to the equation depends on the relative size of the coefficients b, k and m. For a small damping b there is an oscillatory behaviour, while for larger b values the position changes without any oscillation to a new value as the force changes. Systems like this one are often characterized by the relative damping, oscillation frequency, bandwidth or gain.

Newton's law for **rotational systems** is:

$$\frac{d(J\omega)}{dt} = T \tag{3.5}$$

where **T** is the sum of all torques on the body, J is the moment of inertia and ω the angular velocity (Figure 3.2). Often J is not constant (like in an industrial robot or a rolling mill), so that its time dependence has to be taken into consideration.

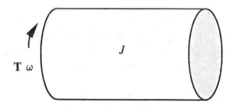

Figure 3.2   Newton's law for rotation.

Defining the angular position ϵ, the rotational dynamics can be written in state/space form. We assume that the rotating direction is given and that J is constant. Then the differential equations are written in the form $d\epsilon/dt = \omega$ and $d\omega/dt = T/J$.

Example 3.3 Electric motor torque

Consider an electric motor connected to a load with a stiff axis. The net torque **T** is the difference between the driving torque $\mathbf{T_m}$ and the load torque $\mathbf{T_L}$. The motor torque $\mathbf{T_m}$ is primarily a function of the rotor current, the field magnetic flux and — in some motor types — the angular velocity and position. The current depends on the electrical transients in the rotor circuit.

The combined load torque $\mathbf{T_L}$ has many sources. Coulomb friction causes a load torque with a magnitude d_0 which depends not on the rotational velocity but on the direction of the

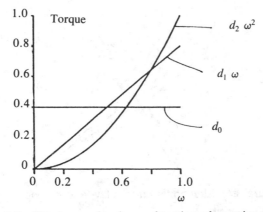

Figure 3.3 The torque is often a function of angular speed.

rotation (notated as sgn(ω)) and is always opposite the rotation (Figure 3.3). In some systems there is a viscous damping where the torque is described as $d_1\omega$ with d_1 as parameter. In a compressor or pump the load torque depends on the turbulent character of the air or water, so the torque varies quadratically with the speed, i.e. like $d_2\omega^2$, where d_2 is a parameter function of the operating conditions.

In summary, a load torque can be described as a sum of these named phenomena, plus some external load T_{L0},

$$T_L = d_0(\text{sgn}(\omega)) + d_1\omega + d_2\omega^2 + T_{L0} \tag{3.6}$$

where d_0, d_1, and d_2 are coefficients. The symbol sgn(ω) describes a relay function that is $+1$ for positive ω and -1 for negative ω. The torque balance on the motor can be described as:

$$\frac{d(J\omega)}{dt} = T_m - T_L \tag{3.7}$$

where J is the total (motor and load) moment of inertia.

An industrial robot is a complex mechanical system, consisting of linked stiff arms. The dynamic description is based on Newton's laws. In generalized form for more general motion they are called the **Lagrange equations**. Other mechanical system structures are elastic, like airplane wings or flexible robot arms. Highly oscillatory modes may appear in an elastic mechanical system. Generally, such dynamic systems are difficult to control.

3.2.2 Electrical and Magnetic Circuits

The dynamic behaviour of most electrical and magnetic circuits is governed by only a few basic laws. **Kirchhoff's laws** describe the relationships between voltages and currents in a circuit. Kirchhoff's current law (Figure 3.4) states:

the net sum of all the currents into any node is zero

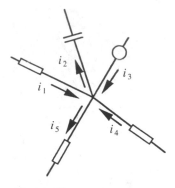

Figure 3.4   Kirchhoff's current law and its application.

A node is defined as a point where three or more connections to elements or sources are made.

Currents directed away from the node are here denoted positive. In Figure 3.4, the equation at the node is:

$$- i_1 + i_2 - i_3 - i_4 + i_5 = 0 \tag{3.8}$$

Kirchhoff's voltage law (Figure 3.5) is stated as:

the net sum of the voltage drops around any closed path is zero

The voltage law is a consequence of the energy conservation principle. In writing the voltage balance, one may go around the path in either direction and sum the voltage drops provided that the voltage across each element in the path is accounted for only once. For the loop in Figure 3.5 the voltage equation is:

$$v_1 - v_2 + v_3 - v_4 + v_5 + v_6 = 0 \tag{3.9}$$

The fundamentals of electromagnetic theory are formulated in the **Maxwell equations**. From a dynamic system point of view there are two elements that contribute to the time

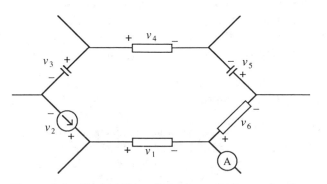

Figure 3.5   Kirchhoff's voltage law and its application.

dependence, the **capacitor** for storing electric charges and the **inductor** for storing magnetic energy.

A capacitor in an electrical circuit represents charge storage, i.e. energy stored in the electric field. The current through the element is proportional to the derivative of the voltage across it:

$$i = C \frac{dv}{dt} \tag{3.10}$$

where C is the capacitance. The unit of capacitance is coulomb/volt or *Farad*.

Example 3.4 A simple *RC* circuit

Consider the simple resistance–capacitance (*RC*) circuit of Figure 3.6. We want to describe how the voltage over the capacitor depends on an input voltage source. The Kirchhoff's voltage law applied to the circuit gives

$$v_i - Ri - v_0 = 0 \tag{3.11}$$

where R is the resistance and the capacitor voltage v_0 is expressed by:

$$\frac{dv_0}{dt} = \frac{1}{C}i \tag{3.12}$$

Eliminating the current i from the circuit differential equation

$$RC \frac{dv_0}{dt} = -v_0 + v_i \tag{3.13}$$

This first-order differential equation is characterized by its **time constant**:

$$T = RC \tag{3.14}$$

Since the dimension is volt ampere^{-1} for R and ampere (volt s^{-1})$^{-1}$ for C then $T = RC$ has the dimension of time (in seconds). Assuming the initial capacitor voltage is zero, a sudden change in the input voltage v_i will cause an exponential change in the capacitor voltage,

$$v_0(t) = v_i(1 - e^{-t/T}) \tag{3.15}$$

Figure 3.7 shows transients for different values of $T = RC$. The response is slower for a larger value of T.

In electronics and communication it has been common practice to examine systems with *sinusoidal inputs*. Assuming that the input voltage is:

$$v_i(t) = V_i \sin(\omega t) \tag{3.16}$$

where V_i is the peak amplitude, the output (capacitor) voltage (after a little while) becomes

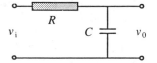

Figure 3.6   A passive first-order low pass RC filter.

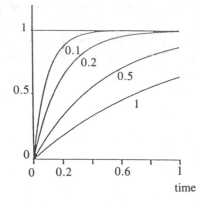

Figure 3.7   Capacitor voltage of the *RC* circuit for different values of *T* = *RC* when a step of the input voltage is applied.

sinusoidal with the same frequency but with a different amplitude and phase,

$$v_0(t) = V_0\sin(\omega t - \phi) \qquad (3.17)$$

where $V_0 = V_i \sqrt{[1 + (\omega RC)^2]}$ and $\phi = \arctan(\omega RC)$.

The amplitude of the output voltage decreases and lags in phase more and more for increasing frequencies. A network with these properties is called a **low pass filter**, since it lets the low frequencies pass but cuts off the high frequencies.

The capacitor network illustrates the two major ways of describing a linear system, the **time domain** and **frequency domain approaches**. When a magnetic field varies with time, an electric field is produced in space, as determined by **Faraday's law** or the **induction law**, which is one of the Maxwell equations. In magnetic structures with windings (and no resistance) there is an induced voltage *e* at the winding terminals, and the induction law is formulated,

$$\frac{d\Psi}{dt} = e \qquad (3.18)$$

where Ψ is the flux linkage. When there is a current in the inductor, then $\Psi = Li$, where *L* is the inductance and *i* the current in the winding. In other words, an inductance is used to represent energy stored in a magnetic field.

The inductance and capacitance differential equations are the basic dynamic elements in electric and magnetic circuits. Other relations are algebraic in nature. The relation between flux density *B* (tesla) and magnetic field intensity *H* is a property of the material:

$$B = \mu H \qquad (3.19)$$

where μ is the permeability. In a ferromagnetic material the total permeability is not constant and for large values of *H* the value of the flux ϕ (proportional to *B*) and the current generating the field intensity is shown in Figure 3.8.

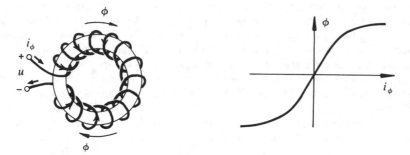

Figure 3.8   Simple magnetic circuit and a typical magnetization curve
without hysteresis.

The phenomenon of hysteresis also has to be taken into consideration. It means that
the flux density is a function of the previous history of the magnetization.

Example 3.5 Separately excited direct current (d.c.) motor

The d.c. motor is the earliest form of an electric motor. It converts direct current electrical
energy into rotational mechanical energy. A sketch of the motor is shown in Figure 3.9.
    There are two magnetic fields in the motor. The stator field is generated either by a
permanent magnet or by an electromagnet, where a separate voltage source (field supply) is
connected to the coils around the stator poles. For simplicity, we assume that the stator field is
constant in time. When a voltage is applied to the rotor circuit a *rotor* magnetic field is
generated.

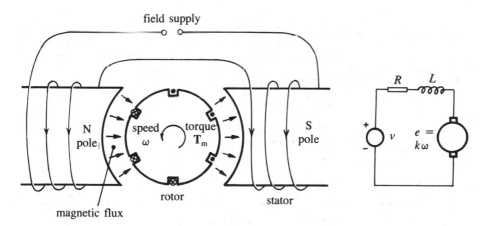

Figure 3.9   Schematic diagram of a d.c. motor and the electric diagram
of the rotor circuit.

The windings are arranged so that the rotor field is always orthogonal to the stator field. Whenever two magnetic fields form an angle between each other there is a resulting torque that tries to make them parallel. (Compare a compass needle: if it is not parallel to the earth's magnetic field lines it turns until it becomes parallel.) As the rotor rotates in order to bring its magnetic field parallel to the stator field the windings of the rotor are switched mechanically by so-called **commutators**. The result is that the rotor field orientation is kept fixed in space and orthogonal to the stator field. Consequently, due to the commutators, the torque is the same for all rotor angles.

The torque generated by the motor is proportional to the stator magnetic flux density and the rotor current i. Since the former is constant in our example, the motor torque $\mathbf{T}_m$ is

$$\mathbf{T}_m = k_m i$$

where k_m is a motor constant. With the load torque $\mathbf{T}_L$ the mechanical part is described by (see Section 3.2.1):

$$\frac{d(J\omega)}{dt} = k_m i - \mathbf{T}_L \tag{3.20}$$

where J is the total (motor and load) moment of inertia.

As a result of the rotation windings through the stator magnetic field, an induced voltage e is formed. With constant stator field it is proportional to the rotational speed ω:

$$e = k_g \omega \tag{3.21}$$

where k_g is a constant. If the units are consistent then $k_g = k_m = k$. By **Lenz's law** it follows that the magnetic flux due to the induced voltage e will be opposing the flux due to the original current through the conductor.

The electrical circuit of the rotor is represented by its resistance R and inductance L. Assuming L constant, the induction law defines the voltage across the circuit as:

$$L\frac{di}{dt} = v - Ri - k\omega \tag{3.22}$$

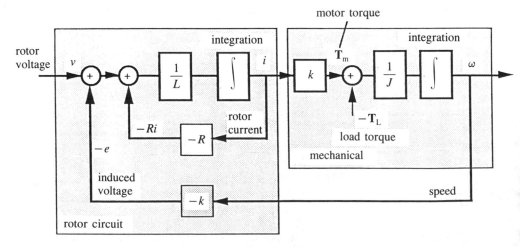

Figure 3.10   Block diagram of the d.c. motor.

where i is the rotor current and v the applied voltage. The motor dynamics is illustrated by Figure 3.10. It is shown how the applied voltage results in a rotor current that generates a motor torque. The torque drives the mechanical part so that a rotating speed is obtained. Note how the induced voltage forms a feedback from the mechanical part to the rotor circuit.

3.2.3 Mass Balances

In many process industries it is of basic importance to model the mass balances of different components. All mass balance equations have the same structure

accumulated mass = input mass − output mass

and can be formulated for each individual component as well as for the total mass. The input mass may come from both an inflow channel or pipe, from chemical reactions or from biological growth. Similarly, the output may be both outflow and consumed mass in chemical reactions or the decay of organisms in a biological reactor. Some examples will illustrate the balance equation principles.

Example 3.6 Total mass balance

Consider a completely mixed tank (Figure 3.11) with an incompressible fluid. The input and output mass flow rates are q_{in} and q_{out} (kg s^{-1}), respectively. A simple balance equation is given by:

$$\frac{dM}{dt} = q_{in} - q_{out} \tag{3.23}$$

where M is the total mass (kg).

Example 3.7 Component mass balance

Consider a mixing tank with a liquid. We will formulate the mass balance of a component

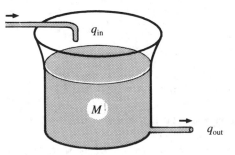

Figure 3.11 A completely mixed tank with a single component.

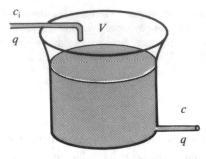

Figure 3.12   Concentration dynamics in a simple mixing tank.

with a homogeneous concentration c (Figure 3.12). The influent concentration c_i (kg m^{-3}) can be manipulated. In this way the mixing tank concentration can be controlled. The influent and effluent flow rates are assumed constant q (m^3 s^{-1}). The total mass of the component in the tank is Vc, with V being the volume. The effluent concentration is assumed to be the same as in the tank. The component mass balance then is written:

$$\frac{d(Vc)}{dt} = qc_i - qc \tag{3.24}$$

Since the volume V is constant,

$$\frac{V}{q}\frac{dc}{dt} = -c + c_i \tag{3.25}$$

Note that the form of this differential equation is the same as for the electrical circuit in Example 3.4. The time constant is defined by $T = V/q$. A sudden change in c_i will change the tank concentration exactly as in Figure 3.7. The differential equation has a solution of the form:

$$c(t) = c_i\,(1 - e^{-t/T}) \tag{3.26}$$

It is intuitively clear that the concentration will change more slowly if the flow rate becomes small relative to the volume V (a large T). Note that the component mass balance has the same dynamic properties as the low pass filter.

In principle the input concentration can be varied like a sinusoidal function and the frequency response of the effluent concentration can be studied in a fashion similar to the electric filter. However, this is often not very practical in chemical engineering processes, where the time constants may be of the order of hours. Such an experiment would then last for many days.

Example 3.8   Aeration of a wastewater treatment tank

The balance of dissolved oxygen (DO) in the aerator of a wastewater treatment plant (or a fermentor) is a *non-linear* dynamic system. The tank is assumed to be a **batch** reactor, i.e. there is no continuous flow of water into and out of the tank. Oxygen is supplied from a compressor with the air flow rate u.

The transfer rate from gaseous oxygen to dissolved oxygen is determined by a transfer rate coefficient $k_L a$. For simplicity we consider it proportional to the air flow rate, i.e. $k_L a = \alpha u$, where α is a proportionality constant. The DO oxygen transfer is zero when the concentration saturates $(c=c^s)$ and is at its maximum when the DO concentration is zero; this is modelled by $\alpha u(c^s - c)$ (mg l^{-1} s^{-1}). Microorganisms consume the DO due to organism growth and decay with a respiration rate R. A simple mass balance of the DO concentration c can be written as:

$$\frac{dc}{dt} = \alpha u(c^s - c) - R \tag{3.27}$$

Because of the product between u and c, the system is non-linear.

Example 3.9 Continuous wastewater treatment plant — simple microorganism and substrate interaction

The basic features of a biological wastewater treatment plant are illustrated in Figure 2.15. The influent wastewater is characterized by its substrate concentration s_i and contains no living organisms. In the aerator (assumed to be completely mixed), a mixture of substrate (concentration s) and microorganisms (concentration c_x) — measured in mg l^{-1} or kg m^{-3} — are kept in suspension. The flow rates are marked in Figure 3.13. The mass balances of substrate and microorganisms in the aerator are written in the form

accumulated mass = influent mass − effluent mass + growth − consumption

Microorganisms with the concentration c_{xr} are recycled (index r) from the settler unit. The growth rate of the organisms is modelled by μc_x where the specific growth rate μ (h^{-1}) depends on the substrate concentration

$$\mu = \hat{\mu}\frac{s}{K+s} \tag{3.28}$$

where K is a parameter. The growth is limited for small values of s and approaches a maximum value $\hat{\mu}$ for high concentrations of substrate. The microorganism concentration decreases due to cell decay, and is proportional to the organism concentration bc_x. The

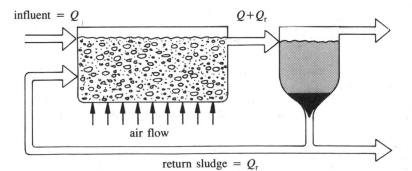

influent = Q

$Q+Q_r$

air flow

return sludge = Q_r

Figure 3.13 Simple model of an aerator in a wastewater treatment plant.

microorganism mass balance is as follows

$$V\frac{dc_x}{dt} = Q_r c_{xr} - (Q + Q_r)c_x + V(\mu c_x - bc_x) \tag{3.29}$$

The substrate enters the aerator via the influent flow rate and the recycle stream. Since it is considered dissolved it has the same concentration throughout the aerator and the settler. The substrate in the aerator is consumed due to the microorganism metabolism. The corresponding substrate utilization rate is μ/Y, where Y is called the yield factor. The substrate mass balance then can be written in the form:

$$V\frac{ds}{dt} = Qs_i + Q_r s - (Q + Q_r)s - V\frac{\mu}{Y}c_x \tag{3.30}$$

The dynamics are non-linear. The specific growth rate μ depends on the substrate concentration and the flow variables are multiplied by the concentrations.

3.2.4 Energy Balances

Several process control systems involve the regulation of temperature. Dynamic models of temperature control systems must consider the flow and storage of heat energy. In many systems, the heat energy flows through substances at a rate proportional to the temperature difference across the substance, i.e.

$$q = \frac{1}{R}(T_1 - T_2) \tag{3.31}$$

where q is the heat energy flow (W), R the **thermal resistance** and T the temperature. The heat transfer is often modelled proportional to the surface area A and inversely proportional to the length l of the heat flow path, i.e. $1/R = kA/l$, where k is the **thermal conductivity**. A net heat energy balance can be formulated as:

$$C\frac{dT}{dt} = q \tag{3.32}$$

where C is the **thermal capacity** (J °C^{-1}) and q the net sum of heat flows into the body.

Example 3.10 Heat balance of a liquid tank

The heat balance of a tank filled with liquid (Figure 3.14) can illustrate this idea. The liquid has a homogeneous temperature T, and the outside temperature is T_o. The thermal capacitance for the tank is C_i. The thermal resistance is R_1 at the top and bottom parts, and R_2 at the walls. A heat element supplies the liquid with the heat energy u_q(W). The heat balance is written as:

$$C_i\frac{dT}{dt} = u_q - \left(\frac{1}{R_1} + \frac{1}{R_2}\right)(T - T_o) \tag{3.32a}$$

A large temperature difference across the walls will cause a rapid temperature change. If the thermal resistances are large then the temperature rate of change will be dampened.

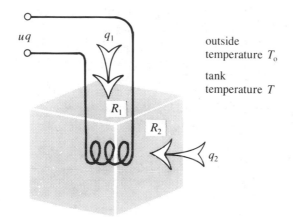

Figure 3.14   A thermal balance system.

3.3 CONTINUOUS TIME DYNAMIC SYSTEMS

3.3.1 Time and Frequency Domain Approaches — a Historical Perspective

The **ordinary differential equation (ODE)** form of dynamics goes back at least to Newton. Maxwell made probably the first systematic study of the stability of feedback systems by studying the centrifugal pendulum governor, developed by Watt in about 1788 to control his steam engine. In his paper from 1868, Maxwell developed the differential equations of the governor, linearized them about equilibrium and found that the stability depends on the roots of the **characteristic equation** having negative real parts.

The development of the electronic amplifier made long distance telephone calls possible after World War 1. To compensate for the loss of electrical energy over long distances, a large number of amplifiers had to be used. Many amplifiers in series caused large distortions, since the non-linearities of each amplifier were also amplified by the cascaded devices. The development of the feedback amplifier by Black solved the problem of distortion. As we will see in Chapter 5, operational amplifiers with feedback have the desired linearity properties. To analyse a system of 50 or more amplifiers the characteristic equation analysis was not suitable. The communication and electronic engineers turned to complex analysis and further developed the frequency response idea. In 1932 H. Nyqvist published his famous theorem on how to determine stability from a graph of the frequency response.

Industrial processes are also very complex and non-linear. Feedback of processes started to become standard in the 1940s and a controller based on the proportional-plus-integral-plus-derivative (PID) concept was developed. Again the frequency response proved to be powerful when used on linearized versions of process dynamics.

During the 1950s several researchers returned to the ODE description as a basis for control. The US and Russian space programmes stimulated this development, since an ODE is a natural form to write a satellite model. The development was supported by digital computers, so that calculations that were not practical before could now be performed. The engineers did not work directly with the frequency form or the characteristic equation but with the ODE in state form. New fundamental issues could be addressed, such as controllability, observability and state feedback. Variational calculus was extended to handle the optimization of trajectories.

In chemical and mechanical engineering it is natural to derive ODE models from the physical modelling approach. It is realistic to use for advanced control, even if PID control is still valid in many systems. Therefore the use of both ODE and frequency descriptions are common.

In electrical and electronic engineering the frequency approach is still common and quite natural in many applications. Many complex systems, however, are preferably described in terms of ODEs.

3.3.2 State/Space Form

The so-called **state/space representation** is just a standard way of representing a set of ordinary differential equations. When the equations are written as a set of first-order differential equations they are said to be in state/space form. The major attraction of this format is that computer algorithms can easily be implemented in this form. Some theoretical developments are also readily seen from the state/space approach. One example is the relation of the internal variables to the external inputs and outputs. In another example the study of control systems with more than one input and/or more than one output can readily be treated in the state/space form. The background mathematics for the study of state/space equations is mainly linear algebra. If one is willing to use vector and matrix notations the descriptions can be greatly reduced. However, a basic understanding of the dynamics does not depend on the linear algebra.

Most physical processes can be modelled by building blocks like the ones described in Section 3.2. Generally the balance equations are non-linear in character and coupled to each other. The dynamic description of the process can thus comprise a number of first-order (coupled) non-linear differential equations denoting energy balances, total mass balances, component mass balances, force or torque balances.

The state/space concept was mentioned in Section 3.2 as a practical way of describing dynamic systems. The **state** is the collection of all the variables (**state variables**) that appear in the first-order derivatives of the dynamic system. It has a fundamental importance. If the present state is known, then it is possible to predict all future values once the input signals are given. Consequently one does not have to know the previous history, i.e. how the state was reached. In other words, the state is the minimum amount of information about the system that is needed to predict its future behaviour. The state $\mathbf{x}$ can be described as a column vector with the state variables being its components:

$$\mathbf{x} = (x_1 x_2 \ldots x_n)^T \tag{3.33}$$

It is seldom possible to measure all the state variables directly. They are internal variables that have to be observed via sensors. Therefore the state/space description is called an

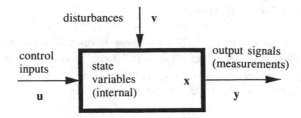

Figure 3.15 Block diagram of a controlled system.

internal description. The measurements are denoted by $y_1, y_2, \ldots, y_p$, and are described by a vector **y**:

$$\mathbf{y} = (y_1 y_2 \ldots y_p)^T \tag{3.34}$$

Usually the number of sensors (p) connected to the system is less than the number of state variables (n). Therefore it is not trivial to calculate **x** from **y**.

We consider systems that are influenced by input signals. There are two kinds of inputs. The signals that can be manipulated are called **control signals** or **control variables** (u_1, $u_2, \ldots, u_r$) and are denoted by the vector **u**,

$$\mathbf{u} = (u_1 \ u_2 \ \ldots \ u_r)^T \tag{3.35}$$

Other input signals can influence the system but are not possible to manipulate. They define the influence from the environment such as disturbances or load changes. Environmental impacts on a system come from surrounding temperature, radiation, undersized magnetic couplings, etc. These signals are collectively indicated with a vector **v**:

$$\mathbf{v} = (v_1 \ v_2 \ \ldots \ v_m)^T \tag{3.36}$$

The system can be symbolized by a block diagram (Figure 3.15) indicating the control inputs, disturbances and outputs. The concepts are illustrated in the following simple example.

Example 3.11 Mechanical systems

The system in Example 3.2 has the two states position **z** and velocity **v**. The input u is the force **F**. Assume that position **z** is measured. In vector form, the system is described by:

$$\mathbf{x} = (z \ v)^T \quad \mathbf{u} = \mathbf{F} \quad \mathbf{y} = \mathbf{z} = (1 \ 0)\mathbf{x}$$

The state equations are then written in the form:

$$\frac{d\mathbf{x}}{dt} = \begin{pmatrix} 0 & 1 \\ 0 & 0 \end{pmatrix} \mathbf{x} + \begin{pmatrix} 0 \\ 1/m \end{pmatrix} \mathbf{u} \quad \mathbf{y} = (1 \ 0)\mathbf{x} \tag{3.37}$$

The ultimate purpose of a control system is, using available measurements **y**, to calculate such control signals **u**, that the purpose of the system is fulfilled despite the influence of the disturbances **v**.

3.3.3 Linear State/Space Systems

Most of the examples shown in Section 3.2 are *linear* dynamic systems and can be modelled by linear ODEs. Typically no product between the states, inputs and outputs appears, such as x_1^2, xu and $x_1 x_2$. A linear system with n state variables, r input variables and constant coefficients is described by the state equations:

$$\frac{dx_1}{dt} = a_{11} x_1 + \ldots + a_{1n} x_n + b_{11} u_1 + \ldots + b_{1r} u_r$$

$$\vdots$$

$$\frac{dx_n}{dt} = a_{n1} x_1 + \ldots + a_{nn} x_n + b_{n1} u_1 + \ldots + b_{nr} u_r$$

where the parameters a_{ij} and b_{ij} are constants. Since the equations are linear differential equations with constant coefficients they have many attractive properties. It is always possible to find an analytical solution of $x(t)$, given arbitrary control signals $\mathbf{u}(t)$. The initial conditions are defined by n constants:

$$\mathbf{x}(0) = (x_{10}\, x_{20}\, \ldots\, x_{n0})^T \tag{3.38}$$

If one is willing to use matrix notation the effort can be greatly reduced

$$\frac{d\mathbf{x}}{dt} = A\mathbf{x} + B\mathbf{u} \tag{3.39}$$

where A and B are matrices, containing the constant parameters:

$$A = \begin{bmatrix} a_{11} & a_{12} & \cdots & a_{1n} \\ a_{21} & & & a_{2n} \\ \vdots & & & \vdots \\ a_{n1} & \cdots & \cdots & a_{nn} \end{bmatrix} \qquad B = \begin{bmatrix} b_{11} & \cdots & b_{1r} \\ b_{21} & \cdots & b_{2r} \\ \vdots & \vdots & \vdots \\ b_{n1} & \cdots & b_{nr} \end{bmatrix}$$

With only one control input, the B matrix is a single column.

There is a linear relationship between the internal state variables $\mathbf{x}$ and the measurements $\mathbf{y}$. Sometimes there is also a direct coupling from the inputs $\mathbf{u}$ to the sensors $\mathbf{y}$,

$$y_1 = c_{11} x_1 + \ldots + c_{1n} x_n + d_{11} u_1 + \ldots + d_{1r} u_r$$

$$\vdots$$

$$y_p = c_{p1} x_1 + \ldots + c_{pn} x_n + d_{p1} u_1 + \ldots + d_{pr} u_r$$

or in vector−matrix notation,

$$\mathbf{y} = C\mathbf{x} + D\mathbf{u} \tag{3.40}$$

where

$$C = \begin{bmatrix} c_{11} & c_{12} & \cdots & c_{1n} \\ \vdots & \ddots & & \vdots \\ \vdots & & \ddots & \vdots \\ c_{p1} & \cdots & \cdots & c_{pn} \end{bmatrix} \qquad D = \begin{bmatrix} d_{11} & \cdots & d_{1r} \\ \vdots & \ddots & \vdots \\ \vdots & & \ddots & \vdots \\ d_{p1} & \cdots & d_{pr} \end{bmatrix}$$

If there is only one output variable, C is a single row. Usually there is no direct coupling from the inputs to the outputs so the D matrix is zero.

A linear system has many attractive properties and obeys the **superposition** principle (in Section 2.3.7 some properties of non-linear systems were mentioned). This means that if the output signal is Δy for a specific input amplitude, then the output will be $2\Delta y$ if the input is doubled. Moreover the contributions from different input signals are additive, i.e. if the input signal u_1 contributes with Δy_1 and u_2 with Δy_2, then the total output change will be $\Delta y_1 + \Delta y_2$. As a consequence the influence from the control input (the manipulating variable) and a disturbance signal can be analysed separately.

It is quite attactive to look for linearized descriptions of dynamic systems. One may ask whether this is a realistic approach, since most processes are basically non-linear. If the non-linearities are 'smooth' (a relay is **not** smooth but exhibits jumps) then a non-linear system can behave like a linear one under certain conditions. A linear description is then valid for small deviations around an equilibrium point.

Many industrial control systems are meant to be held around some steady-state value. The purpose of the control system is to bring the variables back to a reference value. As long as the deviations are reasonably small the linear description is adequate. If, however, the deviations are too large, then more elaborate models may be needed since the non-linear terms will be significant.

3.3.4 Input/Output Descriptions

The frequency response method (Example 3.4 and Section 3.3.1) leads to the complex analysis and the **Laplace transform**. The main concepts are the transfer function, block diagrams and their manipulations, poles and zeros. A special advantage of the frequency response approach is the fact that often this data can be obtained experimentally and leads directly to a useful model. For this reason, frequency response is often used to describe complex systems such as feedback amplifiers (Chapters 4 and 5) and many electromechanical devices and systems. In Chapter 9 the Laplace transform will be used to describe controllers.

If only the relation between the inputs and the outputs is described, some of the internal couplings are hidden and the system representation becomes more compact with fewer parameters than the state/space approach. Since only the inputs and outputs are included in the model it is called an **external** description, as opposed to the internal state/space form. Many controllers (e.g. a PID controller, Chapter 9) are tuned based on a model of the input/output relationship of the process.

In the internal description (Equations 3.39, 3.40) the state variables **x** can be eliminated and the dynamics can be written in the form

$$\frac{d^n y}{dt^n} + a_1 \frac{d^{n-1}y}{dt^{n-1}} + \ldots + a_n \, y = b_0 \frac{d^n u}{dt^n} + b_1 \frac{d^{n-1}u}{dt^{n-1}} + \ldots + b_n \, u \quad (3.41)$$

where the coefficients a_i and b_i can be derived from the A, B, C and D matrices. The input/output relations of a linear system can also be expressed in terms of its transfer function. In systems with many inputs and outputs there is an input/output relation between every input/output pair. We will restrict our discussions to systems with only **one** input u and **one** output y. From the nth order differential equation we obtain the Laplace transforms:

$$(s^n + a_1 \, s^{n-1} + \ldots + a_n) \, Y(s) = (b_0 \, s^n + b_1 \, s^{n-1} + \ldots + b_n) \, U(s) \quad (3.42)$$

where s is the Laplace variable and $Y(s)$ and $U(s)$ the Laplace transforms of $y(t)$ and $u(t)$, respectively. The linear differential equation now can be analysed by *algebraic* methods.

The **transfer function** $G(s)$ is defined as the ratio between the Laplace transforms of the output and the input,

$$G(s) = \frac{Y(s)}{U(s)} = \frac{b_0 \, s^n + b_1 \, s^{n-1} + \ldots + b_n}{s^n + a_1 \, s^{n-1} + \ldots + a_n} \quad (3.43)$$

It can also be calculated directly from the internal description Equations 3.39 and 3.40. It can be shown that:

$$G(s) = \frac{Y(s)}{U(s)} = C(sI - A)^{-1} B + D \quad (3.44)$$

where I is an identity matrix of order n. The derivation is quite straightforward and can be found in control textbooks. Note, that in a system with one input and one output the C matrix is a single row and the B matrix a single column, while A is an $n \times n$ matrix. Usually D (which then is a 1×1 matrix) is zero.

Example 3.12 Transfer function of mechanical system

The transfer function of the mechanical system in Example 3.2 is

$$G(s) = \frac{Z(s)}{F(s)} = \frac{1}{ms^2} \quad (3.45)$$

where $Z(s)$ and $F(s)$ are the Laplace transforms of the position **z** and the force F, respectively. The state equations were derived in Section 3.2.1. The transfer function is also calculated directly from the state equations (Equation 3.44):

$$G(s) = C(sI - A)^{-1} B = (1 \; 0) \begin{pmatrix} s & -1 \\ 0 & s \end{pmatrix}^{-1} \begin{pmatrix} 0 \\ 1/m \end{pmatrix} = \frac{1}{ms^2} \quad (3.46)$$

Example 3.13 Low pass filter

The RC filter, (Example 3.4) can be characterized by its transfer function. Assuming that the voltages are initially zero, the input/output relation can be written as:

$$G(s) = \frac{V_o(s)}{V_i(s)} = \frac{1}{1 + RCs} \tag{3.47}$$

The amplitude gain and the phase shift for a sinusoidal input is obtained by replacing s with $j\omega$ in the transfer function.

Since the input/output description contains fewer coefficients than the internal description, it is always possible to change the description from the internal to the input/output form but not uniquely in the other direction. This is quite natural, since the state variables x can be expressed in different units and coordinate systems, while y and u are defined by the physical nature of the process.

The transfer function denominator forms the **characteristic equation**. The roots of the characteristic equation are called **poles** and have a crucial importance. The values of the poles are identical with the **eigenvalues** of the A matrix. The roots of the transfer function numerator are called the **zeros**. Calling the zeros $z_1, \ldots, z_m$ and the poles $p_1, \ldots, p_n$ $(n \geq m)$ the transfer function (Equation 3.43) can be written in the form:

$$G(s) = \frac{K(s-z_1) \ldots (s-z_m)}{(s-p_1) \ldots (s-p_n)} = \frac{\alpha_1}{s-p_1} + \ldots + \frac{\alpha_n}{s-p_n} \tag{3.48}$$

where α_i are (real or complex) constants. This means that the output y can be written as a sum of exponential functions (**modes**):

$$y(t) = c_1 e^{-p_1 t} + \ldots + c_n e^{-p_n t} + [\text{terms that depend on } u(t)] \tag{3.49}$$

A real pole corresponds to a real exponential function, while two complex conjugated poles can always be combined into one term. If two poles are located in:

$$p_{k,k+1} = -\sigma \pm j\omega \tag{3.50}$$

then the pole pair corresponds to a term of the form:

$$c_k e^{-\sigma t} \sin(\omega t)$$

in the transient. If the poles have negative real parts, then it is clear that the transient is limited if u is limited, i.e. the system is **stable**. In other words, the poles (or eigenvalues) of a linear system completely determine whether or not the system is stable.

The zeros do not influence the stability. Instead they determine the size of the coefficients of the exponential functions in the transient. Intuitively it is clear that if a zero is located closer to a pole, then the corresponding mode is small. If the zero coincides with the pole the corresponding mode is cancelled.

3.3.5 The Validity of Linear Models

In practice there are several dynamic phenomena that can not be described by linear differential equations with constant coefficients. We will illustrate the consequence of non-linearities in some examples. The systems below behave like linear systems for small signals. The non-linearities show up for large amplitudes.

Example 3.14 Signal limitations

In practical systems all signals are limited. In processes it is common to use a valve as the final control element. Since the valve cannot be more than fully opened a desired control signal cannot always be realized (Figure 3.16). This causes certain problems in control, called windup, and are further discussed in Chapter 9.

Another example of signal limitation is the rotor current in an electrical motor. The current has to be limited otherwise the motor would burn. Consequently an electrical drive system does not behave linearly, particularly for fast accelerations or large torques when the currents need to be large.

Example 3.15 Aeration process

Consider the wastewater treatment plant (Example 3.8 in Section 3.2.3). The superposition principle is not valid for the aeration process. The air supply u and the respiration rate R are assumed constant to keep the dissolved oxygen (DO) concentration at an equilibrium of 3 mgl^{-1}. Figure 3.17 shows that when the air flow is changed by a step value (2 per cent, 4 per cent, etc.), the DO concentration approaches a new steady-state value within an hour. A 4 per cent change will quite accurately double the concentration change with respect to a 2% variation, i.e. the behaviour looks quite linear. Note that these changes are symmetrical around the steady-state value.

For an 8 per cent change, the asymmetry of the response is obvious. If the air flow rate is changed $\pm$ 20 per cent, the upward and downward changes are not symmetrical, and furthermore are not 5 times larger than the 4 per cent change. These curves, derived from actual measurements, show how non-linearities appear in practice.

The systems above had 'smooth' non-linearities, i.e. the systems behave linearly for small inputs. Many systems need a more accurate description than linear differential equations

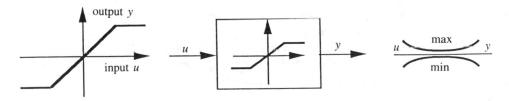

Figure 3.16   An actuator with limitation (and its symbols).

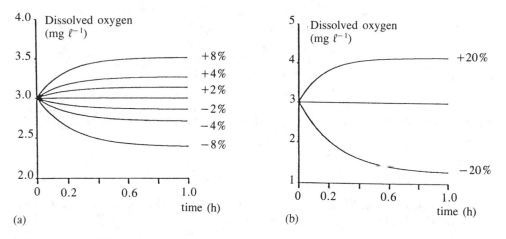

Figure 3.17 Changes of the dissolved oxygen concentration in an aerator system when the air flow has been changed as a step at time 0. Different air flow step sizes have been applied. (a) ±2 per cent, ±4 per cent and ±8 per cent; (b) ±20 per cent.

for large deviations from some equilibrium point, so that non-linear terms have to be added. It is the purpose of the model that ultimately warrants the adequacy of a linear description.

3.3.6 Non-linear Systems

The systems in Section 3.3.5 are non-linear systems that, under some assumptions, can be approximated by linear equations. Other types of non-linearities can not be reduced to a linear description, however simplified. Relay systems are a common example. A relay delivers an *on/off* signal; the ideal relay has a positive constant output for any positive input and a constant negative output for any negative input. Naturally, such a system does not satisfy the superposition principle. Examples of systems with significant non-linearities are:

- different kinds of relays (with dead bands, hysteresis, etc.)
- valves (dead band, saturation, etc.)
- non-linear deformations in mechanical springs
- pressure drops in constrictions in pipes
- friction forces
- aerodynamic damping (e.g. in fans)
- steam properties
- d.c. motors with series field windings (the torque is a function of the square of the rotor current)
- alternating current (a.c.) motors

A non-linear system (cf. Examples 3.8 and 3.9 in Section 3.2.3) can be written in the form

$$\frac{dx_1}{dt} = f_1\,(x_1,\ x_2,\ \ldots,\ x_n,\ u_1,\ \ldots,\ u_r)$$

$$\vdots$$

$$\frac{dx_n}{dt} = f_n\,(x_1,\ x_2,\ \ldots,\ x_n,\ u_1,\ \ldots,\ u_r)$$

where n states and r inputs have been defined. Also this system can be written in a compact vector form:

$$\frac{dx}{dt} = f(x,\ u) \tag{3.51}$$

where the state vector x and the control vector u are defined in Section 3.3.2. The function f is a vector where each component is a function, i.e.

$$f = (f_1\ f_2\ \ldots\ f_n)^T$$

When the system is in the steady state the derivatives of f are zero. Assuming that the state in equilibrium is $\bar{x}$ with the corresponding constant control signal $\bar{u}$ the condition at steady state is:

$$f(\bar{x},\ \bar{u}) = 0 \tag{3.52}$$

Note that Equation 3.51 corresponds to n equations. There may be several solutions to these equations, where each solution corresponds to some equilibrium point.

A sensor may also behave non-linearly (see Chapter 4). Thermistors or pressure sensors have a non-linear relationship between the physical variable and the sensor output signal. The measurement characteristics may be linear for small signals, but have to be described by non-linear relations for large signals. Thus Equation 3.40 has to be written in a more general form:

$$y_1 = g_1\,(x_1,\ x_2,\ \ldots,\ x_n,\ u_1,\ \ldots,\ u_r)$$

$$\vdots$$

$$y_p = g_p(x_1,\ x_2,\ \ldots,\ x_n,\ u_1,\ \ldots,\ u_r)$$

In matrix notation this can be written as:

$$y(t) = g(x(t),\ u(t)) \tag{3.53}$$

where the vector g consists of the functions $g_1,\ g_2,\ \ldots,\ g_p$, i.e.

$$g = (g_1\ g_2\ \ldots\ g_p)^T \tag{3.54}$$

Usually there is no analytical solution to non-linear systems. However, the solutions can be obtained numerically, which in most cases is sufficiently adequate. Note that it is sufficient to obtain the state equations of a system to get the model. Once the model is given in differential equation form there are always methods to find the solution.

3.3.7 Numerical Simulation of Dynamic Systems

In order to solve the non-linear differential equations we refer mostly to numerical methods. An elementary solution to the differential equation is obtained by approximating the time derivative with a simple difference equation (the **forward Euler approximation**), i.e.

$$x(t+h) \approx x(t) + h\mathbf{f}(\mathbf{x}(t), \mathbf{u}(t)) \tag{3.55}$$

Knowing the initial conditions $\mathbf{x}(0)$ one can compute the states $\mathbf{x}(h)$, $\mathbf{x}(2h)$, $\mathbf{x}(3h)$, ..., that are close to the true solution at times h, $2h$, $3h$, etc. It is crucial to choose a step size h that is sufficiently small. Too short a step size will give unreasonably long solution times, while h that is too large will cause numerical problems. These problems may be significant, particularly if the plant dynamics contain both fast and slow dynamics together.

Example 3.16 The problem of too long a step size

To illustrate the problem of too long a step size consider the simple first-order system:

$$\frac{dx}{dt} = -ax \tag{3.56}$$

where $x(0) = 1$ and $a>0$. The system has the analytical solution $x(t) = e^{-at}$. Let us solve the differential equation numerically by a forward Euler approximation. Approximating the derivative with a finite difference:

$$\frac{dx(t)}{dt} \approx \frac{x(t+h) - x(t)}{h} \tag{3.57}$$

we obtain:

$$x(t+h) = x(t) - ha\,x(t) = (1 - ha)\,x(t) \tag{3.58}$$

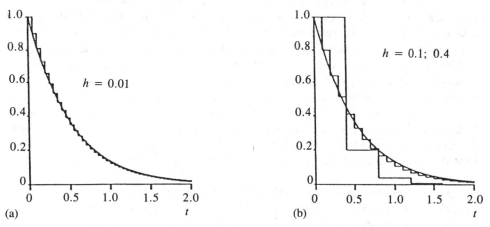

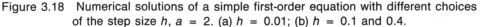

Figure 3.18   Numerical solutions of a simple first-order equation with different choices of the step size h, $a = 2$. (a) $h = 0.01$; (b) $h = 0.1$ and 0.4.

Figure 3.18 shows what happens for different choices of the stepsize h. For larger values of h such that $|1 - ha| > 1$, i.e. $h>2/a$, the solution x will oscillate with alternating sign and with an increasing amplitude. This instability has nothing to do with the system property but is only caused by a too crude approximation in the solution method.

The problem of oscillations due to too long an integration step is called **numerical instability**. Today there are several commercial simulation packages available for the solution of non-linear differential equations. By solution we mean that the transient response of the state variables can be obtained by numerically integrating the differential equations, given the appropriate initial conditions and the inputs specified as functions of time. There are many integration methods that have their merits and drawbacks. A popular class of integration methods are the **Runge–Kutta** methods. Most of the integration techniques have a variable step length that is automatically adjusted to fit an error criterion.

Simulation programs are today in common use. With such programs, the user has to formulate the equations, along with some conditions like the total integration period, numerical integration method, variables to be printed out or plotted, etc. The simulation program takes care of:

- Checking out the equations to examine if they are consistent
- Sorting the equations into an appropriate sequence for iterative solution
- Integrating the equations
- Displaying the results in the desired format (tables or graphical outputs)

Modern simulators have easy commands for parameter or initial value changes and have several integration routines to choose from. They also have advanced output features to present the results in easy readable graphic formats. There are several powerful simulation software packages available in the market. The packages Simnon, Simulab, Easy-5 and ACSL are all commercially available for personal computers or workstations. The Matlab package has rapidly gained enormous popularity as an analysis tool, not only for matrix calculations, but for all kinds of linear algebra analysis, parameter identification, time series analysis and control system synthesis. The simulated curves in this book were obtained with Simnon, a package developed at the Department of Automatic Control in Lund.

Simnon and ACSL are equation-oriented simulators, i.e. the systems are defined in ordinary differential equation form. Other types of simulators (such as Easy-5 and Simulab) are supplied with prewritten modules to describe parts of a process unit. Users can add their own dynamic modules. Using such a simulator means connecting together a number of modules of unit processes. The package then contains the same type of numerical integration tools and interaction tools as the equation-oriented simulator. The module-oriented simulator trades ease of use for less flexibility compared with the equation-oriented simulator.

Special simulators have been developed for specific applications, for instance, flight simulators or nuclear reactor simulators. These are meant to simulate a well defined system in order to train an operator. Naturally they are less flexible than the general packages, but are meant to be easily operated by the user.

3.4 DISCRETE TIME DYNAMIC SYSTEMS

A computer works in discrete time and therefore cannot process data that varies continuously in time. When a computer is used to collect data and to generate control signals, this will necessarily take place at defined time instances. A faster processor does not operate according to a concept, it will just collect more data in the same amount of time, while the data remains discrete.

In the following we develop a model of the physical process suitable for computer control. According to this model, the process measurements will be collected at regular intervals. These intervals do not need to be constant; however, in order to develop a simple form for the discrete dynamic model, we will assume a constant interval length. This interval is known as **sampling time** (there is more on sampling in Section 5.2).

Another simplification useful for the development of discrete models is for both measurements and control signals to be constant during the sampling interval. In fact, this is the way the sample-and-hold circuits in the computer interface operate.

3.4.1 State Descriptions

Since the measurements are made only at intermittent intervals the systems dynamics is formulated only in these instances. The non-linear dynamics (Equation 3.51) can be approximated as a difference equation:

$$\mathbf{x}(kh+h) \approx \mathbf{x}(kh) + h\mathbf{f}(\mathbf{x}, \mathbf{u}) \tag{3.59}$$

where h is the sampling time and k the number of the sampling interval. The approximation is valid if h is sufficiently small and the derivative is 'smooth'. A linear system with constant coefficients (Equation 3.39) is discretized as:

$$x_1(kh+h) = (1+ha_{11}) x_1(kh)+ \ldots +ha_{1n} x_n(kh) + hb_{11} u_1(kh)+ \ldots +hb_{1r} u_r(kh)$$

$$\vdots$$

$$x_n(kh+h) = hu_{n1} x_1(kh) + \ldots +(1+ha_{nn}) x_n(kh) + hb_{n1} u_1(kh)+ \ldots +hb_{nr} u_r(kh)$$

This can be written in matrix notation as:

$$\mathbf{x}(kh+h) \approx \mathbf{x}(kh) + hA\mathbf{x}(kh) + hB\mathbf{u}(kh) = (I+hA) \mathbf{x}(kh) + hB\mathbf{u}(kh) \tag{3.60}$$

A linearized system does not need to be approximated by a finite difference as in Equation 3.59. Since there is an analytical solution to linear differential equations the exact corresponding discrete time equation can be derived from Equation 3.39. It is assumed that the control signal $\mathbf{u}(t)$ is constant between the sampling instances (the system includes a sample-and-hold circuit). Then the discrete time system can be written in the matrix form:

$$\mathbf{x}(kh+h) = \Phi\mathbf{x}(kh) + \Gamma\mathbf{u}(kh) \tag{3.61}$$

where Φ is an $n \times n$ matrix and Γ an $n \times r$ matrix. The exact relations between the

A and *B* matrices and the Φ and Γ matrices are:

$$\Phi = e^{Ah} = I + hA + \frac{(hA)^2}{2!} + \dots \tag{3.62}$$

$$\Gamma = \left(Ih + \frac{Ah^2}{2!} + \dots \right) B$$

The transformation between the continuous and the discrete form matrices can be done with standard software. Note that the exact solution approaches the finite difference approximations $\Phi \approx I+hA$ and $\Gamma \approx hB$ for small sampling intervals h.

The measurements are made only intermittently and therefore Equation 3.40 is valid at the sampling instances:

$$\mathbf{y}(kh) = C\mathbf{x}(kh) + D\mathbf{u}(kh) \tag{3.63}$$

Note that we do not have to worry about the solution of a discrete system. The solution is automatically generated from the formulation of the difference equation, and the computer generates the solution of $\mathbf{x}$ step by step (recursively).

3.4.2 Input/Output Relations and the Shift Operator

As in continuous systems, it is convenient to directly relate the input u to the output y, particularly when the controller is written in the same form, i.e. acting on the process output in order to produce a control signal. The analysis becomes easy to handle by using the **shift operator** q. The definition of q means that, acting on a time dependent variable $z(t)$, the operator q shifts the time by one sampling interval:

$$qz(kh) = z(kh+h) \tag{3.64}$$

Its inverse q^{-1} shifts one sampling interval backwards:

$$q^{-1} z(kh) = z(kh-h) \tag{3.65}$$

In general, q can operate several times on a variable:

$$q^n z(kh) = qq \dots qz(kh) = z((k+n)h)$$

Using the q operator on a vector $\mathbf{x}(kh)$ simply means that it acts on each vector component.

By eliminating the state vector $\mathbf{x}$ in the internal description (Equations 3.61 and 3.63) the relation between the input and output can be expressed as:

$$y((k+n)h) + a_1 y((k+n-1)h) + \dots + a_n y(kh) = b_0 u((k+n)h) + \dots + b_n u(kh) \tag{3.66}$$

The shift operator makes a more compact description possible:

$$(q^n + a_1 q^{n-1} + \dots + a_n) y(kh) = (b_0 q^n + b_1 q^{n-1} + \dots + b_n) u(kh) \tag{3.67}$$

The **transfer operator** $H(q)$ is defined as:

$$H(q) = \frac{y(kh)}{u(kh)} = \frac{b_0 q^n + b_1 q^{n-1} + \ldots + b_n}{q^n + a_1 q^{n-1} + \ldots + a_n} \qquad (3.68)$$

The use of the backward shift operator (Equation 3.65) is illustrated if the time is translated backwards by n sampling intervals. Then the input/output relation becomes:

$$y(kh) + a_1 y((k-1)h) + \ldots + a_n y((k-n)h) = b_0 u(kh) + \ldots + b_n u((k-n)h) \qquad (3.69)$$

Using the backward shift operator, we obtain:

$$(1+a_1 q^{-1} + \ldots + a_n q^{-n}) y(kh) = (b_0 + b_1 q^{-1} + \ldots + b_n q^{-n}) u(kh) \qquad (3.70)$$

Note that this expression can be obtained by dividing the previous difference equation with q^n. The corresponding transfer operator is:

$$H(q^{-1}) = \frac{y(kh)}{u(kh)} = \frac{b_0 + b_1 q^{-1} + \ldots + b_n q^{-n}}{1 + a_1 q^{-1} + \ldots + a_n q^{-n}} \qquad (3.71)$$

By multiplying the numerator and denominator in Equation 3.71 with q^n we obtain Equation 3.68 and confirm that $H(q^{-1}) = H(q)$.

The transfer operator can be derived directly from the state equations (Equations 3.61, 3.63). We just state the result and refer the reader to a control textbook for proof. The transfer operator is:

$$H(q) = H(q^{-1}) = \frac{y(kh)}{u(kh)} = C(qI - \Phi)^{-1} \Gamma + D \qquad (3.72)$$

This calculation is made as if q were a complex number, although it is formally an operator. We are mostly interested in systems with one input u and one output y so the C matrix becomes a single row and Γ a single column, while Φ is an $n \times n$ matrix. Usually D is zero, i.e. there is no algebraic (i.e. instantaneous physical) coupling from the input to the output.

Again we note, that the input/output coefficients can be uniquely defined from the internal description. However, since the state vector **x** can be expressed in any coordinate system there are many possible Φ, Γ, C and D calculated from $H(q)$.

3.5 CONTROLLABILITY, ESTIMATION AND OBSERVABILITY

In any system there are some fundamental issues that have to be addressed. The first such issue is whether there are sufficient control variables to manipulate the complete system in a desired way. This is known as the **controllability** issue. The second has to do with instrumentation and observations. Does the available instrumentation give sufficient information about the state of the system? Is it possible to indirectly calculate the current state vector $\mathbf{x}(t)$, if current and previous values of the available output signal $\mathbf{y}(t)$ are known? This property is called **observability**.

3.5.1 Controllability

Roughly speaking, a system is controllable if its state $\mathbf{x}$ can be moved to any desired point by a suitable choice of control signals $\mathbf{u}$. Controllability is the property that makes it possible to place the poles of a closed loop system in any desired position or, equivalently, to move the eigenvalues of the system to any desired location by feedback. This is discussed further in Chapter 9.

If a process is 'uncontrollable' it usually means that parts of the system are physically disconnected from the input (here we assume only one input u). To see that, we demonstrate the case when the system contains 'distinct' eigenvalues or poles (that is, no repeated poles), then it is always possible to select the states so that they are decoupled. This means that the system can be represented as:

$$\frac{dx_1}{dt} = \lambda_1 \, x_1 + \beta_1 \, u$$

$$\vdots$$

$$\frac{dx_n}{dt} = \lambda_n \, x_n + \beta_n \, u$$

where λ_i and β_i are constant parameters; or written in matrix form:

$$\frac{d\mathbf{x}}{dt} = \begin{bmatrix} \lambda_1 & 0 & . & . & 0 \\ 0 & \lambda_2 & 0 & . & 0 \\ . & . & . & . & . \\ . & . & . & . & . \\ 0 & . & . & 0 & \lambda_n \end{bmatrix} \mathbf{x} + \begin{bmatrix} \beta_1 \\ \beta_2 \\ . \\ \\ \beta_n \end{bmatrix} u \qquad (3.73)$$

This is called the **diagonal form**. The states in a diagonal system are also called **natural modes**. We see that the input will influence the different state variables separately. All the elements in the B matrix are non-zero if the system is controllable. Otherwise, the state variables corresponding to zero entries in B are not influenced by control, and the associated states will follow their natural behaviour. The same reasoning applies for time-discrete systems.

There are mathematical methods to test the controllability of linear systems, both time-continuous and time-discrete. However, no mathematical test can replace the control engineer's physical understanding of the process. In particular, it often happens that every mode is controllable only to some degree (this corresponds to small values of the β-coefficient). While the mathematical test tells that the system is controllable certain modes may be so weakly controllable that control design becomes useless for all practical purposes.

3.5.2 Estimating the State from the Measurements

In most cases all the states are not measured directly, i.e. there are fewer sensors than number of state variables. Still, in many situations it is interesting to know all the states **x** even if adequate sensors are not available or simply the cost is prohibitive. It is possible — under certain conditions — to estimate the state **x** from the measurements **y**. To denote that the estimated value of **x** differs from the true value we use the notation **x̂**.

A so-called **full-order estimator** can be derived for either the continuous or the time-discrete case. Here we discuss the time-discrete estimator, since it is directly suitable for process computer implementation. The structure of the estimator is similar to the real system description (Equation 3.61). A correction term is added that takes the real measurements **y** into consideration:

$$\hat{\mathbf{x}}(kh+h) = \Phi\hat{\mathbf{x}}(kh) + \Gamma\mathbf{u}(kh) + \mathbf{K}\,[\mathbf{y}(kh) - C\,\hat{\mathbf{x}}(kh)] \tag{3.74}$$

The D matrix (Equation 3.63) is most often zero. If there is only one sensor, then **k** is a vector, otherwise it is a matrix. It has to be chosen appropriately. If **x** were equal to **x̂** then the last term in Equation 3.74 would be zero, since $\mathbf{y} = C\mathbf{x}$. The estimated value then would obey the same dynamic equation as the true state (Equation 3.61). As long as **x̂** is different from **x**, the last term is an error correction term, the difference between the true measurement **y** and the estimated measurement $C\hat{\mathbf{x}}$. The error corrects the equation and compensates the estimated value. The estimator is illustrated in Figure 3.19, which shows that the model output $C\hat{\mathbf{x}}$ is corrected all the time with the true measurement **y**. If **K** is properly chosen, then **x̂** can approach **x**.

The 'plant' (upper box) represents the physical process, while the estimator is an algorithm that implements Equation 3.74 in a computer. In other words, the estimator

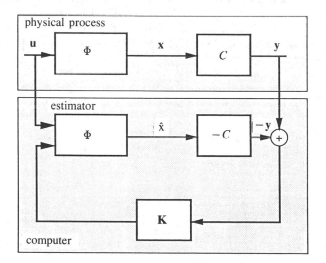

Figure 3.19 The full order estimator.

is nothing more than a computer simulation of the physical process, continuously corrected with the measured real data. To examine how the estimator converges, consider the estimation error $\tilde{\mathbf{x}}$

$$\tilde{\mathbf{x}}(kh) = \mathbf{x}(kh) - \hat{\mathbf{x}}(kh) \tag{3.75}$$

Subtract Equation 3.74 from Equation 3.61 and insert Equation 3.63:

$$\tilde{\mathbf{x}}(kh+h) = \Phi\tilde{\mathbf{x}}(kh) - \mathbf{K}[\mathbf{y} - C\hat{\mathbf{x}}(kh)] = \Phi\tilde{\mathbf{x}}(kh) - \mathbf{K}C\tilde{\mathbf{x}}(kh)$$
$$= (\Phi - \mathbf{K}C)\,\tilde{\mathbf{x}}(kh) \tag{3.76}$$

If $\mathbf{K}$ (hopefully) can be chosen so that $\tilde{\mathbf{x}}(kh)$ converges to zero sufficiently fast, then the estimator is satisfactory. This means that $\hat{\mathbf{x}}(kh)$ will converge to $\mathbf{x}(kh)$ regardless of the initial conditions. Furthermore, the error convergence can be made much faster than the system dynamics that is determined by Φ.

The fact that $\mathbf{K}$ can be chosen so that the estimator works in the desired way depends on the system property, **observability**. This property depends only on the Φ and C matrices. Roughly speaking observability refers to the ability to deduce information on all the states $\mathbf{x}$ of the system by monitoring the sensed outputs $\mathbf{y}$. It also guarantees that $\mathbf{K}$ can be found so that $\tilde{\mathbf{x}}(kh)$ converges to zero *arbitrarily* fast. Unobservability results from some state or subsystem being disconnected physically from the output and therefore not appearing in the measurements.

Example 3.17 Estimation of angular velocity in a mechanical system

Consider a motor that is connected to a heavy load by a resilient shaft. In a paper machine or in a rolling mill, the elasticity of the axis may cause the load to oscillate (Figure 3.20). Since the primary goal is to control the angular velocity of the load it has to be measured. In deriving an estimation, we first assume that the angular position of the load can be measured (Figure 3.20). The dynamics of the load can be described by Newton's law. The resilient shaft acts like a spring with a force proportional to the angular difference ϵ between the motor and the load. We also assume that the velocity difference ω between the motor and load causes a damping torque that is proportional to the velocity. Assuming that the motor torque is $\mathbf{T}_m$ and the moment of inertia is J_L then Newton's law for the load is:

$$J_L \frac{d^2\epsilon}{dt^2} = \mathbf{T}_m - k\epsilon - c\frac{d\epsilon}{dt} \tag{3.77}$$

The dynamics is written in state format:

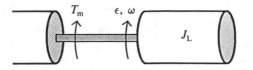

Figure 3.20   Electric drive consisting of motor and load which are coupled by a resilient shaft.

$$x_1 = \epsilon \qquad \frac{dx_1}{dt} = x_2$$

$$x_2 = \frac{d\epsilon}{dt} = \omega \qquad \frac{dx_2}{dt} = -\frac{k}{J_L}x_1 - \frac{c}{J_L}x_2 + \frac{1}{J_L}\mathbf{T}_m$$

$$y(t) = x_1$$

The derivatives are approximated by finite differences with the sampling interval h:

$$x_1(kh+h) \approx x_1(kh) + h\,x_2(kh) \qquad (3.78a)$$

$$x_2(kh+h) \approx x_2(kh) - \frac{hk}{J_L}x_1(kh) - \frac{hc}{J_L}x_2(kh) + \frac{h}{J_L}\mathbf{T}_m(kh) \qquad (3.78b)$$

where $k = 0, 1, 2, \ldots$.

The estimator has the form:

$$\hat{x}_1(kh+h) = \hat{x}_1(kh) + h\hat{x}_2(kh) + K_1[x_1(kh) - \hat{x}_1(kh)] \qquad (3.78c)$$

$$\hat{x}_2(kh+h) = \hat{x}_2(kh) - \frac{hk}{J_L}\hat{x}_1(kh) - \frac{hc}{J_L}\hat{x}_2(kh) + \frac{h}{J_L}\mathbf{T}_m(kh) + K_2[x_1(kh) - \hat{x}_1(kh)] \quad (3.78d)$$

The purpose of the estimator is to find $x_2(kh)$ from the measurements $y = x_1(kh)$. The initial conditions of the real system are unknown and therefore the estimator starts arbitrarily at zero. In Figure 3.21, the angular velocity is varying in an unknown way. At time $t=5$ it

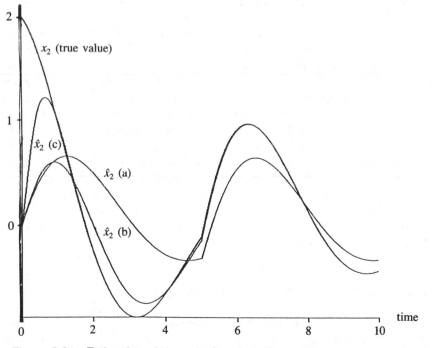

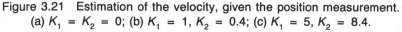

Figure 3.21 Estimation of the velocity, given the position measurement.
(a) $K_1 = K_2 = 0$; (b) $K_1 = 1$, $K_2 = 0.4$; (c) $K_1 = 5$, $K_2 = 8.4$.

gets another kick, since the motor torque suddenly increases. The figure shows how $\hat{x}_2(kh)$ approaches $x_2(kh)$. The real velocity $x_2(kh)$ is unknown and starts with the value 2. Even if K_1 and K_2 are 0 the estimated value will finally approach the real value, since the estimation model is perfect. If the values of K_1 and K_2 are positive, however, the estimator manages to approach the true value faster. This is because we have used the measurements and have a more efficient estimator. For increasing values of K_1 and K_2 the estimation is quite fast.

Once the estimator has reached the true value (the estimation error has converged to zero) the estimated and the true remain identical, even if the input torque $\mathbf{T}_m$ changes, since the estimator 'knows' the input function.

In the actual system the velocity can be obtained simply by taking the derivative of the position. However, taking a derivative of a measurement is often tricky, since the signal is noisy and the derivative of a rough signal usually gives poor information.

If the model of the physical system is not accurate, the dynamics of the error is no longer governed by Equation 3.76. However, **K** can often be chosen so that the error system is stable and the error is acceptably small, even with small errors in the model or disturbances acting on the measurements.

3.6 SYSTEMS WITH UNCERTAINTY

A mathematical model is hardly ever a perfect representation of reality. There are several imperfections in system description. In many systems the model does not include all the phenomena that take place, and some states are simply neglected. Other systems are difficult to quantify by mathematical expressions. Instead semantic information may represent the system better. This is particularly true in systems where a person is included in the control loop. Also many biological systems are too complex or unknown to be described quantitatively. For the control of systems we should always ask what is an *adequate* representation of the uncertainty.

Stochastic processes are used to model both disturbances to the process and random errors in the sensors. A stochastic process is a sequence of stochastic variables. In principle this means that some random variable with a certain probability distribution is added to each process variable every sampling interval. Similarly, measurement noise added to a sensor signal is modelled as a random variable. Other aspects of measurement inaccuracy are discussed in Section 4.2.

3.6.1 State Estimation with Stochastic Disturbances

In the previous section we assumed that the measurement information in the estimator was perfect. This is hardly ever the case since every sensor has some imperfection. For instance, the electrical noise of a sensor can be described as an additional random variable **e** in the output equation. Since each sensor obtains a random error term they can be written in compact form as a vector **e** added to output Equation 3.63:

$$\mathbf{y}(kh) = C\,\mathbf{x}(kh) + \mathbf{e}(kh) \tag{3.79}$$

Each component of the noise vector **e**(*kh*) is modelled as a sequence of stochastic variables, in other words as random numbers. If they are independent, the amplitude of the noise at time *kh* does not depend on the amplitudes at previous times. Often the random number amplitude can be assumed to be normally distributed so the mean value and the standard deviation completely characterize the noise.

When measurement noise is present, the estimation described in Section 3.5.2 has to be made more cautious. Equation 3.79 is used instead of Equation 3.63 to calculate the error. The estimator structure is changed to:

$$\begin{aligned}\hat{\mathbf{x}}(kh+h) &= \Phi\hat{\mathbf{x}}(kh) + \Gamma\mathbf{u}(kh) + \mathbf{K}[\mathbf{y}(kh) - C\hat{\mathbf{x}}(kh)] \\ &= \Phi\hat{\mathbf{x}}(kh) + \Gamma\mathbf{u}(kh) + \mathbf{K}[C\mathbf{x}(kh) + \mathbf{e}(kh) - C\hat{\mathbf{x}}(kh)]\end{aligned} \tag{3.80}$$

Now the choice of **K** has to be a compromise. If **K** is large the estimation error tends to zero fast. However, the noise term **e** is amplified which will cause an error. The value of **K** therefore has to be sufficiently large so that $\hat{\mathbf{x}}(kh)$ approaches $\mathbf{x}(kh)$ as fast as possible, and yet sufficiently small so that the noise term does not corrupt the result.

The mechanical system of Example 3.17 is considered once more.

Example 3.18 Estimation with measurement noise

Now assume that the position is measured with a noisy measurement,

$$\mathbf{y}(t) = x_1 + \mathbf{e} \tag{3.81}$$

The result of estimating the velocity using the same **K** values as for perfect measurements is shown in Figure 3.22. It indicates how speed and estimation accuracy have to be weighted against each other. With small **K** the convergence is poor, but the ultimate accuracy is quite good. A large **K** makes the error converge faster, but the final accuracy is poor. It is obvious that estimating the velocity by differentiating the angular position would give a very noisy signal.

In order to find the best **K** values with noisy measurements more sophisticated methods have to be used. The best choice of **K** is often time varying. Typically **K** can be large as long as the difference between the **real** measurement **y**(*kh*) and the *estimated* measurement $\hat{\mathbf{y}}(kh) = C\hat{\mathbf{x}}(kh)$ is large compared to **e**(*kh*). As the error gets smaller its amplitude is comparable with the noise **e**(*kh*) and **K** has to be decreased accordingly.

The process variables themselves may contain disturbances that cannot be modelled in any simple deterministic way. Thus, the difference Equation 3.61 can be complemented with some additional noise term that will describe either the modelling error or some real process noise. For example, the liquid surface of a large tank may not be smooth due to wind and waves, causing random variations of the level. Another example is the torque of an electrical motor that may have small pulsations due to the frequency converter

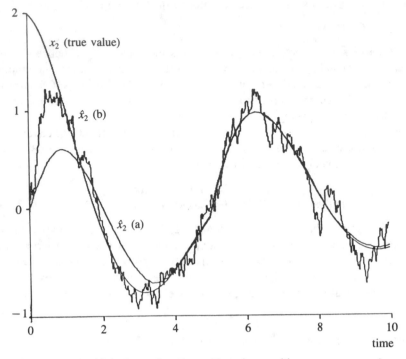

Figure 3.22   Velocity estimation with noisy position measurement.
(a) $K_1 = 1$, $K_2 = 0.4$; (b) $K_1 = 5$, $K_2 = 8.4$.
Note that the approach is faster in (b) but the estimate is more noisy.

properties. Such variations can be modelled as random numbers $v(kh)$ being added to the
state:

$$x(kh+h) = \Phi x(kh) + \Gamma u(kh) + v(kh) \tag{3.82}$$

The random variable can be considered in a similar way as the measurement noise
e. Given the noise description, this means that a controller can take the uncertainty into
consideration. It seems reasonable that any controller action would be more cautious as
soon as the state variables are distorted by noise: this means a smaller gain of a controller.

There is an optimal choice of **K** in a noisy situation. A **Kalman filter** has the structure
given in Equation 3.80 and is based on the system description in Equations 3.79 and 3.82.
The **K** that is obtained from the Kalman filter is time varying and represents the optimal
compromise between system and sensor disturbances and the estimation error.

3.6.2 Fuzzy Systems

Many systems are not only non-linear and time variant but are generally ill defined. They
can not easily be modelled by equations, nor even be represented by straightforward logic

such as *if-then-else* rules. This is the background against which Lofti A. Zadeh developed the **fuzzy logic**. The name fuzzy is a misnomer, since the logic is firmly grounded in mathematical theory.

Fuzzy logic can be regarded as a discrete time control methodology that simulates human thinking by incorporating the imprecision inherent in all physical systems. In traditional logic and computing, sets of elements are distinct; either an element is an element of a set, or it is not. The conventional (binary) logic considers only opposite states (fast/slow, open/closed, hot/cold). According to this logic a temperature of 25 °C may be regarded as 'hot' while 24.9 °C would still be 'cold', to which a temperature controller would react consequently.

Fuzzy logic, on the contrary, works by turning sharp binary variables (hot/cold, fast/slow, open/closed) into 'soft' grades (warm/cool, moderately fast/somewhat slow) with varying degrees of **membership**. A temperature of 20 °C, for instance, can be both 'warm' and 'somewhat cool' at the same time. Such a condition is ignored by traditional logic but is a cornerstone of fuzzy logic. The 'degree of membership' is defined as the confidence or certainty — expressed as a number from 0 to 1 — that a particular value belongs to a fuzzy set.

Fuzzy systems base their decisions on inputs in the form of linguistic variables, i.e. common language terms such as 'hot', 'slow' and 'dark'. The variables are tested with a small number of *if-then* rules, which produce one or more responses depending on which rules were asserted. The response of each rule is weighted according to the confidence or degree of membership of its inputs.

There are some similarities between the *if-then* rules of **artificial intelligence (AI)** and fuzzy logic. Yet AI is a symbolic process while fuzzy logic is not. In AI, **neural networks** represent data and decisions in special structures. Each data input is assigned a relative discrete weight. The weighted data are combined in the network in a precise way to make decisions. The weighting functions in fuzzy logic, on the contrary, are defined as continuously valued functions with their membership values.

Fuzzy logic often deals with observed rather than measured variables of the system. Traditional control modelling requires a mathematical model of the system, which requires a detailed knowledge of all the relevant variables. Fuzzy modelling deals with input/output relationships, where many parameters are lumped together. In fuzzy control a 'preprocessing' of a large range of values into a small number of membership grades helps reduce the number of values that the controller has to contend with. Because fewer values have to be evaluated fewer rules are needed, and in many cases a fuzzy controller can solve for the same output faster than an expert system with its set of *if-then* rules. In some prototyping cases fuzzy logic has proven to be a good way of starting with little information.

There is no guarantee that fuzzy logic can deal with complex systems successfully. A controller based on fuzzy logic is in practice an estimator of the system not based on a particular model; it is very difficult to prove the stability of such a controller.

An automatic controller for a train vehicle provides a simple illustration of the application of fuzzy set theory. The criterion for the controller is to optimize travel time within certain constraints. The distance from the destination, speed and acceleration are measured, while the controller output is the motor power.

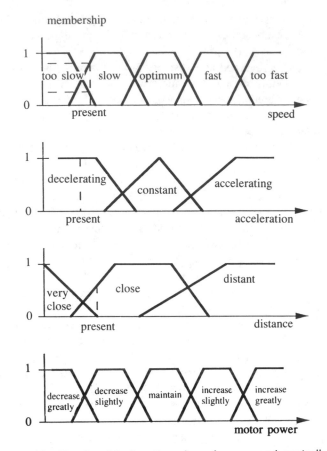

membership

Figure 3.23   Membership functions for a fuzzy speed controller.

The membership functions assign linguistic variables to the measurements (Figure 3.23). In the present state the acceleration is 'decelerating' due to a steep hill. The velocity is a member of the membership 'slow' (weight 0.8) and 'too slow' (weight 0.2), while the distance is close with a weight 0.5.

A few rules can give a flavour of the control logic:

- If speed is *too slow* and acceleration is *decelerating*, then *increase power greatly*
- If speed is *slow* and acceleration is *decelerating*, then *increase power slightly*
- If distance is *close*, then *decrease power slightly*

Which rule should be selected? The output has a confidence level depending on the confidence of the inputs. In this case the final selection is to increase the power slightly. Even if the speed is almost too slow the vehicle is close to the destination.

Particularly in Japan, fuzzy logic has become extremely popular for control system design, while — ironically — it has not caught the same interest in the United States where it was first devised, or in other western countries. Products based on fuzzy logic (all of them Japanese) include autofocusing cameras, air conditioners, washing machines, vacuum cleaners, elevator control and subway system speed control.

3.7 SEQUENTIAL SYSTEMS

Many industrial processes are based on binary measurements and control signals. In Sections 2.2 and 2.4 simple examples of sequencing networks were given. Many automation problems refer to two types of systems with binary inputs and outputs. One is a combinatorial network that can be considered as a collection of logical expressions. The second is a sequencing network.

In a **combinatorial network** the output condition y (*true* or *false*) depends on a number of input conditions u that have to be satisfied simultaneously. The system has *no memory*, i.e.

$$y(t) - f[u(t)] \tag{3.83}$$

This *network* can be used to check if a manual control action is allowed. During a manual start-up of a complex process the computer may check all logical conditions that have to be satisfied before a certain actuator is turned *on* or *off*. In a **sequencing network** the output depends on both the present values and earlier states or inputs. A sequencing system has a memory function and the concept of **state** is used.

In a simple sequencing network the execution proceeds as:

Step 1 → Step 2 → ... → Step n

When the transition from one step to the next is determined by logical conditions the sequence is called **asynchronous**. In a **synchronous** sequence the state transition is triggered by a clock pulse. In industrial applications the asynchronous type is more common.

Some concepts can be illustrated by a simple configuration in discrete manufacturing. Consider two machines M_1 and M_2 in a **transfer line** (Figure 3.24). The buffer B between them can contain 0 or 1 part.

Each machine is defined to be in one of two discrete states, *operational* or *broken*. During a short time interval Δt there is a probability ($f_1\Delta t$ or $f_2\Delta t$) that one of the two machines will fail. There is another probability ($r_1\Delta t$ or $r_2\Delta t$) that a broken machine will be repaired. The buffer has two discrete states, *full* or *empty*. An *empty* buffer becomes *full* if M_1 produces some parts (with a production rate $\mu_1\Delta t$). A *full* buffer can only

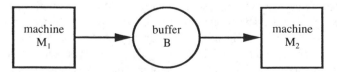

Figure 3.24 Two machines with an intermediate buffer.

become *empty* if M_2 produces, i.e. consumes the stored parts (with a production rate μ_2 Δt). Since the states are discrete we can describe them with binary digits (0 or 1),

1 — machine operational	0 — machine broken
1 — buffer full	0 — buffer empty

The system is defined by eight states (Table 3.1).

Table 3.1   Definition of discrete
states in the transfer line

State	B	M_1	M_2
s_{000}	0	0	0
s_{001}	0	0	1
s_{010}	0	1	0
s_{011}	0	1	1
s_{100}	1	0	0
s_{101}	1	0	1
s_{110}	1	1	0
s_{111}	1	1	1

Machine M_1 can finish its operation only if the buffer is *empty*; otherwise it is said to be *blocked*. Machine M_2 can produce only if the buffer contains a part; otherwise it has to wait (it is said to be *starved*). Thus, each machine can be idle due to blocking or starving.

The operation of the machines can be illustrated by a state graph or an **automaton** (Figure 3.25) described by the eight states $s_{000}, \ldots, s_{111}$. The system can be in only one state at a time. The transfer rate between the states is defined by the probabilities that a machine fails or is repaired or will complete its operation within a specified time.

Assume that the system is in state s_{101}. In this state M_1 is idle and M_2 can produce, since the buffer is full. The system can leave the state along three routes.

- If M_2 fails: go to state s_{100}
- If M_1 becomes repaired: go to state s_{111}
- As M_2 produces: go to state s_{001}

By modelling the machine system like this it is possible to simulate how the states will be changed. In fact, the model can estimate the probability that a certain state will be reached. Repair rates or production rates will influence how well the system can produce. Naturally, an optimal system will not let the machines be blocked or starved and will have a small failure rate. State graph modelling of this type is a tool for systematic analysis of such systems and is commonly used in the design of industrial applications.

A process where the transition depends only on the current state and input signal is called a **Markov process**. In that sense it resembles the differential equations in Section 3.4. Note, however, the fundamental difference between the time-discrete dynamic systems

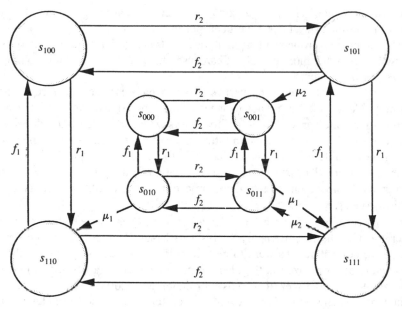

Figure 3.25   A state graph of the transfer line.

and the sequential systems. In the former case, each state variable has a continuously varying amplitude (such as temperature and pressure) and is defined in discrete times. A sequential system described as a Markov process 'jumps' between a finite number of distinct states.

The state graph can be considered to be an information structure of the automation process. It does not say anything about the implementation, and in Chapter 8 we will show how switching elements can be used to realize such systems.

3.8 SUMMARY

Models are essential descriptions of the physical process that is to be controlled. In this chapter we have considered four classes of mathematical models:

- Continuous dynamic systems described by linear or non-linear differential equations
- Time-discrete (sampled) dynamic systems represented by linear or non-linear difference equations
- Discrete event, or sequential systems, described by finite states
- Systems with uncertainty expressed either by statistical or by linguistic methods

It is necessary to formulate any dynamic system in time-discrete form to make it suitable for computer control. The systems can be represented in either state/space form (internal representation) or in an input/output form (external representation). Which one is selected has to do with the controller structure that will be designed. Linear models are very attractive from an analysis point of view, but in computer control we are not restricted to linearity. Examples of non-linear systems have been presented.

For linear dynamic systems many analytical tools are available for the analysis. Non-linear systems are much more complex to analyse, and simulators are important tools for the analysis of the systems.

Two important structural properties have been described, controllability and observability. Controllability relates to whether the input signals are sufficient to reach anywhere in the state/space. Observability marks whether there is an adequate set of measurement variables by which to find the values of the internal state of the system. If a process is observable it is possible to indirectly measure the state variables that are not supplied with sensors. This process is called estimation. If the process or the disturbances are corrupted by noise, then the estimation procedure needs to contain a statistical model of the disturbances.

Many industrial processes are controlled without any quantitative mathematical models. The operator often has a sufficiently sophisticated mental model for the control of the process. If the control actions were to be automated in a process computer, the control schemes had to be formalized in linguistic terms. Fuzzy algebra is shown to be a useful methodology to describe mental models with inherent uncertainty.

Sequential systems and discrete events are extremely common in industrial process control. Some systems are controlled by relatively simple combinatorial networks, while others may be extremely complex. We have seen a few ways to systematically model some classes of sequential systems.

FURTHER READING

There are many books on different aspects of modelling dynamic systems. Luenberger (1979) presents a very readable introduction to modelling and dynamic systems.

The areas of *process identification* and *parameter estimation* have been mentioned briefly, for which there are several good textbooks. Söderström and Stoica (1989) and Ljung (1987) are excellent introductions to the subject, while Ljung and Söderström (1983) provide a comprehensive text on recursive estimation.

Robot kinematics and dynamics is an essential part of *mechanical* systems and Craig (1989) gives quite a short yet excellent introduction on industrial robots. Spong and Vidyasagar (1989) and Asada and Slotine (1986) present more comprehensive treatments of robot control.

Electrical systems are covered in numerous textbooks, and of these Fitzgerald, Kingsley and Umans (1983) is a standard text on electrical machinery. The dynamic aspects are emphasized more in Leonhard (1985) while Elgerd (1982) offers a good introduction to electric power systems.

Buckley (1964) is a classical book on *chemical process* modelling and control. An appreciation of this subject may also be gained through Luyben (1973) who presents several illustrative applications of chemical modelling while Seborg, Edgar and Mellichamp (1989) give quite a comprehensive description of chemical process control.

Some aspects on modelling of *biological wastewater treatment* are found in Henze *et al.* (1987) and Olsson (1985). Cannon (1967) discusses *heat transfer processes* in detail.

There are numerous good books on *control theory*. Franklin, Powell and Emami-Naemi (1986) give an excellent introduction to continuous control systems. A recent book by Kuo (1991) contains large introductory sections to modelling of physical systems and the necessary mathematics for their analysis. Time-discrete control systems, both deterministic and stochastic, are thoroughly covered by Åström and Wittenmark (1990). Non-linear systems control is described well by Asada and Slotine (1986).

Numerous journals and papers deal with *fuzzy logic* and its applications. A good overview of fuzzy set theory is provided by Yager (1987) and by Klir and Folger (1988). The relation between neural networks and fuzzy systems is well documented in Kosko (1990).

Sequential systems are described by Fletcher (1980) and modelling in the manufacturing industry is treated in Desrochers (1990), in which Markov chains and Markov processes are introduced as modelling tools. Some of the *historical notes* on Watt and Maxwell are found in Fuller (1976) and the development of the feedback amplifier is described in Bode (1964). The Nyqvist criterion is nowadays a minor part of introductory control courses, but was considered a revolution in control system design at the time of introduction. In fact, it was considered so important for the military that it was kept secret in the United States until after World War II. Some of the early work on state/space theory and optimization are reported in Bellman and Kalaba (1964).

Finally Kreutzer (1986) presents a general view on modelling and shows the application of different modelling styles in simulation.

The *simulation* packages mentioned in this chapter are marketed by the following:

Simnon   SSPA Maritime Consulting AB, S-40022 Göteborg, Sweden
Matlab, Simulab   MathWorks, Inc., South Natick, MA 01760, USA
Easy5   Boeing Computer Services, Seattle, WA 98124, USA
ACSL   Mitchell and Gauthier Assoc., Concord, MA 01742, USA

4

Physical Process Input and Output

Aim: To acquire a basic understanding of the most important issues to connect a physical process to a computer, including primary sensor and actuator technology, signal transmission, conditioning and modification

This chapter is dedicated to the interface between the physical process and computer inputs and outputs. In order to physically connect the computer to the process many technologies have to be applied. Even if it is impossible to become an expert in all the related fields it is important that the process computer engineer is aware of the different interfacing problems in order to communicate adequately with the specialists.

The chapter begins with an introduction to the components of the process−computer interface. Measurement and sensor technology is of vital importance for control. Sensors have to accurately represent the physical variables under both steady-state and transient conditions. Instrumentation technology is a vast topic in itself and a few examples are presented here to give the flavour of the measurement problems and possibilities. General properties to characterize sensors are described in Section 4.2. A brief survey of different analog sensor types is made in Section 4.3. Measurement devices that transmit only on/off signals, binary sensors, are discussed in Section 4.4.

The computer output will be converted to mechanical movement or power so that a control action can take place in the process. This is achieved with actuators — again there are many specialist areas of which electric drive systems are important. Electrical drive systems as actuators for electrical to mechanical energy conversion are treated in Section 4.5, and other actuators — for on/off control — are described in Section 4.6.

The transmission of signals between the computer, the measuring system and the actuators is another large subject. Signals have to be conditioned to fit both the measurement device and the computer input. Similarly the actuator must receive a signal that corresponds accurately to the computer output. Many different kinds of electrical disturbances can distort the signal and precautions have to be made for these. The disturbance pattern often determines whether the transmission will be by a voltage, current or optical signal. These questions are discussed in Section 4.7.

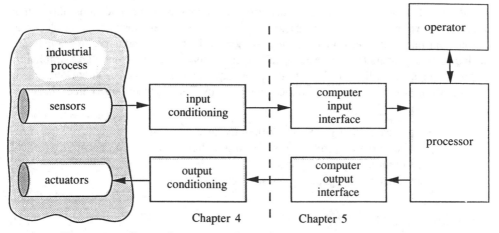

Figure 4.1   General structure of the process computer input/output

4.1 COMPONENTS OF THE PROCESS INTERFACE

The general structure of the process computer input/output is shown in Figure 4.1. There are many types of sensors, actuators and input/output conditioning circuits, but the basic structure need not be more complicated than shown. The conceptual simplicity does not mean that practical realization is also simple. As Murphy would put it, 'if everything seems to be going well, you have obviously overlooked something'.

4.1.1 Measurements

For most physical systems a wide variety of measurement technologies exist, each producing its own characteristic signal as a function of the physical variable being measured. The measuring device consists of two parts (Figure 4.2), the **sensor** and the **transducer** — these terms are sometimes erroneously used interchangeably.

The signal being measured is felt as the 'response of the sensor element' and is converted by the transducer into the transmitted quantity. The output of a measuring device

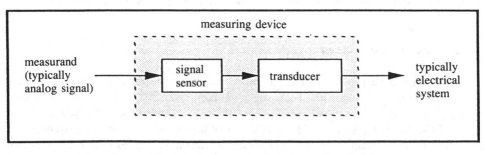

Figure 4.2   Block diagram of a measuring device.

is interpreted as the 'response of the transducer'. In control system applications, this output is typically (and preferably) an electrical signal. The sensors described here give electrical output, but pneumatic sensors are quite common in the process industry. In comparison with electrical sensors they are usually of low cost, small size, robustness and simplicity. In applications with inflammable material, pneumatic sensors provide the safety demanded.

In this book sensors will refer to the complete sensor—transducer unit. Three categories of sensors can be recognized:

- **analog** sensors that generate continuous analog signals
- **digital** sensors that produce a pulse train or binary word from a sensed physical pulse duration or pulse interval
- **binary** sensors that produce on/off digital signal levels

4.1.2 Actuators

The actuator is the device that mechanically drives a process and converts electrical energy to the required control output. In an industrial robot the motors of the mechanical joints are the actuators. In a chemical process, the final control elements may be servo-controlled valves.

The requirements for an actuator — such as power, motion resolution, repeatability and operating bandwidth — can differ significantly, depending on the particular system. A proper selection of actuators is of crucial importance.

An actuator system can be divided into two elements, a *transducer* and an *amplifier*. The transducer of an actuator performs the electromechanical power conversion, such as an electric motor. The amplifier portion of the actuator system amplifies the low level control signal from the computer output interface to a level high enough to drive the transducer. In some cases both the transducer and the amplifier portion of the actuator system are mounted in a single unit. Thus, some of the final control elements are control systems themselves and the computer output is the reference value (setpoint) to the final control element.

Compressed air is often used to set valves as final control elements. When large forces are needed hydraulic actuators are used. The electrical signal from the computer then has to be converted to compressed air or oil pressure. The physical control variable can also be an on/off value produced by electromechanical relays or electronic switches.

4.1.3 Bandwidth and Noise

Two fundamental issues, bandwidth and noise, influence the interfacing of a computer to a physical process. Bandwidth is an important parameter in a variety of contexts such as signal transmission, computer buses and feedback control, where it means different things. In signal transmission and feedback control bandwidth is the range of frequencies amplified above a certain threshold. In computer bus technology bandwidth is synonymous with data carrying capacity (see also Chapter 6). Here bandwidth is defined as the useful frequency range of a sensor or an actuator. Only signals with significant frequency content limited to the bandwidth are measured. This means that the sensor is sufficiently fast to faithfully measure the signal without sensor dynamics corrupting the signal. Likewise, an actuator has to have an adequate bandwidth in order to realize the desired control signal. The larger the bandwidth, the faster the sensor or actuator response will be. Unfortunately, the larger the bandwidth, the more susceptible the device will be to unwanted high frequency disturbances.

Any measurement signal will be corrupted by **noise**, and part of the signal transmission problem is to reduce the influence of the noise. The sources of noise have to be eliminated or minimized. Noise that corrupts the information content in signals or messages is a problem not only in a process interface but appears in all kinds of communication (see Chapter 10). Modelling of noise was discussed in Section 3.6.1. Controllers can be designed to deal with noise in a systematic way (Chapter 9).

4.1.4 Signal Transmission

Analog signals generated by the measuring device often must be processed in some way before being sampled by the computer. A voltage signal has to be amplified in order to match the sensor voltage range to the computer interface range. Moreover, the voltage level has to be shifted to align the minimum sensor output voltage with the minimum voltage for the input interface. This is called **signal conditioning**.

The transmission of analog signals presents special problems due to electrical disturbances. Noise that corrupts the signal may be generated by resistance, inductance or capacitance couplings to the environment. A possible solution is the conversion to a pulse rate or pulse duration, proportional to the voltage level. This is useful when noise is influencing the same frequency band as the original signal. Pulse trains can be conducted either electrically or optically (i.e. by using optical fibre technology).

The analog signals are converted to numerical (digital) form in the input interface. The computer processes the input data to generate a control signal or other output signal. Usually this digital value has to be converted to analog form in the computer output interface. The signal level has to provide the right amplitude to drive an actuator.

4.2 SENSOR PROPERTIES

A sensor should represent a physical property as fast and as accurately as possible. Often the choice of a sensor is based on its reliability and maintenance properties. Yet sensor accuracy is a primary factor. For process instrumentation operating stability and reproducibility are crucial factors, so that process product quality will not vary too much.

Most of the sensor characteristics provided by manufacturers are in the form of **static** parameters. Such parameters do not describe how quickly and accurately the sensor can measure a rapidly varying signal. The properties that describe the sensor properties for varying input signals are called **dynamic** characteristics and are essential for control applications. A sensor should ideally react immediately to a change in the observed process and present the new value at its output. In practice, all sensors need some time to adjust and indicate a new state. Obviously the response time of a sensor has to be sufficiently small in order to adequately represent the true variations of the interesting signals. Do not forget, however, the bandwidth compromise between sensor speed and sensitivity to noise (see Section 4.1.3).

4.2.1 Accuracy and Precision

Accuracy (or rather inaccuracy) marks the difference between the measured and the real value, and can be assigned to a particular reading or to a sensor. Note that **resolution** is something else; it is the smallest change in measurement value that can be indicated. This is often better than the accuracy which has to be observed in sales material. It is important to note that sensor accuracy depends not only on the physical hardware but also on the measurement environment. The **measurement error** is defined as the difference between the measured value and the true value. Of course the true value is not known, so in a given situation we can only make estimates for accuracy by providing further analysis or precision measurements.

Errors can be classified as deterministic (or systematic) and random (or stochastic). **Deterministic errors** may be caused by malfunctions of the sensor or by the measurement procedure itself. Consequently, they are repeated at every measurement. A typical systematic error is reading **offset** or **bias**. In principle, systematic errors can be corrected by calibration. **Random errors** may be caused by several parameters such as temperature, humidity or electrical noise, and cause a variability of the measurement value. Some of them are known and can be compensated for, while others are unknown. Often the disturbance influence can be quantified by mean errors, standard deviations or variances.

The difference between systematic and random uncertainty is illustrated by Figure 4.3. The centre of each target represents the true measurement value and each dot is a measurement. The measurements are primarily characterized by their bias and variance. Both bias and variance have to be small in order to obtain good accuracy. Figures 4.3(a) and (c) represent biased measurements. The variance (or standard deviation) between individual measurements is a measure of **precision**. A sensor with good reproducibility (or small random errors) has good precision but not necessarily good accuracy, since a bias may distort the result. The measurements in Figures 4.3(c) and (d) are said to be precise, while Figure 4.3(d) is also accurate.

4.2.2 Dynamic Characteristics

There are several parameters that can characterize sensor dynamics but it is uncommon to receive more sophisticated dynamic information from the vendors. The sensor can be tested by a step response where the sensor output is recorded for a sudden change of the

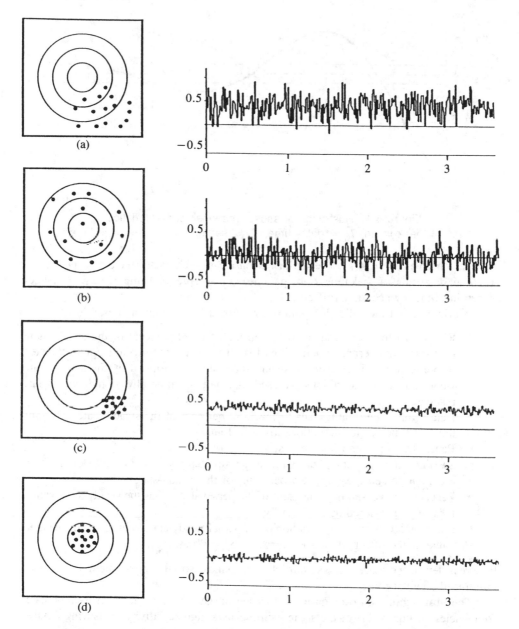

Figure 4.3   Illustration of bias, precision and accuracy. In the target symbol the centre is the true value and each dot is a measurement. In the signal display the true value is a constant marked with the straight line. Precision is seen to be equivalent to the signal variance, i.e. low variance = high precision. (a) Large bias + low precision = low accuracy; (b) small bias + low precision = low accuracy; (c) large bias + high precision = low accuracy; (d) small bias + high precision = high accuracy.

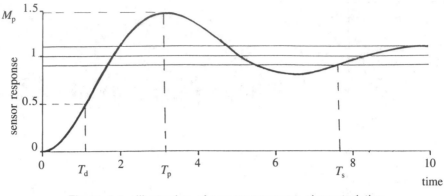

Figure 4.4   Illustration of sensor response characteristics.
(M_p = peak value; T_d = delay time; T_s = setting time; T_p = peak time.)

physical variable (Figure 4.4). The sensor response can be characterized by its speed (e.g. rise time, delay time or peak time) or by parameters that represent damping (e.g. overshoot or settling time) or accuracy (offset).

In principle, all the following parameters should be as small as possible:

- **Rise time**   the time it takes to pass from 10 to 90 per cent of the steady-state response. An alternative definition of rise time is the reciprocal slope of the step response at 50 per cent of the steady-state value, multiplied by the steady-state value. Still other definitions are used. A small rise time always indicates a fast response.
- **Delay time**   the time it takes to reach 50 per cent of the steady-state value for the first time. Other definitions are also found.
- **Peak time**   the time for the first peak or overshoot.
- **Settling time**   the time it takes for the sensor step response to settle down within a certain percentage (e.g. ±5 per cent) of the steady-state value.
- **Percentage overshoot**   a measure of the peak value minus the steady-state value. It is often measured as a percentage.
- **Steady-state error**   the deviation of the actual steady-state value from the desired value or the offset. It can be corrected by calibration.

In a real device there are always conflicting demands, so all the parameters cannot be minimized simultaneously.

The **static gain** (or **d.c. gain**) is the sensor gain at very low (or steady-state) frequencies. A high value of the static gain means a high sensitivity measuring device.

4.2.3 Static Characteristics

The static accuracy of a sensor indicates how much the sensor signal correctly represents the measurand *after* a transient period. Some important static parameters are sensitivity, resolution, linearity, zero drift and full scale drift, range, repeatability and reproducibility.

Sensitivity of a sensor is measured by the magnitude of the output signal corresponding to a unit input of the measured variable. More elaborate definitions are found in sophisticated instruments. **Resolution** is the smallest change in a signal that can be detected and accurately indicated by a sensor. **Linearity** is not defined analytically, but is determined by a calibration curve. The static calibration curve shows the output versus the input amplitudes under static conditions. Its closeness to a straight line determines the degree of linearity.

Drift is defined as the deviation from the null reading of the sensor when the measured value is kept steady for a long period. At a zero drift test the measured value is kept at zero or a value that corresponds to zero reading, while at a full scale drift the measured value is kept at its full scale value. The drift of a sensor may be caused by instability in amplifiers, ambient changes (e.g. changes in temperature, pressure, humidity and vibration), changes in power supply or parameter changes in the sensor itself (aging, wear-out, non-linearities, etc.).

The **range** of a sensor is determined by the allowed lower and upper limits of its input or output so as to maintain a required level of measurement accuracy. **Repeatability** is characterized as the deviation between measurements in a sequence when the object under test is the same and approaches its value from the same direction every time. The measurements have to be made under such a short period that long term drift is not noticeable. Repeatability is often given as a percentage of the sensor range.

Reproducibility resembles repeatability but implies a long time lapse between measurements. The sensor has to be in operation between the measurements, but calibration must be performed. Reproducibility is denoted as a percentage of the sensor range per unit of time (e.g. per month).

4.2.4 Non-linearities

Many sensors have a non-linear behaviour. Typically some sensors exhibit saturation so the output is limited even if the input increases. Examples include:

- Non-linear deformation in mechanical springs
- Coulomb friction
- Magnetic saturation in transformer parts
- Flow rate measurements (in open channel measurements there is a non-linear relation between the measured level h and the flow rate F such as $F = (constant)h^\alpha$
- A thermistor has a non-linear resistance change to temperature, $R = R_0 e^{\beta(1/T - 1/T_0)}$, where T is the temperature in Kelvin and R_0, T_0 and β are constants

Particular problems appear for backlash in gears and other loose components, and for hysteresis in magnetic circuits. The output signal is a multivalued function of the input and depends on the direction of motion.

4.2.5 Impedance Characteristics

When two or more components are interconnected, the behaviour of individual components in the system can deviate significantly from their behaviour when each component operates separately. For example, a heavy accelerometer can introduce an additional dynamic load that will modify the actual acceleration being measured. Similarly, a voltmeter can modify the currents and voltages in a circuit, and a thermocouple junction can modify the temperature under measure. This is called a **loading effect**. Loading errors can exceed other types of measurement errors and have to be avoided when different sensors and signal transmission devices are connected.

The term **impedance** has fundamental importance. The mechanical analogy of electrical impedance can be defined if *velocity* is interpreted as *voltage* and *force* as *current*. In electrical systems, a device with high **input impedance** draws less current for a given voltage. Consequently the power consumption is low. A low input impedance device requires a higher current for a given voltage. Therefore it extracts more power from the preceding device which can explain the loading error. A device with a high **output impedance** generates a low output signal and will require conditioning to step up the signal level. A device with a high input impedance and low output impedance is used to condition the signals. This is called **impedance matching** and has to be done properly at each interface.

Consider a standard electrical device with two ports. The output impedance Z_o is defined as the ratio of the open-circuit (no-load) voltage at the output port (no current floating at the output port) to the short-circuit current at the output port. The input impedance Z_i is defined as the ratio of the rated input voltage to the corresponding current through the input terminals while the output terminals are kept open (Figure 4.5).

When a load is connected at the output port, the voltage across the load will be different from v_o, due to the presence of a current through the output impedance. It is possible to relate the input and output voltages under open-circuit conditions by a gain G:

$$v_o = Gv_i \tag{4.1}$$

where v_i is the input voltage.

4.2.6 Choosing the Proper Input/Output Impedances

Consider two electrical two-port devices connected in cascade (Figure 4.6). Defining the gains G_1 and G_2, respectively, between the input and output voltages, it is straightforward

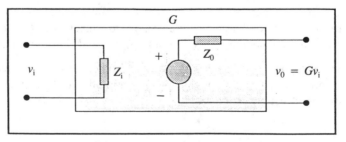

Figure 4.5   Schematic representation of input and output impedances. The gain is G, and v_o is the open-circuit voltage.

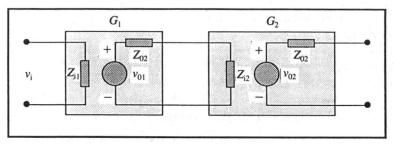

Figure 4.6   Impedance matching of cascaded devices.

to verify the following relations:

$$v_{o1} = G_1 v_i \tag{4.2}$$

$$v_{i2} = \frac{Z_{i2}}{Z_{o1} + Z_{i2}} v_{o1} \tag{4.3}$$

$$v_{o2} = G_2 v_{i2} \tag{4.4}$$

The combination of the gains gives the overall gain:

$$v_{o2} = G_1 G_2 \frac{Z_{i2}}{Z_{o1} + Z_{i2}} v_i \tag{4.5}$$

This expression can come close to the gain $G_1 G_2$ if the output impedance of device 1 is much smaller than the input impedance of device 2. In other words, the frequency characteristics of the device with the two amplifiers in cascade will not be significantly distorted if the proper impedance matching is done, i.e. the output impedance of the first device is much smaller than the input impedance of the second. In order to obtain a proper impedance matching some amplifier circuitry has to be introduced between the sensor output and the data acquisition unit input. Such impedance matching filters are based on **operational amplifiers** (see Section 4.6).

4.3 ANALOG SENSORS

The sensor–transducer unit connects to an input conditioning circuitry (see Figure 4.2) consisting of any electronics between the sensor and the computer interface. The dividing line between 'sensor electronics' and 'signal conditional electronics' is sometimes open to individual interpretation.

Most sensor–transducer devices used in feedback control applications are components that generate analog signals. Common classes of physical values that are measured in computer-controlled systems are the following:

- Electrical and magnetic properties
- Motion

- Force, torque, pressure and touch
- Temperature
- Levels
- Flow rates
- Density, viscosity and consistency
- Concentrations (gas, liquid, dissolved, suspended)
- Chemical or biochemical activity measurements

Here we will present a selection of analog sensor devices commonly used in control systems. Electrical measurements — including current, voltage, resistance, magnetic field, radiation and power — are the cornerstones of measurement technology. Even if voltage and current are analog in themselves, they are mostly measured using digital electronics. Then an analog–digital converter (see Section 5.3.2) is built into the device itself.

4.3.1 Motion Sensors

By **motion** we mean the four kinematic variables:

- Displacement (including position, distance, proximity, size)
- Velocity (including angular velocity)
- Acceleration
- Shock or impact

Each variable is the time derivative of the preceding one. In theory, it should be possible to measure only one of them, and then obtain the others by differentiation or integration. In practice, however, many factors make this approach highly unsuitable due to the nature of the signal (steady, transient, etc.), the frequency content, noise corrupting the signal, and available processing capabilities.

Motion measurements are extremely common in many applications involving mechanical equipment such as servo systems, robots, electrical drive systems or other manipulators. Displacement measurements are used for positioning valves in process applications. Plate thickness is continuously monitored by gauge control systems in steel rolling mills, while strain gauges are devices that measure strains, stresses and forces, but can be adapted to measure displacements.

Motion sensors include the following types of devices:

- **Potentiometers**   resistively coupled displacement sensors
- **Variable inductance sensors**   employ the principle of electromagnetic induction, e.g. differential transformers, resolvers and synchro-transformers
- **Variable capacitance sensors**   used to measure small displacements, rotations and fluid levels
- **Piezoelectric sensors**   based on the piezoelectric effect and constructed to measure pressure and strain as well as acceleration, velocity, force and torque. Piezoelectric materials deform when subjected to potential difference
- **Laser sensors**   useful for very accurate measurements of small displacements

- **Ultrasonic sensors** used in many applications to measure distances, including medical systems, ranging systems for autofocusing capability, level and speed sensing

Example 4.1 Resolver

Resolvers are used in accurate servo and robot systems to measure angular displacement. Their signals can be differentiated to obtain the velocity. A resolver is a sensor based on the principle of mutual induction. The rotor is connected with the rotating object and contains a primary coil (Figure 4.7) supplied by an alternating current (a.c.) source voltage v_{ref}. The stator consists of two windings separated by 90°, with induced voltages of:

$$v_{01} = K v_{ref} \sin \theta \tag{4.6a}$$

$$v_{02} = K v_{ref} \cos \theta \tag{4.6b}$$

respectively, where θ is the angular position of the rotor.

The resolver can be said to have modulated signals, i.e. v_{ref} is modulated by the angle θ. By using only one signal, angles between 0° and 90° can be measured. By using both signals it is possible to measure between 0° and 360° without any ambiguity.

The resolver output is a trigonometric function of the angle. This non-linearity is not always a drawback. For example, in robotics trigonometric functions of angles are needed for torque control so the resolver output gives a favorable signal that reduces the computer load.

A resolver gives fine resolution and high accuracy. It has a low impedance output signal (high signal level) and is available in reliable constructions. The bandwidth depends on the frequency of the supplied voltage. The brushes to the rotor cause different kinds of problems, for instance, wear-out, noise and mechanical load.

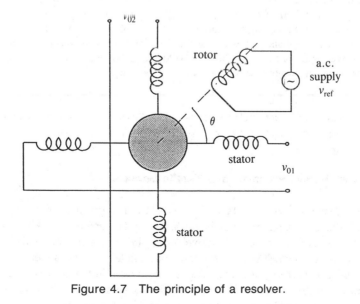

Figure 4.7 The principle of a resolver.

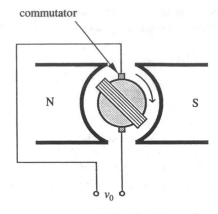

Figure 4.8   The principle of a d.c. tachometer.

Both angular and translatory velocities are of fundamental interest in drive systems and robot manipulators. An interesting application of acceleration and jerk measurements is found in active suspension control in vehicles.

Example 4.2 Tachometer

As an example of angular velocity measurement, consider a tachometer. Figure 4.8 illustrates the principle of a permanent magnet direct current (d.c.) tachometer. The permanent magnet generates a steady and uniform magnetic field. Relative motion between the field and an electrical conductor induces a voltage proportional to the rotating speed of the conductor. The rotor is connected directly to the rotating object. The output voltage is generated by the motion itself and is picked up via the commutator device, usually a pair of low resistance carbon brushes. The inductance gives the tachometer a certain time constant so the induced voltage is not exactly proportional to the velocity if the tachometer is to measure transient velocities. Therefore it cannot measure fast transients accurately.

Accelerometers are very common in the on-line control of machinery and in machine protection systems to detect faults and monitor the equipment condition.

4.3.2 Torque, Force, Pressure and Tactile Sensors

Many types of **force/torque** sensors are based on strain gauge measurements. The change of electrical resistance in material that is mechanically deformed is used in resistance-type strain gauges. The sensitivity of the relative change in resistance with strain depends very much on the material used. It is $1-2$ orders of magnitude lower for a metallic strain gauge than for a semiconductor strain gauge. The strain element in the latter is made of a single crystal of piezoresistive material. The resistivity in a semiconductor strain gauge is also higher, providing reduced power consumption and heat generation.

Torque and force sensing is useful in many applications including control of fine motions (such as manipulation and gripping in robotics) and measuring transmitted mechanical power in rotating devices. Instead of measuring strain, the torque can also be measured by actual deflection. If the twist angle of an axis is measured by an angular displacement sensor then the torque can be determined. In Example 3.5 we saw that the torque of a d.c. motor is proportional to the rotor current.

Pressure can be measured by mechanical deformation, for example, by bending a tube, or the deformation of a membrane that is transmitted to an electrical signal. The deformation can be measured by a **differential transformer** or strain gauge. The differential transformer is a sensor used to indicate position changes. It basically consists of a ferromagnetic core moving inside two coils of a transformer. The differential transformer is built in such a way, that when the core is in the mid position, the transformer output is a zero voltage, while any displacement brings a voltage change in the output. A couple of principles are shown in Figure 4.9. By measuring the pressure at the bottom of a vessel the liquid height can be determined.

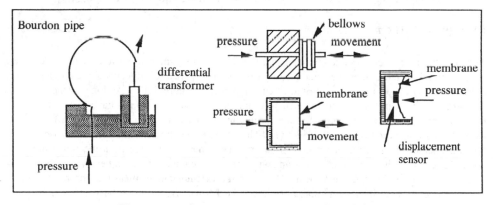

Figure 4.9 Some pressure sensor principles.

Tactile sensing is usually interpreted as touch sensing, but tactile sensing is different from a simple touch where very few discrete force measurements are made. In tactile sensing, a force 'distribution' is measured, using a closely spaced array of force sensors and usually exploiting the skin-like properties of the sensor array. Tactile sensing is used in grasping and object identification in robotics.

4.3.3 Temperature Sensors

The temperature sensitivity of many components is not necessarily only a nuisance, it can also be used for temperature measurements. Often the temperature dependence is non-linear and not always reproducible. There are three common types of sensors:

- thermoelements
- resistance temperature detectors (RTDs)
- thermistors

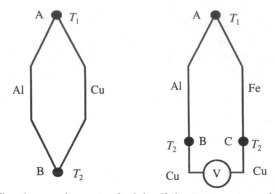

Figure 4.10   The thermoelement principle. If the temperatures in A and B are different there is a current in the circuit. The right diagram shows a practical circuit for measuring the generated voltage. Point A is called the hot junction and B and C the cold junction. Note that B and C should have the same temperature!

Example 4.3 Thermoelement

The first thermoelement was constructed in 1887 by the Frenchman le Chatelier. Consider two contact points A and B that are joined by two wires of two different metals (for example, aluminium and copper) so that a closed circuit is formed (Figure 4.10). As long as the temperatures in A and B are the same there is no current, but if A and B have different temperatures an electric current is formed in the circuit. This is called the Seebeck effect, having been discovered by Seebeck in 1821. This 'thermo-electromotoric force' increases when the temperature difference becomes larger. The generated voltage is in the range of about 10 μV to a few millivolts so the analog circuitry is comparatively expensive. Cabling and signal transmission become crucial. Note that the thermoelement measures the temperature *difference* so the cold junction temperature has to be known. Different metal combinations are used for different temperature ranges.

Thermoelements are inexpensive, have a small heat capacity, are robust and have a wide temperature range. The International Electrotechnical Commission has defined some standard types of thermoelements (norm IEC 584-1). The elements are typically called R, S, B, K, J, E, T according to the different temperature ranges.

The resistance of metals has a positive temperature coefficient. It is this feature that is used in **resistance temperature detectors (RTDs)**.

Example 4.4 Resistance temperature detectors

RTD devices are often made of platinum wire. The resistance R is an almost linear function of temperature T (°C) from a reference temperature $T_0 = 0$ °C. Expressing the ratio of the resistance R at temperature T to the resistance R_0 at the reference temperature T_0, we have:

$$R/R_0 = 1 + aT + bT^2 + \dots \tag{4.7}$$

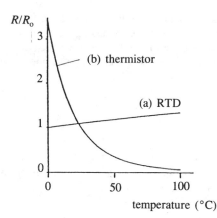

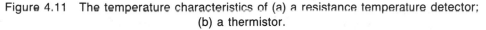

Figure 4.11 The temperature characteristics of (a) a resistance temperature detector; (b) a thermistor.

where a is the resistance temperature coefficient and b a positive or negative constant (Figure 4.11(a)). For platinum, typical parameter values are $a = 0.004$ per °C and $b = 0.59 \times 10^{-6}$ °C^{-2}.

The RTD is available for a number of standardized resistances. The dominating type has a resistance of 100 Ω at the reference temperature and is so common that it has got its own name **Pt-100**. The sensitivity is very small for the RTD sensor and any current i used to sense the resistance change will heat up the sensor, thus changing the measurement by an amount proportional to i^2. The resistance output is often measured in a bridge configuration.

Example 4.5 Thermistor

A thermistor ('thermally sensitive resistor') is made of semiconductor material having negative temperature coefficient and large sensitivity. As mentioned in Section 4.2, its resistance depends on temperature in a non-linear fashion:

$$R = R_0 e^{\beta(1/T - 1/T_0)} \tag{4.8}$$

where T is the temperature in Kelvin, R_0 is the resistance at a reference temperature T_0 (usually 298 K) and β is a constant (typically 3000–5000 K). The slope of the R–T curve (Figure 4.11(b)) is the temperature coefficient a and is a function of the temperature:

$$a = \frac{1}{(R/R_0)} \frac{d(R/R_0)}{dT} = \frac{-\beta}{T^2} \tag{4.9}$$

The value of coefficient a is typically between -0.03 and -0.06 per °C at 25 °C.

Due to the resistance of the thermistor, a current passing through it will generate heat. The power dissipated in the thermistor at 25 °C is typically of the order 0.002 mW. With a dissipation constant of about 1 mW °C^{-1}, the device temperature will rise by 1 °C (in air) for each milliwatt of power dissipated.

The thermistor is not an accurate temperature sensor. Due to its sensitivity, however, it is used to measure small temperature changes. It is also quite robust, both mechanically and electrically. The non-linear output voltage of the thermistor has to be linearized to a linear function of temperature. This can be done either by analog electronics or with software, using a calibration table or an inverted function of the thermistor characteristic function. Better electronic couplings have achieved good linearity and the thermistor has useful versions up to 500–600 °C.

4.3.4 Flow Rate Sensors

Flow rate measurements are of vital importance in many process industries, but it is still difficult to measure flow rates with great accuracy. There are three common principles for **volumetric** flow rate measurements,

- Difference pressure
- Turbine sensors
- Magnetic flow sensors

By making a constriction in a pipe and measuring the pressure difference it is possible to obtain the fluid velocity. At the constriction the velocity will increase and a pressure drop $\Delta p = p_1 - p_2$ appears (Figure 4.12). The flow rate is proportional to $\sqrt{\Delta p}$, where the proportionality constant depends on the constriction geometry. The pressure difference can be translated to motion which is monitored by a differential transformer.

There are friction losses in the constriction so the pressure losses can be significant, but the hydrodynamic design of the constriction can be made more streamlined so that the pressure losses are minimized. This is done in the **Venturi** pipe which is commonly used to measure large flow rates in open channels (see Figure 4.12).

A liquid flow can be measured with a turbine. Its rotation is transmitted by mechanical or magnetic means to a counter or some electronic equipment. A turbine can only be used for clear liquids because the mechanical movement will be distorted by particles in the fluid.

Magnetic flow sensors are built on the induction principle which states that a voltage is induced in a conductor that is moved in a magnetic flux. In magnetic flow measurements, the conductor is the flowing electrically conductive liquid (Figure 4.13). It can be shown that the voltage generated between two opposite points on the inner pipe wall

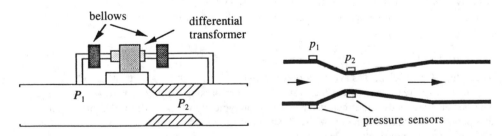

Figure 4.12   Flow measurement by difference pressure. A general constriction and a Venturi pipe principle.

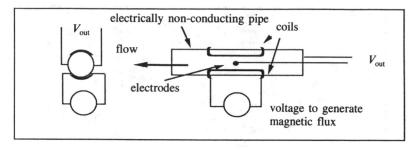

Figure 4.13 The magnetic flow rate sensor principle.

is largely independent of the velocity profile. The magnetic field is generated by two coils which are energized by either a.c. power line voltage or a pulsed d.c. voltage. The induced voltage is received by two insulated electrodes and becomes proportional to the liquid flow velocity. Since the flow cross-section area is assumed *constant* the voltage becomes proportional to the flow rate.

In many applications it is important to know the **mass** flow (measured in kg s^{-1}) rather than the volumetric flow rate (in m^3 s^{-1}). If the density of an incompressible fluid is known then the mass flow rate can be calculated directly from the volumetric flow rate. In practice, however, it often happens that the density is unknown for both compressible and for incompressible fluids. Ever-increasing demands for product quality means more and more motivation to measure mass flows directly in many processes, and even quite expensive sensors may be acceptable.

When measuring the mass flow of liquids in pipes it is very difficult to rapidly weigh the liquid. For this reason, mass flow in many cases is calculated from the flow velocity, density or pressure, and temperature. Many attempts to measure mass flow by measuring force and acceleration have failed. One principle, however, has been accepted for industrial applications, mass flow measurement using the gyrostatic principle and the influence of the **Coriolis acceleration** or the **Coriolis force**.

Example 4.6 Mass flow measurement based on the Coriolis principle

For all systems which rotate about an axis, a Coriolis force is generated when a mass moves radially in the system. This force is directed along the tangent and is proportional to the rotating speed and the radial velocity of the mass particle. Coriolis mass flow meters based on this principle work very well and deliver direct in-line accurate mass flow measurements without pressure and temperature compensation. The sensor is by no means simple and is quite expensive, but it has no moving parts and causes very limited pressure drop in the system. It is, however, sensitive to vibrations. A simplified explanation of the principle follows.

By excitation an absolutely straight pipe will be sent into resonance. At two precisely defined points, the inlet and the outlet of the pipe, infrared sensors are placed to detect the phase of the oscillation (Figure 4.14). When no mass flows through the pipe, the phases of the two signals coincide. Any particle in the liquid flowing into the pipe system will undergo a

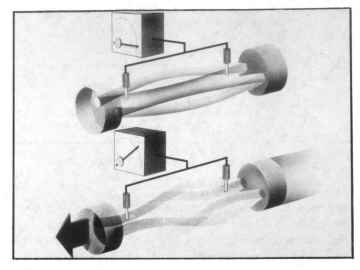

Figure 4.14   A flow sensor based on the Coriolis principle
(Courtesy of Endress and Hauser, Switzerland)

lateral (sidelong) acceleration. Its inertia will attenuate the vibration at the inlet of the pipe. As the particle passes through the pipe, it transmits the absorbed energy into the pipe and the oscillation at the output will be intensified. The phase difference between the two optical sensors is directly proportional to the mass flow rate.

4.3.5 Concentration and Biochemical Measurements

There is a vast number of measurements of different variables in the chemical or biochemical industry. Some on-line measurements can be made on a more routine basis, such as:

- conductivity
- salt content
- redox (oxygen−reduction potential, ORP)
- pH
- dissolved oxygen
- suspended solids

For these types of measurements, sensors with acceptable performance are commercially available in most cases. Of course, even if a device is deemed to be reliable it still requires proper maintenance which can vary greatly between different applications.

There are many types of sensors for measuring quality variables, such as different concentrations of organic carbon, nitrogen components, different phosphorus components, etc. Many of those instruments are based on some type of automated wet chemistry principle and are usually very complex. Maintenance becomes a vital part of functioning, and reliability is not always satisfactory.

4.4 BINARY AND DIGITAL SENSORS

In sequential control, measurements are of the *on/off* type and depend on binary sensors. In a typical process or manufacturing industry there are literally thousands of on/off conditions that have to be recorded. Binary sensors are used to detect the position of contacts, count discrete components in material flows, detect alarm limits of levels and pressures, or find end positions of manipulators. Here we will present some types of sensors that produce binary information. The sensors indicate positions, alarm conditions and levels.

4.4.1 Electric Position Sensors

Limit switches have been used for decades to indicate positions. They consist of electrical contacts that are mechanically actuated. A contact opens or closes when some variable (position, level) has reached a certain value. There are hundreds of different types of limit switches. We will emphasize their importance in process control applications by way of a couple of examples. Limit switches represent a crucial part of many control systems, and the system reliability depends to a great extent on these. Many process failures are due to the limit switches. They are located 'where the action is' and are often subject to excessive mechanical forces or too large currents.

A normally open (n.o.) and a normally closed (n.c.) switch contact are shown in their normal and actuated positions in Figure 4.15. A switch can have two outputs, called change over or transfer contacts. In a circuit diagram it is common practice to draw each switch contact the way it appears with the system at rest.

The simplest type of sensor consists of a **single-pole single throw (SPST)** mechanical switch. The device is shown in Figure 4.16 connected to a 'pull-up' resistor. When the switch is open, the resistor pulls the output to 5 V, which is interpreted by a TTL (transistor—transistor logic) level compatible gate in the computer interface as one logical state. When the switch is closed, the output is pulled to ground potential, which is interpreted as the other logic state.

To close a mechanical switch causes a problem since it 'bounces' for a few milliseconds. When it is important to detect only the first closure, such as in a limit switch

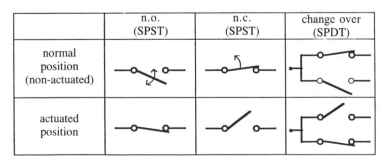

	n.o. (SPST)	n.c. (SPST)	change over (SPDT)
normal position (non-actuated)			
actuated position			

Figure 4.15 Limit switch symbols for different contact configurations.

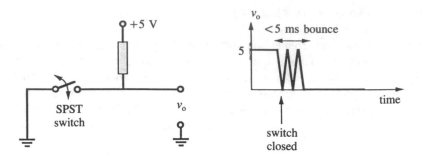

Figure 4.16   Simple SPST switch.

or in a 'panic button', the subsequent opening and closing bounces need not be monitored. When the opening of the switch must be detected after a closing, the switch must not be interrogated until after the switch 'settling time' has expired. The use of a programmed delay is one means of overcoming the effects of switch bounce.

A **transfer contact** (sometimes called **single-pole double-throw, SPDT**) can be classified as either 'break-before-make' (BBM) or 'make-before-break' (MBB) (Figure 4.17). Note that in a BBM contact, both contacts are open for a short moment during switching. In a MBB contact there is a current in both contacts briefly during a switch.

Contact debouncing in a SPDT switch can be made in hardware. When the grounded moving contact touches either input, the input is pulled low and the circuit is designed to latch the logic state corresponding to the first contact closure and to ignore the subsequent bounces.

There are several other methods to indicate positions by binary sensors. A few principles will be mentioned.

- **Mercury switches** consist of small hermetically sealed glass tubes containing two contact terminals and sufficient mercury to bridge these. The switch is opened or closed by tilting the tube.
- **Reed switches** consist of two leaf springs (the reeds) sealed in a small glass tube. The two free ends overlap and almost touch. When a magnet approaches the tube the reeds assume opposite magnetic polarity and attract each other to make contact.

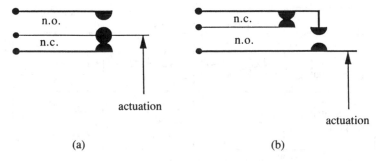

Figure 4.17   (a) Break-before-make (BBM); (b) make-before-break (MBB) contact.

- **Photoelectric sensors** consist basically of a source emitting a light beam and a light-sensing detector receiving the beam. The object to be sensed interrupts or reflects the beam. Sensing thus is made without physical contact.
- **Ultrasonic and microwave sensors** are used to sense objects in industrial applications, within a range varying from a few centimetres to several metres. They can be used in either reflective mode (with emitter and detector in the same unit) or in through-beam mode (with emitter and detector in separate units).
- **Inductive proximity sensors** operate by generating a high frequency electromagnetic field that induces eddy currents in a metal target. The sensor inductance is part of an oscillator circuit. When the target (a conducting material) nears the sensor (typically 2−30 mm), the oscillations are dampened and the change in the oscillator current actuates a solid-state switch.
- **Capacitive proximity sensors** contain a damped *RC* oscillator. When the target (that needs not be conducting) nears the sensor (typically 5−40 mm) the resulting change in capacitance causes the circuit to actuate a solid-state circuit.
- **Magnetic proximity sensors** use magnetism without any moving parts to detect an object near the sensor (a reed relay has moving parts). The principle is based on inductance, reluctance (magnetic 'resistance') or the **Hall** effect. A sensor based on the Hall effect has a resistance that can be controlled by an external magnetic field.

4.4.2 Point Sensors

There are many kinds of measurement sensors that switch whenever a variable (level, pressure, temperature or flow) reaches a certain point. Therefore, they are often called 'point sensors'. They are often used as devices that actuate some kind of alarm signal or shut down the process whenever some dangerous situation occurs. Consequently, they have to be robust and reliable.

To give a comprehensive description of point sensors is outside the scope of this book but a few references are listed in the bibliography.

4.4.3 Level Switches

A level switch is used to send a signal whenever the level in a tank reaches a certain value. Depending on the medium (liquid, slurry, granule or powder solids), a limit switch can often be used to detect the proximity of the surface.

A float can be connected to the surface of a liquid and influence a limit switch. A reed switch is an ideal sensor in a liquid, since it is waterproof. If the float contains a magnet it may actuate the reed relay in a nice manner. Photoelectric sensors are common and popular for the same purpose.

Capacity proximity sensors are particularly suitable for solids, where floats cannot be used. As the material rises it displaces air between the capacitance probe and the vessel wall which changes the dielectric between the probe and ground, and consequently the capacitance. Finally, the level can be measured by placing a pressure sensor at the bottom of the vessel. As the level reaches a certain height, the pressure produced by the liquid head actuates a pressure switch.

4.4.4 Digital Sensors

Digital measuring devices (digital sensors) generate discrete output signals such as pulse trains or encoded data that can be directly read by the processor. The sensor part of a digital measuring device is usually quite similar to that of its analog counterpart. There are digital sensors that incorporate microprocessors to perform numerical manipulations and conditioning locally, and provide output signals in either digital or analog form.

When the output of a digital sensor is a pulse signal, a counter is used either to count the pulses or to count clock cycles over one pulse duration. The count is first represented as a digital word according to some code and then read by the computer.

Example 4.7 Shaft encoders

Shaft encoders are digital sensors that are used for measuring angular displacements and velocities. They are much used in robotics, machine tools, servo systems and electric drive systems. An encoder generates coded data (or a pulse train) to represent angular position. In an optical encoder a transparent area may correspond to binary 1 and an opaque area to binary 0. A light source illuminates one side of the disc. As the disc rotates a bank of pick-off sensors generate a set of pulse signals. It is possible to obtain high resolution (number of pulses per revolution), high accuracy (due to the noise immunity of digital signals) and relative ease of use in digital systems. Encoders can be either incremental or absolute. Shaft encoders can also be used to measure angular velocities, and can be considered as **digital tachometers**.

4.5 ELECTRICAL DRIVE ACTUATORS

Most actuators used in control applications are continuous drive systems, such as d.c. motors, induction (asynchronous) motors and synchronous motors. Stepping motors are considered a special case of synchronous motors, but are fed with digital information and are therefore known as digital actuators.

4.5.1 Amplifiers

Computer output ports have a very low power and cannot control any physical devices directly. For this reason, computer output signals must be amplified before they can be fed to the actuators that control the physical process. The power amplifier may be a separate unit but may also be part of the actuator. At power levels up to a few hundred watts it is possible to use amplifiers that are not much different from those used in audio equipment; these are known as **power operational amplifiers**. At higher power ratings, the amplifiers are often called either **servo amplifiers** or **programmable power supplies** and are usually designed to fit motors or electrically controlled valves. The output voltage of a programmable power supply can be controlled by an analog or digital signal.

A common technique to control power levels in the order of kilowatts and higher is **pulse width modulation (PWM)**. In this technique, the output voltage is switched between two constant values at a high frequency, typically in the kilohertz range. The average voltage level is controlled by changing (modulating) the width of the pulses. 'Narrow' pulses will result in a low and 'broad' pulses in a high average voltage. A PWM unit is built around a power control section with solid-state switches (such as transistors, power MOSFETs — metal oxide field effect transistors — and thyristors). These electronic switches turn the power fully on or off.

The reason why pulse width modulation is used is that conventional power amplifiers would simply burn at high power levels. The advantage of switching is that the solid-state devices are not continuously loaded with high power and therefore their power dissipation is low. This fact makes PWM amplifiers very efficient. An additional advantage of PWM amplifiers is that the switching can be directly controlled from the digital output ports of a computer.

PWM signals are increasingly being used to drive continuous actuators such as d.c. motors, hydraulic servos and a.c. motors. If the switching frequency of the PWM amplifier is sufficiently high in relation to the actuator time constants, then the signals will average around some value. This is, for instance, the case with the heater system for a water tank (shown in Figure 4.18). The input heating power is switched between zero and a maximum, resulting in a certain average power that allows the water temperature to oscillate slightly around a constant value. If the switching frequency is sufficiently high in relation to the time constant of the tank system, then the temperature variations are negligible.

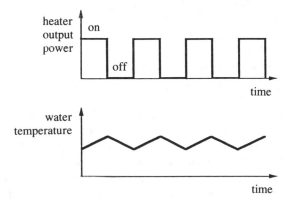

Figure 4.18   Using on/off control for a heater system.

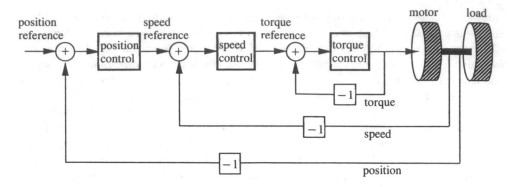

Figure 4.19 Schematic structure of a position servo control system.

4.5.2 Velocity and Position Control

In order to realize a position for a robot arm, a machine tool, a computer disk drive or some other servomechanism, a feedback control system is applied to some motor (see Section 2.3.8). The structure of a position control system is shown in Figure 4.19 and is the same regardless of the type of motor.

The angular position is measured with some sensor (see Sections 4.3.1 and 4.4.4) and is compared with the position reference value (**setpoint**) that the computer has to actuate. From the position error the position controller calculates how much the velocity has to change to correct the position. The desired angular velocity (speed reference) is compared with the actual speed.

The motor has to produce a change in the driving torque (see Example 3.3, Section 3.2.1) in order to get a different angular velocity. The necessary motor torque is calculated by the speed controller. Again, this value is compared with the actual torque (in Example 3.5, Section 3.2.2, we saw that the torque is measured through the rotor current). The torque controller contains a model of motor dynamics and calculates the voltage that is needed to produce the torque. Naturally, this calculation depends on the type of motor. The torque controller is usually equipped with power electronic components that can generate the desired voltage and frequency.

The position control contains three nested loops, the torque, speed and position loops and is called a **cascade control system**. This control structure is discussed in Chapter 9. Many actuators have to produce a velocity instead of a position. Then the outer position loop is removed, and the speed reference is given directly by the control computer.

If the position control system is implemented in analog techniques the computer sends position commands to a conventional analog position servo via a DAC (see Section 5.2). The velocity may be measured with a tachometer (outlined in Example 4.2, Section 4.3.1) and the position with a potentiometer. A second approach combines analog and digital control techniques (Figure 4.20). This has been a standard structure since the early 1960s.

The velocity control is implemented with analog electronics, while the position control loop is implemented in the computer. The motor-position sensor is an incremental shaft encoder (as for Example 4.7) which generates a fixed number of pulses per motor

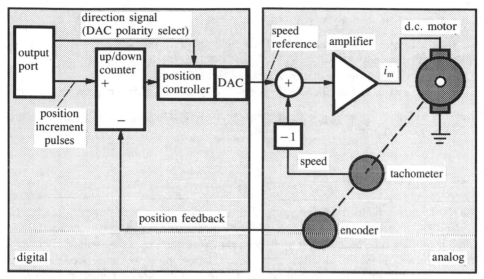

Figure 4.20   Mixed analog and digital servo control loops.

revolution. When the motor rotates, a train of pulses is generated and the pulse rate is proportional to the motor speed. The computer sends out position pulses that are added to the counter, while the encoder pulses are subtracted. The position error corresponds to the pulse difference and is sent to a DAC which converts it to a voltage that becomes the speed reference value. To the computer this interface resembles that of a stepping motor.

Another interface option is a completely digital servo loop. The shaft encoder provides the feedback that can be used for both position and velocity measurement. Circuit boards implementing the mixed analog and digital servo loop or the completely digital servo loop are available for most common microcomputer bus systems. Controllers are also available in units combined with servo amplifiers.

4.5.3 Stepping Motors

Stepping motors are incremental-drive actuators that are driven in fixed angular increments (steps). Each step of rotation is the response of the motor rotor to an input pulse. Since each step of the motor is synchronized with pulses from the computer command, its control is very simple. This, however, assumes that the motor position is well defined by the pulses (no steps are missed) so that no feedback from the motor angular position (**open loop control**) is needed. Pulse missing can be a problem under highly transient conditions near the rated torque. Then feedback control is used to compensate for the motion errors. The position controller produces a suitable pulse train to control the motor.

Some of the advantages of stepping motors are:

- high accuracy of motion, even under open loop control
- easy implementation in digital applications
- no mechanical commutator

Some of the negative sides of stepping motors are:

- low torque capacity compared to continuous drive motors
- limited speed
- high vibration levels due to the stepwise motion
- large errors and oscillations as a result of a missed pulse in open loop control

There are still many advantages with stepping motors so that they are used in a great number of low-power applications.

4.5.4 Direct Current Motors

The mechanical and electrical dynamics of the d.c. motor have been described earlier (in Example 3.3, Section 3.2.1 and Example 3.5, Section 3.2.2). The d.c. motor is very common as a servomotor, even if it is challenged today by alternating current (a.c.) motors. The commutator is the real disadvantage which restricts the power and speed of the d.c. motor. This has been overcome in brushless d.c. motors that are permanent magnet rotor d.c. motors where commutation is accomplished by electronic switching of the stator current. In concept, brushless d.c. motors are somewhat similar to stepping motors and to some types of synchronous a.c. motors (see below). The d.c. motor will be considered in control applications in Section 9.7.2.

The control of d.c. motors is accomplished by controlling the supply voltage to the rotor and (sometimes) the field circuits. To use a variable resistor in series with the supply source has many disadvantages. It is wasteful and the heat generated in the resistor has to be removed promptly to avoid damage. The usual way to control the supply voltage is by using solid-state devices. By varying the on and off times of a switch, the supply voltage to the motor circuit can be 'chopped' so that the average supply voltage is controlled. A common switch in d.c. motor control is the **thyristor** (see Sections 4.5.1 and 4.6.1).

4.5.5 Asynchronous and Synchronous Motors

Alternating current (a.c.) motors are widely used as servomotors. This has been made possible by the development of power electronics in combination with new control methods. With microelectronic drive systems the frequency of the applied voltage can be suitably controlled. The torque cannot be measured as easily as in a d.c. motor, but can be estimated on-line. Due to the time requirements, processors with a special architecture for very fast computations, digital signal processors (DSPs), are used. Alternating current motors are attractive not only for heavy duty applications but are increasingly of interest in industrial automation applications. Some of the advantages of these motors are:

- cost-effectiveness
- robust and simple construction
- high reliability

- convenient power supply
- no commutator needed
- virtually no electric arcing (because of no commutator)

On the minus side the a.c. motor has a lower starting torque than a d.c. motor and needs a more complex control device. However, the merits of a.c. motor systems are such that they can challenge the d.c. motors in robots, manipulators and other industrial servo systems.

In an **induction** (**asynchronous**) motor the stator magnetic field is not constant as for the d.c. motor. In the simplest (two-pole) machine there are three **stator** windings, separated 120° in space around the stator. When they are fed with three-phase alternating voltage, the resulting magnetic flux from the stator will rotate with the same frequency as the voltage. The **rotor** windings are not energized by an external voltage so no commutator brush devices are needed in induction motors.

The rotating field in the stator intercepts the rotor windings and generates a rotor current due to mutual induction. The resulting rotor flux interacts with the rotating stator flux and produces a torque in the direction of the rotation of the stator field. This torque drives the motor. As long as the rotor speed has not reached the stator flux rotating speed (the synchronous speed), there is a driving torque. The velocity difference is called the slip *s*, defined as:

$$s = \frac{\omega_s - \omega_m}{\omega_0} \tag{4.10}$$

where ω_s is the stator field angular velocity and ω_m the rotor (motor) speed. Apparently for $s = 0$ there is no driving torque.

The rotor of a **synchronous** motor runs in synchronism with a rotating magnetic field generated by stator windings, that in principle are like those of an asynchronous machine. In contrast to the asynchronous motor, the synchronous motor has rotor windings that are energized by an external d.c. source. The rotor poles obtained in this manner will lock themselves with the rotating field by the stator and will rotate at the *same* speed. The motors are used frequently for constant speed applications under variable load. With modern frequency converters the synchronous motors can be used with a variable speed. Synchronous motors supplied with permanent magnet rotors are becoming more common today. A stepping motor can be considered a special case of a synchronous motor.

4.6 BINARY ACTUATORS

In many situations sufficient control of a system can be achieved if the actuator has only two states: one with electrical energy applied (*on*) and the other with no energy applied (*off*). In such a system no digital-to-analog converter is required and amplification can be performed by a simple switching device rather than by a linear amplifier.

Many types of actuators can receive digital signals from a controller, such as magnetic valves controlling pneumatic or hydraulic cylinders, electromagnetic relays controlling electrical motors, and lamps. There are two main groups of binary actuators, **monostable** and **bistable** units. For a monostable actuator there is only one stable position and only one signal can be given. A contactor for an electric motor is monostable. As long as a signal is sent to the contactor the motor rotates, but as soon as the signal is broken the motor will stop.

A bistable unit remains in its given position until another signal comes. In that sense the actuator is said to have a memory. For example, in order to move a cylinder, controlled by a bistable magnet valve, one signal is needed for the positive movement and another one for the negative movement.

4.6.1 Switches

The output lines from a computer output port can supply only small amounts of power. Typically, a high level output signal has a voltage between 2 and 5 V and a low level output less than 1 V. The current capacity depends on the connection of the load but is generally less than 20 mA, so the output can switch power less than the order of 100 mW.

A number of switches can be used for low and medium power switching. Integrated circuits containing transistor switches can be used when the actuator voltage is below about 80 V and the current below 1.5 A. When the computer output is larger than 2.4 V a current flows through the actuator, while for outputs smaller than 0.4 V the transistor is *off* and no current flows through the actuator.

For actuators designed for higher power the switch design can be based on discrete power transistors or MOSFETs. Typically, such circuits can carry 5−10 A and block more than 100 V. The transistors have some internal resistance and dissipate some power when carrying current, and they will overheat if not mounted adequately to be cooled.

For higher power handling capability it becomes more important to avoid a direct electrical connection between a computer output port and the switch. The switch may generate electrical noise which would affect the operation of the computer if there were a common electrical circuit between the computer and the switch. Also, if the switch fails it would be possible that a high voltage from the switch could affect the computer output port and damage the computer.

The most common electrically isolated switch used in control applications has been the **electromechanical relay** (Figure 4.21). A current through the relay coil makes the armature move from the normally closed to the normally open contact. Typically the coil may draw 0.5 A with 12 V, so it cannot be directly driven from the computer output, but has to be switched through some other medium power switch. Low power reed relays are available for many computer bus systems and can be used for isolated switching of signals. Relays for larger power ratings cannot be mounted on the computer board.

A relay is a robust switch that can block both direct and alternating currents. Relays are available for a wide range of power, from reed relays used to switch millivolt signals to contactors for hundreds of kilowatts. Moreover, their function is well understood by maintenance personnel. Some of the disadvantages are that relays are relatively slow,

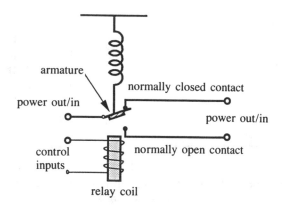

Figure 4.21   Electromechanical relay.

switching in the order of milliseconds instead of microseconds. They suffer from contact bouncing problems which can generate electric noise, that in turn may influence the computer.

The switching of high power is easily made in solid-state switches where many deficiencies in relays are not present. A solid-state switch has a control input which is coupled optically or inductively to a solid-state power switching device. The control inputs to solid-state switches are designed to be driven directly from digital logic circuits so they are quite easy to adapt to computer control.

Thyristors are an important class of semiconductor switches. Important examples are **triacs** and a device consisting of two **silicon-controlled rectifiers (SCR)**, also called solid-state controlled rectifiers or semiconductor-controlled rectifiers.

Once a thyristor is fired (switched *on*), it will remain *on* as long as a current flows through the switch and is unable to turn itself *off*, unlike a power transistor or power MOSFET. The thyristor switches *off* only if current ceases to flow through it. It is not sufficient to bring the firing voltage to zero to turn *off* the thyristor. When the supply voltage is direct current, then it does not drop to zero. It is necessary to bring a slightly negative voltage over the thyristor to turn it *off*. This can be done with a commutating circuit in what is called **forced commutation**. Thyristors are most commonly used to switch alternating current, since it passes through zero at regular intervals, allowing the thyristor to shut *off* if no firing signal is present. Then we have a **natural commutation**.

Thyristors can handle considerably more power than transistors or power MOSFETs. When the thyristor is conducting it offers virtually no resistance and the voltage drop across the thyristor can be neglected for practical purposes.

4.6.2 Switching Off an Inductive Load

To turn an actuator off by opening a switch may create problems if the actuator is inductive like an electrical motor or a solenoid coil. The voltage across the inductor is $v = L(\mathrm{d}i/\mathrm{d}t)$,

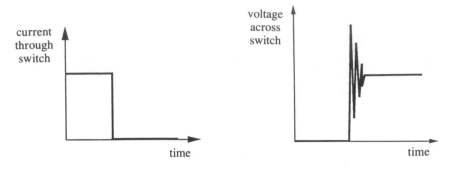

Figure 4.22   Illustration of the voltage spikes that occur in an inductive actuator when a switch is opened.

where L is the actuator inductance and i the actuator current. If the current is switched off rapidly the voltage across the actuator can become very high during the switching phase (Figure 4.22). The switch may be damaged by the voltage spike which is why it is necessary to dampen it. This can be done with a **free-wheeling diode** across the load (Figure 4.23).

When the switch is opened the load current starts to flow through the free-wheeling diode. The voltage across the actuator is limited to less than 1 V, the voltage drop across the diode. The switch must block only the supply voltage. The load current flowing through the free-wheeling diode decays exponentially. The decay rate can be increased by adding another resistor in series with the diode.

4.7 SIGNAL TRANSMISSION

The transmission of information between different units is both a necessary and a critical part of the system. We will discuss different practical methods to deal with this problem.

The signal levels and impedances between the measuring device and computer input have to fit together. This is ensured by **signal conditioning**. All electronic equipment contributes to the **electrical disturbances**. If a current or voltage in one circuit causes

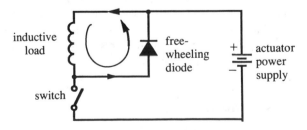

Figure 4.23   The use of a free-wheeling diode to dampen the voltage spikes for inductive loads.

current and voltage in another, the two circuits are coupled to each other. Most often the coupling elements are combinations of resistive, capacitive and inductive elements. Signal **grounding** is done in order to manage electrical noise problems. Some basic rules for grounding are mentioned. Finally, the choice of the **signal form** (voltage, current or optical) depends on several factors but noise sensitivity is of primary importance.

4.7.1 Signal Conditioning Using Operational Amplifiers

In order to obtain faithful measurements the loading effect has to be minimized (see Section 4.2). The sensor output signal has to be sufficiently large and the computer input should have high input impedance in comparison to the output impedance of the sensor−transducer. In order to match impedances and signal levels, an amplifier circuitry is introduced between the sensor output and computer input.

An impedance-matching filter has a high input impedance, a low output impedance and almost a unit gain. The last stage in the circuitry is usually some stable high-gain amplifier to step up the signal level. Impedance matching can be implemented using operational amplifiers with feedback. The operational amplifier is a versatile building block to create analog circuitry.

An **operational amplifier (op-amp)** is an integrated circuit amplifier with a very large voltage gain (typically of the order of 10^6 to 10^{12}), a high input impedance (usually several megohms) and a low output impedance (typically smaller than $100\ \Omega$). The output current is typically limited to 10 mA at ± 10 V. Op-amps are available in hundreds of different types. A schematic representation of an operational amplifier is shown in Figure 4.24.

The output v_o is an amplified version of the difference between the two inputs v_i^+ and v_i^-:

$$v_o = G(v_i^+ - v_i^-) \tag{4.11}$$

where G is the gain. A very small difference in potential between the two inputs is sufficient to cause a large output-voltage swing. In this form, the op-amp is the simplest form of a comparator and goes into positive or negative saturation, depending on the difference between the input voltages.

Since the voltage gain is large but unpredictable, the device is never used as an amplifier

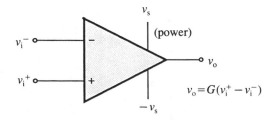

Figure 4.24   Schematic representation of an operational amplifier. The power connections are not usually shown.

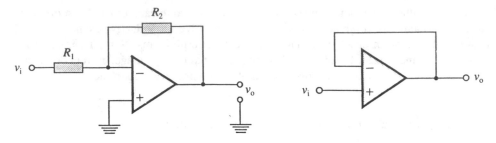

Figure 4.25   Schematic representation of an inverted amplifier and a voltage follower.

without negative feedback being applied. For reasonable signal frequencies (less than about 20 kHz), most of the necessary amplifier couplings can be realized by different passive components (resistors and capacitors) around the operational amplifier. The inverted amplifier, shown in Figure 4.25, is a basic op-amp feedback. The voltage gain (the ratio of the output to the input voltage) is:

$$\frac{v_o}{v_i} = -\frac{R_2}{R_1} \bigg/ \left(1 + \frac{1}{\beta G}\right) \tag{4.12}$$

where $\beta = R_1/(R_1 + R_2)$ and G the op-amp gain. The gain decreases with signal frequency, but as long as $\beta G \gg 1$ the voltage gain depends only on the resistors. Generally, for an ideal op-amp, the gain can be expressed by the ratio of the feedback impedance to the input impedance.

A **voltage follower** (or an **impedance transformer**) is realized by feedback coupling (Figure 4.25). The voltage follower has a gain equal to about 1. By connecting a voltage follower to a high-impedance sensor–transducer device, a low impedance output is obtained. Signal amplification might be necessary before the signal is transmitted or processed.

In data acquisition systems there are often many signals that lie on different potentials. If the signal level is low it is suitable to use a differential amplifier to increase the signal

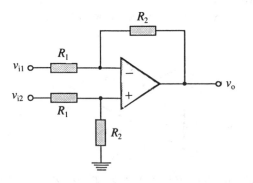

Figure 4.26   Differential amplifier.

level. The amplifier shown in Figure 4.26 has the output voltage:

$$v_o = \frac{R_2}{R_1}(v_{i2} - v_{i1}) \tag{4.13}$$

4.7.2 Electrical Conductors

A signal that is transmitted along a conductor is influenced by the transfer function of the conductor. To transfer a direct current is simple as long as the conductor impedance is small compared to the receiver input impedance. If, however, the transfer time of the signal between the end points of the conductor is not negligible compared to the signal period time or rise time, then other effects become significant. (Remember that a current propagates about 300 m per microsecond.) One of these problems is signal reflection. Since there are capacitances in all electrical circuits, they become a load that easily gets larger than the resistive load at high frequencies.

In order to describe **signal reflection** we have to look closer at the transmission line. It can be considered as two parallel conductors with inductances L (per length unit) in *series* along the line and capacitances C (per length unit) in *parallel* between the conductors. Analysing the transmission line one finds that if a signal is added to one point along the line, it will propagate in *both* directions with the velocity $v = 1\sqrt{LC}$. A signal that propagates along a transmission line terminated with the impedance Z_L will be partially reflected, and the reflection ratio ρ is given by:

$$\rho = \frac{Z_L - Z_0}{Z_L + Z_0} \tag{4.14}$$

$Z_0 = L/C$ is called the **characteristic impedance**. Consequently, the load impedance should ideally be $Z_L = Z_0$ to avoid reflection.

Example 4.8 A pulse signal in a transmission line

For a signal pulse along the transmission line the relation between voltage and current in the pulse holds $v = Z_0 I$. If the line is open at the end ($Z_L = \infty$), the current has no other way to go than back again. It is completely reflected and so is the voltage since it is equal to $Z_0 I$. The other extreme case is for a short circuit termination ($Z_L = 0$). The voltage over the cable end has to be zero all the time. The incoming voltage pulse then has to be compensated by an equally large voltage of negative sign. Therefore, the voltage pulse is completely reflected but with changed sign.

There are similar conditions at the source end. If the source impedance is not Z_0, the pulse is reflected again. Pulses sent over a line with incorrect impedances at the ends will travel back and forth with decreasing amplitudes. At digital transmission there are long pulse trains from the source and reflections will cause the pulses to mix so they are

misinterpreted by the receiver. We will show in Section 6.2.3 how to use a termination
network to match the impedance at the end of bus lines.

Example 4.9 Connecting two cables

If two cables with different characteristic impedances are connected, the line balancing has to
be considered. For example, if a 50 Ω cable is connected to a 300 Ω cable, a resistor net has
to be connected in between. The net must be such that the 300 Ω line together with the
resistors is seen from the 50 Ω line as a 50 Ω load. The same must be true from the 300 Ω
line.

When both wires in a line carry current and the currents have the same amplitude
but are 180° out of phase, the line is said to be **balanced**. In an unbalanced circuit only
one wire carries signals and the other serves as reference ground (which does not exclude
it also carrying a current).

Electrical conductors are basically of two kinds: twisted pair and coax cable. As the
name indicates, a **twisted pair** consists of two insulated copper conductors twisted together.
The twisted pair is a cheap and simple electric communication medium. The wide use
of twisted pair and coax cables in telephony and television distribution has made the cables
and their ancillary components, such as connectors and tools, widely available at low cost.

The electrical circuit of the twisted pair cable can be either balanced or unbalanced.
The advantage of the balanced circuit is that it is less sensitive to external interferences
and conversely irradiates less energy because of the mutual elimination of the induced
electromagnetic fields with the same intensity and opposite polarity. On the other hand,
a balanced circuit is such only if it is properly trimmed and maintained. An unbalanced
twisted pair cable is much more sensitive to disturbances but requires almost no effort
to be installed and operated. A cable with four twisted conductors is even more insensitive
to inductive disturbances (see Section 4.7.3). Figure 4.27 shows a shielded multiconductor
with four twisted pair cables for use in data communication.

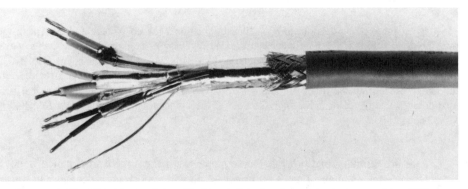

Figure 4.27   Shielded cable with four twisted-pair conductors, used in data
communication. (Courtesy of ELFA AB, Sweden.)

Figure 4.28 Coax cable for baseband data communication. (Courtesy of ELFA AB, Sweden.)

Coax cables are made of an electric conductor isolated with a plastic coat and surrounded by a conducting wire mesh, the shield (Figure 4.28). Because of their geometry, coax cables are intrinsically unbalanced. The shield is held at ground potential and prevents energy from irradiating from the central conductor, an important effect at high frequencies. Conversely, the shield hinders external interferences from reaching the central conductor. Coax cables are denominated by the value of their distributed impedance, common values are 50 Ω (the cable is commonly known as RG-58 type) and 75 Ω (RG-59). The importance of coax cables is due to their use in high-frequency applications and communication technology (see Section 10.3).

4.7.3 Electrical Disturbances

Disturbances are generated in many ways of which the most important are those caused via coupling by:

- resistance (via the conductor)
- capacitance
- induction (magnetic)
- radiation

Resistive (galvanic) coupling via a conductor is independent of the frequency of the disturbance. In a capacitive or inductive coupling, however, the degree of coupling depends on the frequency, whereby at higher frequencies more energy is transmitted. This means in practice that fast circuits (with fast voltage or current changes) are more serious disturbance sources than slow circuits. Usually there is a combination of different types of electrical couplings. Typically, difficult problems arise as soon as signal conductors are located close to power cables.

A lot of research and development is going on to develop a better noise environment for electronic equipment. The goal is to achieve **electromagnetic compatibility (EMC)** within and between electronic and communication systems. A device has to be insensitive to external disturbances and should not generate such disturbances that other equipment will pick up.

4.7.3.1 Resistive couplings

Noise currents flow through conductors. Any conductor joining a sensor with its signal conditioning circuitry is a potential collector of electrical noise. For example, resistive coupling is obtained when a common power unit delivers voltage to different electronic systems, and they in turn are connected to a common ground (see Section 7.3.2). Other common noise sources may be poorly shielded motors, and frequency converters with semiconductor switches. One way to avoid the problem is to supply different sensitive electronic devices with separate power supplies. Another way is to galvanically separate different power units and devices.

Example 4.10   Flying capacitor isolation

A flying capacitor isolation (Figure 4.29) is an example of galvanic isolation. The first pair of switches (S_1) is closed while the second pair (S_2) is opened. The capacitor becomes charged very quickly, then switches S_1 are opened and the S_2 pair closed so that the voltage is transferred to the output (which can be an analog-to-digital converter of the computer, see Section 5.3.2). Thus the input device is never electrically connected to the computer input. The capacitor is said to 'fly' the external voltage to the computer input.

4.7.3.2 Capacitive couplings

Usually there is a capacitance (leaking capacitance) between two conductors or between a noise source and a conductor. **Capacitive coupling** is characterized by the fact that a variable voltage induces a current i in the conductor that is proportional to the voltage time derivative, $i = C(\mathrm{d}v/\mathrm{d}t)$, where C is the capacitance. Capacitive coupling should be minimized, and it becomes smaller if the conductors, i.e. the noise source and the receiver, are kept far from each other.

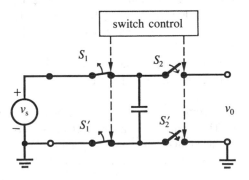

Figure 4.29   Flying capacitor isolation. The capacitor is charged from the voltage source. After the S_1 pair is opened and the S_2 pair closed, $v_0 = v_s$. Note that the two grounds are never connected through the switches.

A good way to decrease capacitive coupling is to supply the measurement conductor with an electrostatic shield to break the disturbance route. The shield has to be grounded so that its potential is zero. Where the measurement conductor to the sensor or to the computer terminates, the shield does not surround the complete conductor. As some small parts at the ends are not protected, there is a little capacitive coupling. Therefore, it is important to make the connections at the termination as short as possible.

4.7.3.3 Inductive (magnetic) coupling

An electric conductor generates a magnetic field around itself with a magnetic flux density dependent on the amplitude of the current. Therefore, **magnetic coupling** is a great nuisance close to power cables. A variable current generates a variable magnetic field and, in accordance with the induction theorem, the varying magnetic field generates a voltage. Considering the mutual inductance, M, the induced voltage v is $v = \mathrm{d}(Mi)/\mathrm{d}t$, where i is the current. If the conductor with induced voltage is a closed circuit, then a current is generated there. This induced current increases with the area that encloses the magnetic flux.

There are several ways to eliminate inductive couplings. The area of the circuit that encloses the magnetic flux can be decreased by twisting the cables. Also, conductors can be kept close to each other so that the area between them is minimal. Furthermore, the small area 'changes sign' at each turn, so that the net magnetic flux becomes very small.

The measurement conductor should be located as far away as possible from the disturbance. In particular, sensitive electronics need to be placed as far as possible away from transformers and inductors. Cables should be placed so that probable disturbance fields propagate along the cable. Two good rules to follow are: to avoid power cables and signal wires in the same cable conduit; and to cross low voltage and high voltage cables at right angles.

The magnetic field can be dampened by shielding. A copper or aluminium shield has very high conductivity and, due to eddy currents in the shield, the magnetic flux is reduced. Shielding can also be realized with high permeability material such as iron. A magnetic shield is often clumsy, since it has to be thick in order to dampen the magnetic flux. Therefore, shielding is used mostly in devices that produce large magnetic fluxes.

4.7.3.4 Some rules of thumb

We summarize some of the basic rules to eliminate or dampen the influence of electrical noise on the measurement environment. Evidently one should first try to:

- Reduce the noise source.

Other important factors are:

- Galvanic couplings.

- The distance between the noise source and the object.
- The noise frequency content.

Capacitive couplings can be reduced by:

- Using shielded cables.
- Minimizing the length of the unshielded wires at the termination panel.

Magnetic couplings are dampened if:

- The cables are twisted, so that the area of the circuit that encloses the magnetic flux is decreased and the orientation is altered.
- Individually twisted pair wires are used, one for each sensor.
- Power cables and signal wires are not placed together.
- Low voltage and high voltage cables are crossed at right angles.
- Low voltage wires are held at some distance (at least half a metre) from interference sources and power cables.

4.7.4 Signal Grounding

Grounding is physical conduction to a common potential. **Signal grounding** creates a voltage or potential reference for measurement signals. In theory, all the points that have to be grounded should be connected to this reference without any resistance or inductance. Unfortunately, this is impossible in practice. Problems caused by poor grounding are among the most common measurement problems and are extremely difficult to detect. The principal problem is the same for an electronic circuit as for a whole industrial plant.

Consider a simple measurement system (Figure 4.30a) where the voltage source v_s is connected to a ground P_1 and the measurement system is grounded at P_2. Two separate grounds seldom or never have the same potential and this causes a leakage current between them. The voltmeter will show a voltage of $v_s + v_g$ instead of the correct value. In large systems it is common to implement several grounding circuits, from analog signals, computer equipment, power units and the chassis. All the different grounds have to be connected in a common ground, as illustrated in Figure 4.30b.

A rule of thumb for analog signal grounding is to ground as close to the source (i.e. the sensor) as possible. Naturally, this is a problem in large distributed systems with long cables. A separate computer equipment grounding is desirable since digital systems can easily pick up high frequency signals. The more traditional analog data acquisition systems detect mostly low frequency signals, while the high frequency disturbances are filtered out by the analog equipment.

A separate **power unit ground** is recommended for relays, motors and other devices that use large currents. Finally, all **chassis** are connected to a separate ground, possibly also to the safety ground.

Usually a cable shield is connected to ground. In order to avoid any ground current loops, it has to be connected to ground in only one point, either to the voltage source (the sensor) ground or to the amplifier ground. The first case, illustrated in Figure 4.31,

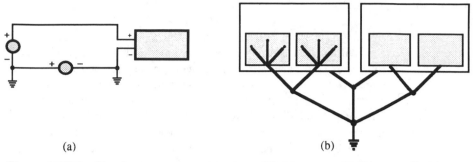

(a) (b)

Figure 4.30(a)   Simple measurement system with two grounds; (b) grounding to a system ground.

gives the best disturbance damping. Multiple shields still have to be connected to ground in the same point.

We can sum up the grounding problems in a couple of rules:

- Think where the currents flow: the current coming from the power supply must return to the same device. The conductors that carry that current may be called 'ground' but may have voltage offsets or spikes due to the impedance of the cable. Use sufficiently heavy wire for the power-supply and ground leads.
- Keep digital and analog grounds separate: as the digital circuits change logic state, the 'digital ground' can carry large voltage spikes. Since analog circuits have poor 'noise immunity' they should have separate ground.

4.7.5 Signal Types — Voltage, Current, Optical

The choice of signal form for transmission from a measuring device to the computer depends on several factors, but it is essential that the transmission is insensitive to electrical disturbances.

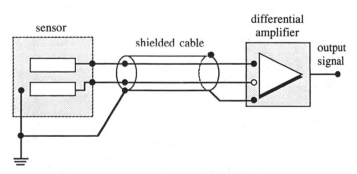

Figure 4.31   A system with a sensor and a differential amplifier. The grounding of the shield is made at the sensor.

4.7.5.1 Voltage signals

Every cable shows a certain resistance per unit length. If the input impedance of the receiver is not infinite, there will be a current in the cable and a resulting voltage drop. As the signal changes there is also a capacitive current between the conductors. It is realistic to always calculate some voltage drop along the transmission. The demand for a high input impedance of the receiver makes it very sensitive to disturbances. Consequently, voltage is not very suitable for transmissions where disturbances may be significant.

Voltage signals are not used very much in industry where distances are long and there are plenty of disturbance sources. The reason that voltage signals are popular in many applications is that there are many devices for amplification, filtering and other signal processing forms. If the same signal has to be sent to several receivers it can easily be coupled in parallel. The most important voltage levels for transmission have been standardized (IEC 381):

$+1$ to $+5$ V
0 to $+5$ V
0 to $+10$ V
-10 to $+10$ V

4.7.5.2 Current signals

For the transmission of signals over long distances current is a better choice than voltage. The reason is that current remains constant along the cable, while voltage drops with the distance due to the cable resistance. At the end of the circuit, the current signal is transformed to a voltage via a high precision shunt resistor (Figure 4.32).

For current transmission, the receiver should ideally have zero input impedance. In reality, the impedance is determined by the shunt resistor and is usually of the order of a few hundred ohms. For a current of 20 mA and a shunt resistance of 250 Ω, the voltage

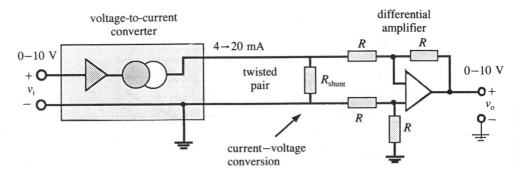

Figure 4.32 Analog transmission with a current loop. The voltage-to-current converter is a commercial unit. The transmission line can be several hundred metres long. The shunt resistance has to be 250 Ω to convert the current to the proper voltage range.

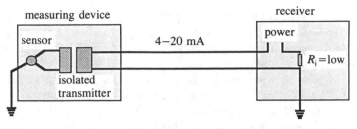

Figure 4.33   Sensor with current output signal and two-wire transmitter which is galvanically isolated from the probe.

drop is 5 V. If the signal source has a high output impedance, then any transmission noise will lead only to a small voltage drop over the shunt resistor, something that can normally be accepted.

Transmission using current is used mostly for low frequency signals (up to 10 Hz). The sensor output voltage is converted (in an operational amplifier circuit) to a current, and because of that it can be constructed so that the sensor probe is galvanically isolated from the signal wires (Figure 4.33). At constant current and ideal isolation, all the current from the source reaches the receiver so the cable resistance has no influence. For changing currents the capacitive effects will become evident and some current is lost along the cable, either to the return conductor or to ground. The international standard IEC 381 recommends a range of 4−20 mA. This means that if the loop is broken (0 mA) the error condition can be detected.

Current transmission has the advantage that both power and signals can be transmitted over the same wires. Only two conductors are needed contrary to voltage transmission which needs at least three conductors. The current in the loop becomes the sum of the current for the power supply and an additional current, so the loop current indicates an adequate measurement value. Since the current is defined only between 4 and 20 mA, a current of less than 4 mA indicates an error.

To sum up, a measuring system based on current signals and a probe which is galvanically isolated from the output signal has several advantages:

- It allows long transmission lines.
- It allows a simple performance check since 0 mA means that the sensor is off-line or not measuring.
- It gives good protection from interference.
- It needs only two transmission wires which reduces costs.

4.7.5.3 Optical transmission

If a voltage signal is converted to a pulse train then a high noise immunity can be obtained. However, the bandwidth of the signal is lower than for the analog techniques. The limit is of the order of hundreds of hertz. Optical fibre transmission has become practical in many measurement (and communication) applications. By using a light-emitting diode (LED), digital electrical signals can be converted to light pulses which are conducted over glass fibres. At the receiving end the light pulses are converted back to electrical signals using phototransistors.

Optical transmission is immune to magnetic and electric field disturbances and also provides total isolation. This form of transmission is suitable not only for long distances (> 1 km) but also in difficult measurement environments, such as close to electric motors and frequency converters. The use of optical transmission is motivated not primarily by capacity but by corrupting disturbances.

4.8 SUMMARY

Measurements of physical variables are performed with **analog**, **digital** or **binary sensors**. The sensor output signal (analog, pulse train, binary, etc.) has to be chosen properly with respect to the application and its demands. Any sensor has to satisfy many demands such as:

- Output ought to vary linearly with the measured value.
- The output signal should be sufficiently large.
- The sensor itself should not distort the measured value.
- Small power consumption.
- Insensitivity to environmental effects.
- The output should remain at the measured value.

In order to faithfully represent transients the sensor must have:

- Adequate response time.

Some examples of sensors for continuously varying physical variables or for on/off conditions have been surveyed.

The **actuator** is the device that mechanically drives a process and converts electrical energy into the required control output. Actuators have to be chosen to be adequate for the purpose: their nature can be binary, digital or continuous. The desired output power and speed determines the type of actuator to choose. We have emphasized electrical—mechanical energy conversion using d.c. motors, induction (asynchronous) motors, synchronous motors and stepping motors. Many actuators are complete control systems in themselves. Velocity and position servo mechanisms have been described where the controllers are implemented either as separate units or in the process computer.

For adequate **signal transmission** the impedances and signal levels along the line have to be matched so that loading effects are avoided and the whole measurement range is accurately represented. The fundamental causes for noise and disturbances have been described, such as:

- resistive couplings
- capacitive couplings
- inductive or magnetic couplings

Different ways of eliminating or dampening these problems have been listed, such as:

- proper shielding
- galvanic isolation
- twisted cables
- separation of power and signal cables

Signal grounding has to be performed properly and is crucial for noise reduction.

Finally, voltage transmission is more sensitive to noise sources than current transmission. The latter is applied in industrial control whenever long distances are involved. To avoid many of the electrical noise problems optical transmission is being used more and more.

FURTHER READING

There are many good general books on measurement and instrumentation technology. Doebelin (1983) is an outstanding reference and contains detailed descriptions of most sensor types. Other good texts are Alloca and Stuart (1984), Barney (1985) and Ott (1976).

For computer control applications, de Silva (1989) is an excellent textbook. Derenzo (1990) describes several instructive laboratory exercises for computer interfacing.

Force measurements are described by Norton (1989) and the principle of Coriolis sensors is described by Vögtlin and Tschabold (1990). Pessen (1989) presents a good survey of *binary sensors*, *actuators* and *switching elements* for both *electric* and *pneumatic* environments and contains several further references to this area. *Photoelectric* sensors are described in detail by Juds (1988).

Actuators represent a whole discipline in themselves. Fitzgerald, Kingsley and Umans (1983) is the standard book on *electrical machinery*, while Leonhard (1985) gives a more specific treatment of the control of electric drive systems. Kenjo (1984) has written a key reference on stepping motors. A modern and comprehensive treatment of *power electronics* and its application for motor control may be found in Mohan, Undeland and Robbins (1989).

Operational amplifiers are described in detail in Glasford (1986), Hufault (1986), Irvine (1987) and Jones (1986). The techniques for *grounding* and *shielding* are described further in Morrison (1967) and Ott (1976). Finally, Wilson and Hawkes (1983) give a good survey on optoelectronics.

5

Signal Processing

Aim: To describe the basic issues in signal conversion from analog to digital
form and the fundamentals of digital signal processing

The generation of the sensor signals and the transmission of analog signals were discussed
in Chapter 4. Adequate sensor technology and analog transmission techniques provide
the computer input with an analog signal, a pulse train or some binary signal that
corresponds to a physical variable. The different components in the computer input and
output (see Figure 5.1) are summarized in Section 5.1. Sampling means to read a continuous
signal at discrete points in time. The components that realize the sampling are described
in Section 5.2. The choice of an adequate sampling rate is of fundamental importance
for a process computer system, and some rules and methods are also explained in Section
5.2. In Section 5.3 the conversions between analog and digital forms are recapitulated.

Before any continuous signal can be sampled one has to ensure that the signal contains
only relevant frequencies and that all irrelevant or undesired components such as high
frequency noise are removed. Analog filters are used for this purpose and are examined
in Section 5.4.

Once the analog signal has been converted to digital form, digital filtering provides
a means to extract interesting information from the signal (Section 5.5). Some preliminary
tests of the digital signal are drawn up in Section 5.6. Such testing is crucial in order
to avoid the measurement value becoming meaningless.

5.1 INTERFACING ANALOG SIGNALS TO A COMPUTER

The computer input and output modules contain several essential links and this chapter
will describe them in some detail. A more detailed look at computer input and output is
given in Figure 5.1. After signal conditioning, an analog filter is usually needed to reject
undesirable frequency components of the original signal. In particular, it is necessary to

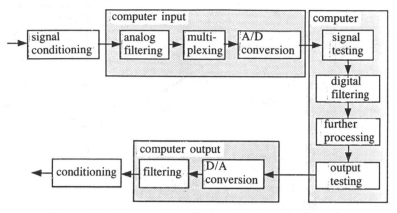

Figure 5.1   An overview of computer input and output.

eliminate high frequency noise before it is time-discretized by the computer (sampled). The sampling rate of a continuous signal is of fundamental interest. Theoretically, the sampling frequency must be equal to or exceed a frequency which is twice as large as the highest frequency component in the analog signal — including both signal and noise. In practice, the sampling rate must be even higher.

The essential components in the sampling procedure are the sample-and-hold circuit and the multiplexer. The former keeps the signal constant between the sampling instances and the latter works like a switch between computer input and the different signal input channels. The signal is now ready to be converted from analog to digital form.

Before the measurement value is used for further processing in the computer, it has to be tested to make sure that it is acceptable. By digital filtering it is possible to dampen irrelevant components of the signal. After computation of a control signal, the computer has to test the value before it is sent to the computer output.

Digital representation of the control signal is converted to analog form in the digital-to-analog (D/A) converter. In system descriptions of time-discrete systems (see Section 3.4.1) we have assumed that the control signal is constant between the sampling instances. This is realized with a sample-and-hold circuit in the computer output.

5.2 SAMPLING OF ANALOG SIGNALS

We have already seen in Section 3.4 how analog signals cannot be read continuously into the computer, but instead are fetched only at intermittent intervals. Thus, a signal is represented in the computer only by a sequence of discrete values. The operation of reading a signal only at determined instances in time is called **sampling**. Sampling is understandably very important; it is performed in the computer by a special unit (Figure 5.2). Sampling is directly related to multiplexing and analog-to-digital conversion; these operations have to be properly synchronized under the control of a clock.

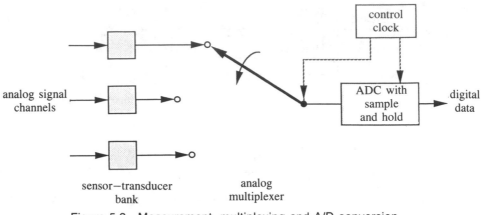

Figure 5.2   Measurement, multiplexing and A/D conversion.

5.2.1 Sampling Units

Sampling itself takes quite a short time. However, any signal transients during the conversion time should not affect the output of the conversion unit. To guarantee this, a **sample-and-hold operation** is required during each sampling period, so that the value of an analog signal is detected (sampled) at the beginning of each sampling period and kept constant throughout the sampling period. This is called a **zero order hold operation** and is illustrated in Figure 5.3. The sampled signal is delayed by about half a sampling period h after the continuous signal.

In a control system, this delay causes a phase lag and makes the stability margins smaller for a time-discrete controller in comparison with a corresponding continuous controller (see also Section 9.6.1). Note that a similar 'sample-and-hold' was assumed

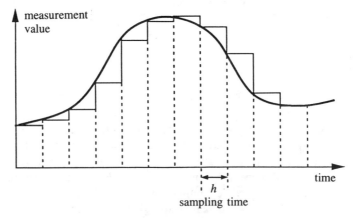

Figure 5.3   Sampling a continuous signal with zero order hold.

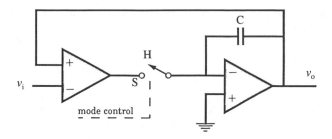

Figure 5.4   Unity-gain sample-and-hold amplifier. In sample mode (S) $v_0 = v_i$. In hold
mode (H) the output is constant.

in the numerical simulation of non-linear systems (Section 3.3.7) and in time discretization
of continuous dynamic systems (Section 3.4.1).

A sample-and-hold circuit is shown in Figure 5.4. The operation of the circuit is
controlled by a switch. It is closed (S) at the sampling instant and the capacitor C is loaded
to the actual value of the input signal. When the switch is opened (H) the output of the
op-amp is ideally constant and the same as the output value when the switch was closed.

5.2.2 Multiplexers

In many situations a limited resource has to be shared among several instances. This is
the case when a series of input signals from different sensors have to be transmitted along
the same physical channel. The signals cannot be fed to the computer at the same time
because they would mix together and lose their meaning. With multiplexing, the computer
selects which signal is read each time. Simply stated, the **multiplexer** is a switch that
connects the computer with one sensor at the time.

Multiplexing is a general principle by which a limited resource (e.g. the input unit
of the computer or a long measurement cable) is assigned to one requesting entity at a
time. Multiplexing is not limited to measuring signals but plays — albeit in different form
— an important role in bus systems (Section 6.2.4) and communication (Section 10.4.2).

A multiplexer can be either electromechanical or electronic and operates roughly like
a switch. The switching order can be either sequential or arbitrary. An electromechanical
multiplexer built with reed switches is a durable, although somewhat slow, system in the
order of a hundred periods per second. The operating life is naturally limited because
of the moving parts, but such a system is robust and has good isolation characteristics
and low cost. Also, the voltage drop over the multiplexer is low, which can be important.

With a solid-state multiplexer, sampling can be completed in a few microseconds.
Combined with an isolation amplifier, it has very good characteristics, but costs more
than the relay multiplexer.

Current leakage and voltage surges at the inputs to a multiplexer can be a problem.
An isolation amplifier between a sensor and the computer input picks up differential voltages
but the potential can 'move' relative to the ground. Possibilities to galvanically isolate

the wires (flying capacitor) from the multiplexer and/or the ADC were discussed in Section 4.7.3.1 (Example 4.10).

5.2.3 Choosing an Adequate Sampling Time

The sampling interval has to be sufficiently short for the continuous signal variations to be truthfully described by the discrete time signal. If the sample time is too long, the computer will get a wrong picture of the original signal. Too short a time interval is not good either because it requires unnecessarily high computer processing capacity; in addition, faster sampling units are usually more expensive. We see, therefore, that the choice of an adequate sampling time is important and not in the least a trivial issue.

Since nothing can be known about the measurement variable between the sampling points, the sampling period has to be sufficiently short so that the continuous signal does not change significantly in the meantime. Moreover, the sampling frequency has to be sufficiently large so that the continuous signal can be reconstructed from the sampled signal. The lower limit of the sampling frequency is apparently related to the process dynamics, i.e. how quickly the original signal varies. By its very nature, with sampling some information about the original signal is invariably lost. The central issue with sampling is therefore to collect all the information that is needed for signal processing.

Example 5.1 Sampling of a sinusoidal signal

Assume that an analog signal has a sinusoidal form (frequency f) and is sampled with a frequency f_s. When the signal is sampled six times per cycle (Figure 5.5) a smooth curve through the sampling points is close to the original continuous curve and the observed frequency f_0 is close to the true frequency f. When the sine wave is sampled only three times per cycle, a smooth curve through the sampling points is a less reliable representation of the original curve, but the observed frequency f_0 is still equal to the true frequency f if we sample for a sufficiently long time.

When the continuous curve is sampled only 5/4 times per cycle (i.e. 5 times in 4 cycles), a smooth curve through the sampled points appears to be a sine wave (Figure 5.6), but the apparent frequency f_0 is only $f/4$, much lower than the original f. We note that the alias (false) frequency is the difference between the sampling frequency ($5f/4$) and the real frequency (f).

If the sampling frequency is too small a false frequency (called an **alias frequency**) appears as shown in Example 5.1. We note that the apparent (alias) frequency f_0 is the difference between the sampling frequency f_s and the real frequency f:

$$f_0 = f_s - f \tag{5.1}$$

The apparent frequency is the same as the real frequency as long as the sampling frequency is sufficiently high, i.e. $f/f_s < 0.5$. When f exceeds $f_s/2$ the apparent frequency drops linearly and reaches zero at $f = f_s$ or one sample per cycle. It is obvious that if the

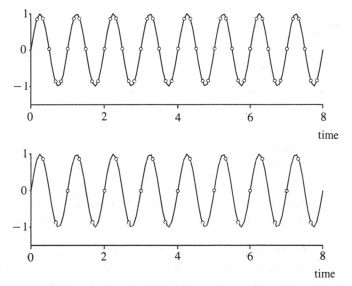

Figure 5.5   When the sine wave is sampled six or three times per cycle the observed frequency is equal to the true frequency.

sampling frequency is exactly 1 sample per 1, 2, 3, 4, ... cycles, then the signal is sampled precisely at the same phase and the apparent frequency becomes zero. The relation between the apparent and real frequencies looks like a sawtooth curve (Figure 5.7).

It appears that for a sample frequency f_s smaller than the double real frequency f, the real frequency can no longer be reconstructed from the sampled values. The frequency $f = f_s/2$ is called the **Nyqvist frequency** f_N,

$$f_N = f_s/2 \tag{5.2}$$

It is important to understand that if the continuous signal contains any frequencies above $f_N = f_s/2$, these high frequency components will appear in the sampled data as waves of

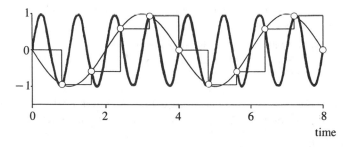

Figure 5.6   When the continuous curve is sampled five times in four cycles then the observed sine wave has a much lower frequency than the original sine wave.

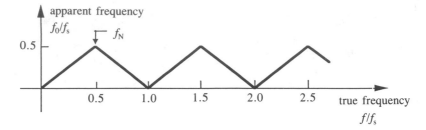

Figure 5.7   The apparent frequency f_0 as a function of the true frequency for a sine signal sampled at the frequency f_s. The apparent frequency is equal to the true one only if $f/f_s < 0.5$.

lower frequency. Frequency aliasing is avoided only if the original signal is sampled at a frequency at least twice as high as the highest frequency component in the signal. This is the essence of the **sampling theorem**.

In practice, the sampling frequency has to be higher than that stated by the sampling theorem. The sampling theorem is based on the assumption that the original signal is periodic and is sampled during an **infinite** time. Since this is obviously not the case in a control system, a higher sampling frequency is needed to collect enough information to adequately describe the signal. There is no theorem that gives a lower limit for the sampling rate when the signal is not periodic. Yet a number of practical rules can be formulated. These will be commented on further in Section 9.6.1.

Usually analog signals contain high frequency noise components. *All* frequencies above the Nyqvist frequency have to be removed before sampling. (Note that it is often incorrectly stated that all frequencies *of interest* must be less than the Nyqvist frequency.) This is done in a low pass analog filter (**anti-alias filter**) which is described in Sections 5.4.1 and 5.4.2.

Figure 5.7 indicates that frequencies equally above or below the Nyqvist frequency $(f_N \pm \Delta f)$ appear to be the same.

Example 5.1 revisited

The alias frequency $f/4$ and the real frequency f lie symmetrically around the Nyqvist frequency $f_N = f_s/2 = 5f/8$, i.e.

$$f_N - \Delta = 5f/8 - 3f/8 = f/4 \qquad \text{(alias frequency)} \qquad\qquad (5.3a)$$
$$f_N + \Delta = 5f/8 + 3f/8 = f \qquad \text{(real frequency)} \qquad\qquad (5.3b)$$

After sampling, there is no way to correct the data. A frequency f cannot be distinguished from its alias frequencies $f + nf_s$. Thus, any of the alias frequencies:

$$f_s - f; f_s + f; 2f_s - f; 2f_s + f; \ldots$$

may appear, if f is higher than the Nyqvist frequency $f_N = f_s/2$.

Example 5.2 Illustration of alias distortion

We will illustrate the alias distortion by a rotating disc with a black spot at the periphery. The disc is rotated with different speeds and illuminated by a stroboscopic lamp which emits short light flashes at given intervals of, say, 1 second. Thus the spot can be seen only intermittently.

If the disc rotates 10° per second clockwise, it is easy to see the black spot in positions 0°, 10°, 20°, ..., etc. Similarly, if the disc rotates counterclockwise, the spot can be seen in positions 0°, 350°, 340°, ..., etc. If the angular speed is increased, the spots will be seen farther apart.

If the disc rotates at 180° per second, the spot will only be visible at 0° and 180° and it is no longer possible to determine the direction of rotation. Assume that the disc is rotated clockwise at 215° per second. It becomes quickly apparent that the spot will appear the same as for a counterclockwise rotation at 135° per second or $135° + n360°$ ($n = 1, 2, ...$) per second.

The angular speed of 180° per second corresponds to the Nyqvist frequency. From the discussion we can see that real frequencies that are equally above or below the Nyqvist frequency ($f_N \pm \Delta f$) appear to be the same (compare with Figure 5.7). The apparent frequency is small near multiples of the sampling frequency (f_s, $2f_s$, $3f_s$, etc.). Thus 10°, 350° and 370° per second appear to have the same frequency after sampling.

The sampling of sine signals in Example 5.1 is analogous to the rotating disc. The sine is the vertical projection of the spot and clockwise or counterclockwise rotation implies different phases for the sine signal.

In old western films, wagon wheels are often seen to slowly rotate backwards. The film 'samples' 24 times per second. If a wheel has N spokes it appears to stand still if it rotates at precisely $1/N$ revolutions in 1/24 seconds. If it rotates slightly faster, then the wheel appears to rotate slowly forward and it will seem to rotate slowly backwards if it rotates slightly slower than $1/N$.

Example 5.3 Alias due to power cable alternating current

High frequency signals are often overlapping low frequency signals. An example is 50 Hz alternating current that is picked up by signal wires from power cables (Figure 5.8). We demonstrate the alias problem by a simulated example. The signal is sampled with $f_s = 60$ Hz. Since the Nyqvist frequency is smaller than 50 Hz, an alias frequency appears, in this case $60 - 50 = 10$ Hz.

Example 5.4 Sampling sludge concentration from a sedimentation unit

This example illustrates that several factors influence the choice of sampling rate. The sedimentation unit in the activated sludge process (Figure 2.15) is used to separate solids from liquid. The thickened sludge is withdrawn from the bottom of the sedimentation unit. Since most of the sludge will be recirculated it is important to know its concentration. This usually

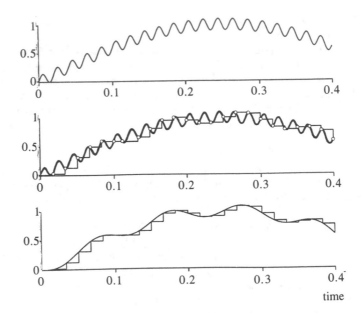

Figure 5.8 Sampling of a measurement signal with a high frequency component. The sampled signal contains an oscillation that is not present in the initial signal.

varies quite slowly, typically on an hourly timescale. Therefore, a sampling interval of about 30 minutes seems to be adequate. Some experimental values from a sedimentation unit are shown in Figure 5.9.

The concentration has significant peaks every 12 minutes. They are caused by a sludge scraper that rotates along the bottom to remove sludge through the bottom valve. The rotation period is 12 minutes. Every time the scraper passes the bottom valve, compacted sludge is brought by the suspended solids sensor. An adequate sampling time has to be in the order of minutes, and the proper process concentration is calculated as some 30-minute moving average value of the measurements (Section 5.5.1).

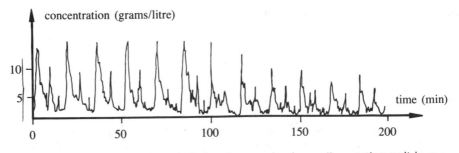

Figure 5.9 Settled activated sludge from a circular sedimentation unit in a wastewater treatment plant.

5.3 CONVERSION BETWEEN ANALOG AND DIGITAL SIGNALS

Once a value is sampled, it is encoded into a digital representation. The analog signals are converted to digital form in an analog–digital conversion (ADC) unit. Similarly, the digital output signals from the computer are converted to analog control signals in a digital–analog converter (DAC).

5.3.1 Digital–analog Conversion

An ideal **digital–analog converter (DAC)** produces an analog output proportional to the digital input. The input is represented in an n bit digital register giving an output of resolution 2^n. Figure 5.10 shows the most common design where a resistor ladder with resistors R and $2R$ is coupled around an operational amplifier.

Switches $s_1, \ldots, s_n$ represent 0 or 1 and are controlled by the digital input. The output voltage is composed by successively smaller terms

$$v_o = -v_{\text{ref}}\left(\frac{s_1}{2^1} + \frac{s_2}{2^2} + \cdots + \frac{s_n}{2^n}\right) \tag{5.4}$$

A DAC can also be designed for current outputs.

There are several DAC properties that have to be considered, such as:

- Linearity, i.e. a linear relationship between output voltage and digital input.
- Offset, the value of the output voltage for a zero input.
- A **glitch** is a transient spike in the output of a DAC that occurs when more than one bit changes in the input code and the corresponding switches do not change simultaneously. A solution is a so-called deglitcher, a sample-and-hold circuit that holds the output constant until the switches settle.
- Settling time, the time for the output voltage to settle.

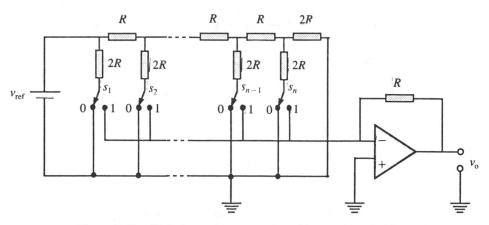

Figure 5.10   Digital–analog converter with a resistor ladder.

- The **slew rate** (expressed in V μs^{-1}) is the maximum rate of change of the output voltage.

5.3.2 Analog–digital Conversion

An **analog–digital converter (ADC)** converts an analog voltage level (the analog input) to a binary number (the digital output). The ADC divides the input voltage range into $2^n - 1$ bands, where n is the number of bits in the output word. The conversion resolution is typically 10–12 bits, i.e. 1023 or 4095 intervals. Like the DAC, the essential parameters are resolution, offset error, linearity and conversion time. There are several different constructions of ADC units and the actual descriptions are usually found in the manufacturers' component catalogues.

It is important to make sure that the analog signal range is such that the ADC accuracy can be fully utilized. During a typical operation a sensor output signal may not vary between 0 and 100 per cent. Assume that the normal variation is only between 20 and 25 per cent of the sensor range. This 5 per cent variation range, after conversion in a 10 bit converter, consists of 5 per cent of 1023 or about 50 intervals. Thus, the digital representation has a resolution of only 2 per cent (Figure 5.11(a)). Instead, the ADC should indicate 0 at 20 per cent sensor signal and 1023 at the 25 per cent level.

To use the full range of the ADC, both the gain and the offset of the input analog signal have to be adjusted. This can be performed with an operational amplifier circuit (Figure 5.11(b)). The offset adjustment is made with the variable resistor R_1 such that the d.c. level is the same as the one required by the ADC. The variable resistor R_2 is used to adjust the gain, so that the maximum analog input will correspond to the maximum converted value in the ADC.

When current transmission is used, a broken connection can be detected by a 0 mA signal (see Section 4.7.5). An ADC can also be used to indicate if a sensor is out of operation. If the ADC is calibrated so that the maximum input signal (e.g. 20 mA) is equivalent to, say, 4000 instead of 4096 in a 12-bit converter, then the highest values of the converter can be used to indicate unpermitted values.

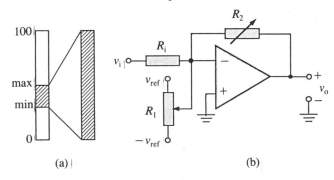

Figure 5.11 (a) Use the full range (0–100 per cent) of the ADC; (b) offset and gain adjustment.

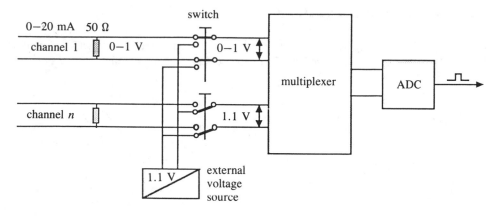

Switch channel 1 on: sensor connected
Switch channel *n* off: sensor disconnected

Figure 5.12   Indicating an off-line sensor. Switch channel 1 on: sensor connected; switch channel *n* off: sensor disconnected.

A switch is placed before the multiplexer (Figure 5.12). In the *on* position the terminal resistor is connected to the sensor. In the *off* position an external voltage source supplies a constant voltage somewhat greater than the voltage equivalent to 20 mA. When the sensor is not connected (during service or calibration) the switch is turned *off* and the computer can identify a measurement value 'sensor off-line'.

The conversion rate for a *typical* ADC is $0.5 - 400 \, \mu s$. A high conversion rate is important if it is desired to measure several signals at the 'same' time, so that the measurement procedure is completed within a fraction of the sampling interval.

5.4 ANALOG FILTERING

Analog filtering is applied to reduce certain frequency components of the signal and is useful whenever the unwanted signals have a frequency content other than the desired signal. Note that analog filtering should be used to reduce electric noise (Section 4.7.3) *only after* proper shielding and differential amplification has reduced the noise as much as possible.

The **passband** and **corner (cut-off) frequency** are two fundamental characteristics of analog filters. The passband is the range of frequencies that are passed unfiltered. At the corner frequency the amplitude has dropped a factor of $1/\sqrt{2}$ from the passband.

A **low pass** filter is designed to pass frequencies below a specified corner frequency and dampen higher frequencies. It is used to eliminate or minimize frequency aliasing and is also called an **anti-alias filter**. A **high pass filter** is designed to pass high frequencies and attenuate low frequencies.

5.4.1 Low Pass Filters — First Order

The simplest form of a low pass filter is a **passive *RC* circuit** (see Example 3.4, Section 3.2.2). The filter is described by the first-order differential equation:

$$T\frac{dv_o}{dt} = -v_o + v_i \tag{5.5}$$

where $T = RC$, v_o the output voltage over the capacitor and v_i the input voltage (Figure 3.6). The filter has unit **static gain**, i.e. $v_o = v_i$ when the derivative is zero. From the Laplace transform of Equation 5.5, the transfer function of the filter is found to be:

$$G(s) = \frac{V_o(s)}{V_i(s)} = \frac{1}{1 + sRC} = \frac{1}{1 + sT} \tag{5.6}$$

In Example 3.4 it was shown that output amplitude for a sine wave input will drop by a factor of $1/\sqrt{[1 + (\omega RC)^2]} = 1/\sqrt{[1 + (\omega T)^2]}$. The static gain is 1 as expected. The cut-off frequency (where the attenuation is $1/\sqrt{2}$) is

$$\omega_c = \frac{1}{RC} = \frac{1}{T} \; (\text{rad s}^{-1}) \quad \text{or} \quad f_c = \frac{\omega_c}{2\pi} = \frac{1}{2\pi RC} = \frac{1}{2\pi T} \; (\text{Hz}) \tag{5.7}$$

Inserting f_c in Equation 5.6 we see that the filter attenuates the amplitude with a factor of:

$$|G| = \left|\frac{1}{1 + j\omega RC}\right| = \left|\frac{1}{1 + j(f/f_c)}\right| = \frac{1}{\sqrt{[1 + (f/f_c)^2]}} \tag{5.8}$$

for sine inputs, i.e. the attenuation above the cut-off frequency is a factor of 10 for every tenfold increase in frequency above the cut-off frequency (Figure 5.13).

Looking in the time domain (compare with Example 3.4), a step change of the input voltage will result in an exponential increase of the output voltage amplitude, with the time constant, RC seconds (if we define R and C in terms of elementary dimensions their product actually has the dimension of time). An **active** low pass filter is obtained by an RC feedback around an operational amplifier (Figure 5.14).

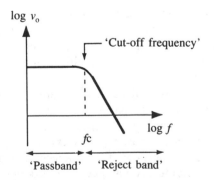

Figure 5.13 Frequency characteristics of a first-order low pass filter.

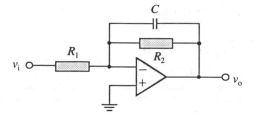

Figure 5.14 An operational amplifier with *RC* feedback, which (under ideal conditions) is a first-order low pass filter.

Generally, the frequency dependent gain $G(j\omega) = v_o/v_i$ for an ideal op-amp with a negative feedback network can be expressed by the ratio of the feedback impedance to the input impedance. For the *RC* filter, the feedback impedance (C in parallel with R_2) is $R_2/(1 + j\omega R_2 C)$ which makes the output voltage gain:

$$|G(j\omega)| = \left|\frac{V_o(j\omega)}{V_i(j\omega)}\right| = \left|\frac{R_2}{R_1}\right| \times \left|\frac{1}{1 + j\omega R_2 C}\right| \tag{5.9}$$

The frequency dependence is the same as for the passive filter, but the amplitude gain can be chosen by resistors R_1 and R_2. In practice, however, op-amp coupling is not a perfect first-order filter. The reason is the limited **slew rate** of the amplifier. The slew rate is the maximum rate of change of voltage of the amplifier output when the input is changed stepwise. This means that very fast signals slip through the filter. A more practical solution is to filter the signal in a passive low pass filter (see Figure 3.6) followed by a voltage follower (Figure 4.25).

Example 5.5 A passive *RL* low pass filter

A resistance–inductance (*RL*) circuit (Figure 5.15) acts as a low pass filter. Kirchhoff's voltage law applied to the circuit gives:

$$v_i - v_L - v_o = 0 \tag{5.10}$$

where the inductor voltage v_L is expressed by $v_L = L(di/dt)$ and the current $i = v_o/R$. Replacing v_L and i we obtain the low pass filter equation (Equation 5.5):

$$T\frac{dv_o}{dt} = -v_o + v_i \tag{5.11}$$

where $T = L/R$.

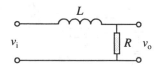

Figure 5.15 A passive first-order low pass *RL* filter.

The transfer function is similar to Equation 5.6:

$$G(s) = \frac{V_o(s)}{V_i(s)} = \frac{1}{1 + s(L/R)} = \frac{1}{1 + sT} \tag{5.12}$$

and the cut-off frequency $f_c = R/2\pi L$ Hz.

The frequency dependence is similar to the RC filter. Components with frequencies much higher than the cut-off frequency are 'choked' by the inductance and appear greatly reduced in amplitude at the output.

5.4.2 Low Pass Filters — Higher Order

Sometimes the slope of the high frequency characteristics of an RC filter is not sufficiently large, which means that the high frequencies are not efficiently cut off. A second-order low pass filter has a frequency slope at high frequencies that is twice as high as that of a first-order filter (see Figure 5.13) and has an attenuation factor of 100 for every tenfold increase of the frequency. Consequently, such a filter is more efficient for removing undesired frequencies.

Example 5.6 A second-order low pass filter

The second-order low pass filter in Figure 5.16 has two independent corner frequencies, f_{c1} and f_{c2}. The frequency dependent voltage gain is:

$$|G(j\omega)| = \left|\frac{V_o(j\omega)}{V_i(j\omega)}\right| = \frac{-R_3/(R_1 + R_2)}{\{[1 + (f/f_{c1})^2][1 + (f/f_{c2})^{-2}]\}^{1/2}} \tag{5.13}$$

where $f_{c1} = R_1 R_2/(2\pi(R_1 + R_2)C_1)$ and $f_{c2} = 1/(2\pi R_3 C_2)$. If the corner frequencies are made equal, $f_c = f_{c1} = f_{c2}$, then the magnitude of G is:

$$|G| = \frac{-R_3/(R_1 + R_2)}{1 + (f/f_c)^2} \tag{5.14}$$

The corner frequency f_c corresponds to an amplitude drop of a factor of 2.

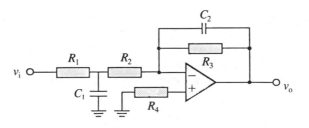

Figure 5.16   A second-order low pass filter.

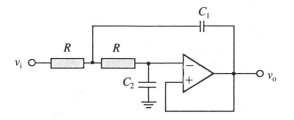

Figure 5.17   A unity-gain second-order Sallen–Key low pass filter.

A **Butterworth filter** has a flat frequency response below a characteristic frequency f_0. As it transmits signal amplitudes faithfully, it is popular as an anti-alias filter. It can be realized as a special case of a filter known as the **Sallen–Key filter**. A second-order version of such a filter is shown in Figure 5.17. For a Butterworth filter the components have to satisfy $2\pi RC_1 f_0 = 1.414$ and $2\pi RC_2 f_0 = 0.707$, respectively. Higher order filters (4, 6, ...) use cascaded stages of second-order filters.

5.4.3 High Pass Filters

One might suspect that by interchanging the capacitance and resistance in Figure 3.6 or the inductance and resistance in Figure 5.15 the result would be a high pass filter (Figure 5.18). Following the same procedure as in Example 3.4 for a high pass RC filter, the output voltage v_o can be expressed as:

$$T\frac{dv_o}{dt} = -v_o + T\frac{dv_i}{dt} \tag{5.15}$$

where $T = RC$. Its transfer function is:

$$G(s) = \frac{V_o(s)}{V_i(s)} = \frac{sRC}{1 + sRC} = \frac{sT}{1 + sT} \tag{5.16}$$

Its frequency dependent voltage gain is expressed by:

$$|G(j\omega)| = \left|\frac{V_o(j\omega)}{V_i(j\omega)}\right| = \left|\frac{j\omega RC}{1 + j\omega RC}\right| = \frac{\omega RC}{\sqrt{[1 + (\omega RC)^2]}} = \frac{\omega T}{\sqrt{[1 + (\omega T)^2]}} \tag{5.17}$$

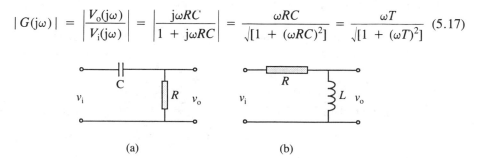

(a)   (b)

Figure 5.18   (a) A passive high pass RC filter; (b) a passive high pass RL filter.

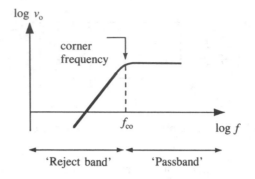

Figure 5.19 Frequency characteristics of a first-order high pass filter.

The circuit will block low frequencies and pass high frequencies, as shown in the frequency diagram (Figure 5.19).

The corner frequency f_{co} is the frequency where the amplitude drop is a factor $1/\sqrt{2}$ of the high frequency gain, $f_{co} = 1/(2\pi RC) = 1/(2\pi T)$ Hz. The voltage gain can be written as:

$$|G| = \frac{(f/f_{co})}{\sqrt{[1 + (f/f_{co})^2]}} \tag{5.18}$$

By arranging the feedback around an *op-amp*, an active high pass filter is obtained (Figure 5.20). Its frequency dependent voltage ratio is the ratio between the feedback and the input impedances:

$$|G| = \left|\frac{-j\omega R_2 C}{1 + j\omega R_1 C}\right| = \frac{\omega R_2 C}{\sqrt{[1 + (\omega R_1 C)^2]}} \tag{5.19}$$

All op-amps have a limited bandwidth which results in a decrease in gain at sufficiently high frequencies. Strictly speaking, all active high pass filters are actually **band-pass filters**, having a passband between two reject bands.

Example 5.7 A passive *RL* high pass filter

The passive high pass filter (Figure 5.18(b)) is defined by its differential equation derived

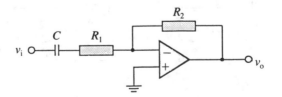

Figure 5.20 An active high pass first-order filter.

from the Kirchhoff voltage law:

$$T\frac{dv_o}{dt} = -v_o + T\frac{dv_i}{dt} \qquad (5.20)$$

which is identical with Equation 5.15 with $T = L/R$. The transfer function is:

$$G(s) = \frac{V_o(s)}{V_i(s)} = \frac{sL}{R + sL} = \frac{sT}{1 + sT} \qquad (5.21)$$

and the frequency dependent gain:

$$|G(j\omega)| = \left|\frac{V_o(j\omega)}{V_i(j\omega)}\right| = \left|\frac{j\omega L}{R + j\omega L}\right| = \frac{\omega(L/R)}{\sqrt{[1 + (\omega L/R)^2]}} = \frac{\omega T}{\sqrt{[1 + (\omega T)^2]}} \qquad (5.22)$$

The corner frequency is $f_{co} = R/(2\pi L) = 1/(2\pi T)$ Hz. With this definition of f_{co}, the voltage gain can be written as Equation 5.18.

5.5 DIGITAL FILTERING

After analog filtering and A/D conversion, further filtering possibilities are offered through digital filtering. Digital filtering provides a great deal of flexibility, since the filter characteristics can easily be changed by tuning a few parameters in the computer. Unlike their analog counterparts, digital filters are easy to use for long time constants.

5.5.1 General Structure

A **digital filter** has the general form:

$$\begin{aligned}\hat{y}(kh) &= -a_1\hat{y}[(k-1)h] - a_2\hat{y}[(k-2)h] - \ldots - a_n\hat{y}[(k-n)h]\\ &\quad + b_0 y(kh) + \ldots + b_m y[(k-m)h]\end{aligned} \qquad (5.23)$$

where h is the sampling interval, $\hat{y}$ the filtered output and y the input measurement value. Note that the argument kh can be considered a simple integer indicator of the parameters. If all the as are zero we will have a **moving average (MA) filter** with a finite impulse response. If some or all of the as are non-zero there is an **autoregressive (AR) filter** which has an infinite impulse response. The filter (Equation 5.23) is in general called an **autoregressive moving average (ARMA)** time series.

There are both **causal** and **non-causal filters**. A causal filter calculates an output value based on old input data. For this reason all *on-line* filters are causal. The filtered time series will be time lagged compared with the original time series. If the data can be processed *off-line*, such as in the analysis phase of a measurement data series, a non-causal filter can be used. Then a measurement value at time t can be smoothed by weighting together both past and future values.

5.5.2 Digital Low Pass Filters

In order to examine slow variations it is necessary to remove individual spikes in the measurement data and other quick disturbances which do not contain relevant information. This is done with a **digital low pass filter**. Constructing a filter which effectively removes the quick variations, and at the same time does not affect the slow variations, is always a compromise. As for the analog filters, a higher order filter dynamics is efficient in removing undesirable high frequencies. The two most important types of low pass filters are **moving averages** and **exponential smoothing**. Low pass filters in the process industry are almost always implemented with these types of simple filters.

Example 5.8 Moving average — the simplest low pass filter

A simple moving average filter is obtained by setting all the a parameters in Equation 5.23 equal to zero. For simple averaging all the weighting factors are the same and the sum of the coefficients is equal to one. For example, a 5-point moving average is calculated by:

$$\hat{y}(kh) = \tfrac{1}{5}\{y(kh) + \ldots + y[(k - 4)h]\} \tag{5.24}$$

When making off-line analysis a non-causal moving average can be calculated using measurements both before and after the current time kh. Then the filtered value is not lagged in relation to the input values. A non-causal simple moving average for five values resembles:

$$\hat{y}(kh) = \tfrac{1}{5}\{y[(k - 2)h] + \ldots + y[(k + 2)h]\} \tag{5.25}$$

If the output is the average of the input over the last n samples, it is also shifted $1 + n/2$ cycles. For increasing n, the filter output becomes smoother but more delayed. The MA has a finite impulse response. If an impulse is given at time $t = 0$ then the MA after time $t = n$ is zero.

The moving average is a simple method but has certain limitations. If equal coefficients are used, the filter can be unnecessarily sluggish and not react adequately to real changes. On the other hand, if the coefficients taper off, it is difficult to analyse how the magnitude of the filter coefficients are related to the filter properties.

An **exponential filter** is a first-order ARMA filter and is defined by:

$$\hat{y}(kh) = \alpha\hat{y}[(k - 1)h] + (1 - \alpha)y(kh) \tag{5.26}$$

The filtered value $\hat{y}(kh)$ is computed by adding a weighted version of the earlier value of the filtered signal $\hat{y}(kh - h)$ to the latest measurement value $y(kh)$. The coefficient α has a value between 0 and 1. Equation 5.26 can be rewritten in the form:

$$\hat{y}(kh) = \hat{y}[(k - 1)h] + (1 - \alpha)\{y(kh) - \hat{y}[(k - 1)h]\} \tag{5.27}$$

which reveals another interpretation. The exponential filter corrects the filtered value as soon as a new measurement arrives. The correction has little gain and becomes small if α is near 1, which means that the filter is sluggish. This will greatly reduce the noise but at a cost of poor agreement with real changes in the measurement signal. If α is small, near 0, the correction gain is large. Consequently, there is a poorer reduction of the noise level but the filter will track real signal changes more easily. For $\alpha = 0$, the filter output is identical with the measurement value. A signal that is changed stepwise and is corrupted by noise (Figure 5.21) is used to illustrate the consequences of different choices of α.

Example 5.9 Interpretation of the exponential filter as an MA filter

The exponential filter can be interpreted as a moving average filter with an infinite number of *b* coefficients and no *a* coefficients in Equation 5.23. The *b* coefficients are gradually smaller for older values. To see this we rewrite Equation 5.26 as:

$$
\begin{aligned}
\hat{y}(kh) &= \alpha\hat{y}[(k - 1)h] + (1 - \alpha)y(kh) \\
&= (1 - \alpha)y(kh) + \alpha(1 - \alpha)y[(k - 1)h] + \alpha^2\hat{y}[(k - 2)h] \qquad (5.26') \\
&= (1 - \alpha)y(kh) + \alpha(1 - \alpha)y[(k - 1)h] + \alpha^2(1 - \alpha)y[(k - 2)h] + \alpha^3\hat{y}[(k - 3)h] \\
&\;\;\vdots \\
&= (1 - \alpha)y(kh) + \alpha(1 - \alpha)y[(k - 1)h] + \ldots + \alpha^n(1 - \alpha)y[(k - n)h] + \ldots
\end{aligned}
$$

Since $0 \leq \alpha \leq 1$, the coefficients are decreasing in an exponential fashion for older values. For example, for $\alpha = 0.5$ the coefficients are 0.5, 0.25, 0.125, ..., while for $\alpha = 0.9$ they are 0.1, 0.09, 0.081, etc. In other words, if α is close to 1 the filter has a longer 'memory' and is more efficient in smoothing the signal. The exponential decay of the size of the coefficients has given the filter its name.

The exponential filter is actually a time-discrete form of a first-order analog low pass filter with a unit static gain (see Section 5.4.1) with the transfer function (Equation 5.6):

$$
G_f(s) = \frac{\hat{Y}(s)}{Y(s)} = \frac{1}{1 + sT} \qquad (5.28)
$$

The time constant T corresponds to RC or L/R in the first-order analog filters. The filter differential equation is:

$$
T\frac{d\hat{y}(t)}{dt} = -\hat{y} + y \qquad (5.29)
$$

When the derivative is approximated by *backward* differences we obtain:

$$
\frac{\hat{y}(t) - \hat{y}(t - h)}{h} \approx -\frac{1}{T}\hat{y}(t) + \frac{1}{T}y(t) \qquad (5.30)
$$

which is a valid approximation for small h, or

$$
\hat{y}(t) = \frac{1}{1 + h/T}\hat{y}(t - h) + \frac{h/T}{1 + h/T}y(t) \qquad (5.31)
$$

which is identical with Equation 5.26 where:

$$
\alpha = \frac{1}{1 + h/T} \quad \text{or} \quad T = \frac{\alpha h}{1 - \alpha} \qquad (5.32)
$$

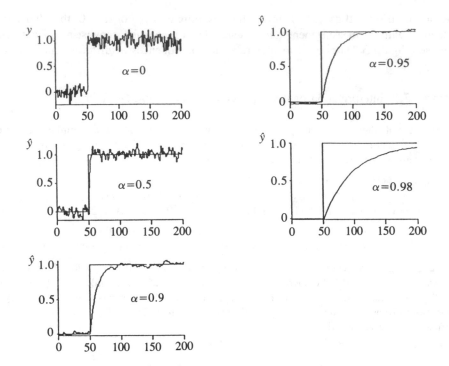

Figure 5.21   Effect of a first-order filter with exponential smoothing. The α parameter has the values, 0, 0.5, 0.9, 0.95 and 0.98. Note that the filter output follows the real signal change rapidly for small values of α, but the noise level is high. At high values of α the filter is slow, but the noise is greatly reduced.

Since it was assumed that h/T is small, the approximation is valid *only* if α is close to 1. Then α can be approximated as:

$$\alpha \approx 1 - \frac{h}{T} \quad \Rightarrow \quad T \approx \frac{h}{1 - \alpha} \tag{5.33}$$

In fact, the exact solution to the differential Equation 5.29 gives Equation 5.26 with:

$$\alpha = e^{-h/T} \quad \Rightarrow \quad T \approx -\frac{h}{\ln(\alpha)} \tag{5.34}$$

which can be approximated by Equation 5.33 for small values of h/T.

The filter response to the step change in Figure 5.21 can illustrate the relation between α and T. It takes one time constant for the step response to reach about 63 per cent of the final value. For $\alpha = 0.95$, T is about 20 sampling intervals and for $\alpha = 0.98$ it is about 50 sampling intervals.

5.5.3 Digital Low Pass Filters — Higher Order

A second-order analog filter is more efficient for eliminating high frequencies (see Section 5.4.2). Such a filter corresponds to a digital filter of the structure expressed by Equation 5.23, with $n = m = 2$. By placing two similar exponential filters in series we get a second-order filter with two equal corner frequencies:

$$\hat{y}_1(kh) = \alpha \hat{y}_1[(k - 1)h] + (1 - \alpha)y(kh) \tag{5.35a}$$

$$\hat{y}_2(kh) = \alpha \hat{y}_2[(k - 1)h] + (1 - \alpha)\hat{y}_1(kh) \tag{5.35b}$$

where y is the real measurement value, $\hat{y}_1$ is the output of the first filter and $\hat{y}_2$ the output of the second one. The filter performance can be tuned with parameter α. Eliminating $\hat{y}_1(kh)$, the second-order filter is written in the form:

$$\hat{y}_2(kh) = 2\alpha \hat{y}_2[(k - 1)h] - \alpha^2 \hat{y}_2[(k - 2)h] + (1 - \alpha)^2 y(kh) \tag{5.35c}$$

The same signal as in Figure 5.21 was filtered through the second-order filter and the result is shown in Figure 5.22.

Note that the second-order filter is more efficient for attenuating high frequencies so that a smaller value of α can be chosen. Therefore, the filter output can follow the real signal changes better than a first-order filter. With a filter of higher order (Equation

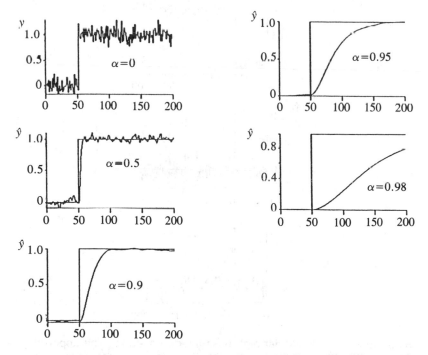

Figure 5.22 Effect of a second-order exponential filter with different values of the parameter α.

5.23) the characteristics can be further improved. The price to be paid lies in greater complexity, but it should be remembered that data processing is not expensive!

5.5.4 Digital High Pass Filters

In some instances it is desirable to highlight the higher frequencies instead of the slow variations. A **difference builder** is a simple example of a digital high pass filter:

$$\hat{y}(kh) = \Delta y(kh) = y(kh) - y(kh - h)$$ (5.36)

The output differs from zero only when a change of the signal occurs.

We will now derive a first-order digital high pass filter from its analog counterpart (Section 5.4.3). The differential equation (Equation 5.15) is repeated:

$$T\frac{d\hat{y}(t)}{dt} = -\hat{y}(t) + T\frac{dy(t)}{dt}$$ (5.37)

where y is the filter input and $\hat{y}$ the filter output. By taking forward differences we obtain a digital high pass filter:

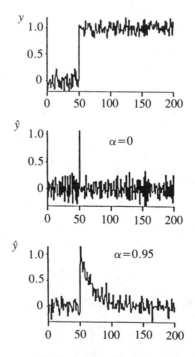

Figure 5.23   Effect of a first-order high pass filter on a signal. The upper diagram shows unfiltered data. The middle diagram shows a high pass filter output with $\alpha = 0$ and the lower diagram with $\alpha = 0.95$.

$$\hat{y}(t+h) = \left(1 - \frac{h}{T}\right)\hat{y}(t) + y(t+h) - y(t) = \alpha\hat{y}(t) + y(t+h) - y(t)\,(5.38)$$

where α is defined in Equation 5.33. The discrete filter equation can also be derived analytically from Equation 5.37 and we find that α is given by Equation 5.34 instead and has to lie between 0 and 1. When $\alpha = 0$, the filter is a pure difference builder. Note again that h/T has to be small if the difference approximation and Equation 5.33 should be valid.

The filter sensitivity for higher frequencies is tuned by the α value. A small α causes greater sensitivity corresponding to a high corner frequency of the high pass filter. Some examples are used to illustrate how a high pass filter works. The same noisy step input as in Figure 5.21 has been used in Figure 5.23. The middle diagram is the output of a pure difference builder ($\alpha = 0$). It contains a spike at $t = 50$ and the filter detects any sudden change. Using $\alpha = 0.95$, the spike at $t = 50$ becomes wider as shown in the lower diagram.

In Figure 5.24 the filter input is a sinusoidal curve with superimposed noise. The high pass filter output retains the high frequency variations and eliminates the slower sinusoidal oscillations. If a step change is added to the noisy sine wave (Figure 5.25) the high pass filter output displays a peak that indicates the step change.

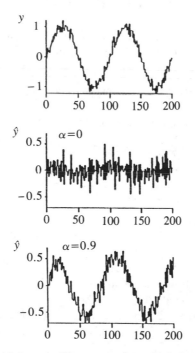

Figure 5.24   Effect of a high pass filter on a sinusoidal signal with overlying noise. The filtered signal (lower diagram) contains only the high frequency variations.

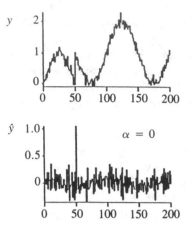

Figure 5.25   Effect of a high pass filter on a sinusoidal signal with overlying noise and a step change at time 200. The filter output (lower diagram) shows a peak at time 200 and contains no low frequency oscillations.

5.6 SIMPLE MEASUREMENT DATA PROCESSING

After a measurement value has been converted to digital form, several tests have to be performed before the data can be accepted, filtered and used for further processing. Initially, to make the A/D conversion the following procedures are needed:

- Addressing the correct measurement location, control of the multiplexer.
- Initiate the A/D conversion.
- Read the ADC register after completing the A/D conversion.

After conversion is completed it is necessary to:

- Test the validity of the measurement value with respect to the minimum and maximum values and the rate of change.
- Scale to suitable units.
- Linearize sensor signals.
- Calculate an average of the 'raw' measurement signals (possibly reject the greatest outliers).
- Calibrate and compensate for drift.
- Filter (digitally) the measurement value.
- Store the measurement value.
- Send alarm messages or other messages to the operator.
- Perform other data processing, registration and reporting.

Commercial data acquisition software usually contains these preliminary tests.

5.6.1 Tests and Alarms

There are many ways to test whether a measurement value is valid. Such tests are important so that incorrect control actions caused by erroneous measurement values can be avoided. The magnitude of the input signal must lie within the sensor range. A minimum requirement is to test for the signals maximum and minimum limits. Small oscillations around a limit can cause a lot of alarm signals. To avoid an unnecessary number of alarms, it is common to define a hysteresis band near the alarm limit. An alarm is generated when the signal passes the second limit. The signal then must have returned within the first limits before a new alarm can be activated at the second limit (Figure 5.26).

The rate of change of a signal is usually tested. This is a good way to detect a sensor error. If the change during the last few sampling intervals exceeds a predetermined value an alarm is generated. It is important that the rate of change check is performed *before* the digital filtering, otherwise it is not a relevant sensor check.

Example 5.10 Testing a dissolved oxygen probe

Dissolved oxygen (DO) in aeration tanks in activated sludge systems is measured with a probe that has a settling time of less than a minute. If the probe is taken out of the water for calibration and cleaning the read-out will increase within a minute from the normal range of about $2-5$ mg l^{-1} to the saturation value of about 10 mg l^{-1}. A real DO concentration increase in the tank cannot occur faster than $10-20$ minutes. Therefore, a significant signal change within only a minute can be detected as an outlier and be used to indicate the calibration and reset the scaling.

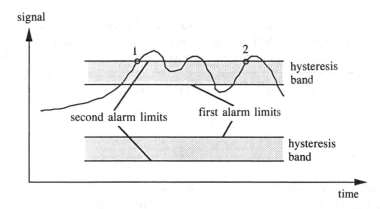

Figure 5.26   Hysteresis band near the alarm limit. An alarm is generated when the signal passes point 1. The signal then must have returned within the first limits before a new alarm is activated at point 2.

5.6.2 Scaling and Linearization

It is important to display the measurement values in suitable engineering units which reduces the risk of misunderstanding. The conversion from an internal representation z to an engineering unit y can often be made by a simple relationship such as $y = k_1 z + k_2$, where k_1 and k_2 are constants. For non-linear sensors (Section 4.2.4, Example 4.1 in Section 4.3.1, Example 4.5 in Section 4.3.3) the relationship is expressed as a function or table. It becomes more complex if the sensor characteristics include dead bands or hysteresis, since the direction in which the measurement signal is changing must be known.

In Section 5.3.2 it was discussed how the analog−digital converter's full range ought to be utilized in order to retain accuracy. If the measurement signal exceeds the A/D converter range, then it has to be assured that the converter output does not 'hop over' and start again from zero. This is incorporated as a standard feature in many data acquisition systems, but one has to check it is included.

5.6.3 Averaging

An initial rejection of erroneous measurement values can be made using simple averaging. Let the ADC sample in the order of 10 times faster than the desired measurement frequency. The 'raw' measurement value is then obtained by averaging. One or two extreme measurement values can be discarded. This reasoning assumes that the signal is constant during the time interval in which the averaging is made and that variations in the signal are noise with a zero average.

5.6.4 Calibration and Drift Compensation

Measurement signals often need to be compensated for drift or calibration errors in sensors or in electronics. Zero point drift or calibration error in the measuring device should be corrected regularly. Corrections can also be made with software (Example 5.10 in Section 5.6.1). The amplifier electronics *after* the multiplexer can be checked by connecting additional inputs to the multiplexer, one with the ground voltage and the others with reference voltages.

5.6.5 Plotting

Simple plots of signals which are functions of time, as functions of each other can reveal interesting details for a process engineer or plant expert, such as:

- Exceptional disturbances
- Missing measurement values
- Tendencies for periodic oscillations

Therefore, plot facilities *in different timescales* are an essential utility in any process computer system.

5.6.6 A Data Structure for Measurement Signals

A number of parameters are associated with each input signal and are used by the program that reads the measurement values. The parameters must be systematically stored so that several routines can access them easily. Parameters associated with the measurement include:

- Pointers to locate the data in the computer system.
- The number of the measurement signal and/or multiplexer address.
- Sampling time — sometimes multiple sampling periods are used.
- Scaling constants.
- Limit value for the variation range (both the sensors and the physical process variables), and first and second alarm limits.
- Rate of change limits.
- Filter constants.
- The measurement value itself (before and after correction).
- Logical variables to indicate operations which may be necessary such as linearization, scaling, alarm action and filtering.

The above parameters have different formats. Some are integers, others are real numbers, logical variables or alphanumeric information. Some principles of database organization to store measurement data are given further treatment in Chapter 12.

5.7 SUMMARY

High frequency components in a signal are usually due to noise and have to be eliminated or dampened before sampling. Analog filters (anti-alias) are used to remove all frequencies above half the sampling rate.

The sampling rate of continuous signals is of fundamental importance in computer control systems. Ideally it has to be at least twice as high as the highest component of the measurement signal. In practice, it has to be higher since the measurement period cannot be infinite. If the sampling rate is not sufficiently high then alias frequency distortion appears, and false frequencies can be seen in the sampled signal. After sampling it is impossible to correct for this false information.

Analog filters can be designed to perform both high frequency elimination (low pass filters) and low frequency elimination (high pass filters). Usually these filters are constructed with op-amp circuits. They have to be used with precaution since the op-amp does not have an infinite bandwidth.

For the conversion of the analog signal to digital form one has to ensure that the converters are sufficiently fast, that the accuracy is relevant and that the range of the converters is used adequately.

Several elementary tests ought to be performed before any signal is accepted for further processing. To extract interesting features of the signal digital filtering offers many possibilities. In this chapter we have demonstrated low order, low pass and high pass filters. Simple moving average and first-order low pass (exponential) digital filters are commonly used. Higher order filters can easily be implemented in a computer.

FURTHER READING

Derenzo (1990) describes many practical aspects on multiplexing and analog—digital conversion and filtering. More details on A/D and D/A conversion are documented in Sheingold (1986). Analog filters are treated in detail in Glasford (1986), Irvine (1987) and Jones (1986).

The sampling theorem is explained in Åström and Wittenmark (1990). Digital filters are special cases of time-discrete dynamic systems that are accounted for in detail in the same book. Digital filtering is further explored by Stearns and David (1988). For noise analysis the two works by Bendat and Piersol (1971, 1980) as well as those by Ljung (1987) and Söderström and Stoica (1989) are recommended.

6

Bus Systems

Aim: To present bus systems as open computer architectures, define their basic operations and describe some current types in order to provide the necessary background to evaluate, select and configure bus applications

Bus systems are in common use in automation applications. Their main advantage is that they allow the selection of the optimal hardware combination for a given control task. The approach to the study and application of bus systems is quite different from that to common computer systems. In the latter, the central processing unit is at the centre of the design and of the operations. In bus systems, the way data is exchanged, i.e. the bus itself, is the centre and the processing details for the single boards bear less importance.

The dozen or so bus systems commonly used in industrial applications cannot be compared with each other directly without qualifications. Each bus reflects the technology and expectations of the time it was designed. A few designs are proprietary of some company, others are supported by standardization bodies and available for everyone to use. The importance of having few bus architectures, independent of a specific processor model and under the control not of a manufacturer but of recognized organizations, is generally agreed.

This chapter begins with a general background about the bus concept (Section 6.1) and then describes bus operations in more detail in Section 6.2. The more in-depth descriptions of VMEbus (Section 6.3) and of Multibus II (Section 6.4) have been included as reference and as extensions of the general bus description in the first sections. The material of Sections 6.3 and 6.4 may be omitted without loss of continuity. Finally, Section 6.5 integrates the theoretical description of Section 6.2 by giving a quick look over other important bus systems, such as the IBM PC bus, Micro Channel and EISA.

6.1 FROM CPU-ORIENTED TO BUS-ORIENTED ARCHITECTURES

The breakthrough in the application of computers for process control took place in 1965,

Figure 6.1   The PDP-11/34 computer system from the mid 1970s. (Courtesy of Digital
Equipment Corporation.)

when Digital Equipment Corporation (DEC) introduced the cabinet-mounted PDP-8 processor, to be followed by the PDP-11 in 1970 (Figures 6.1 and 6.2). The success of the PDP computers was based not only on price (they cost more than $US100,000 per unit but were still much cheaper than other computers of the time), but also on the fact that they were based on open bus architectures. Customers could select the most appropriate

Figure 6.2 Front panel of a PDP-11/45 computer system (mid 1970s). The programming addresses and data were introduced manually via the console switches one byte at a time. (Courtesy of Digital Equipment Corporation.)

hardware for their needs and design their own interface cards on the base of the published (i.e. 'open') bus specifications. The PDP minicomputers became very popular in industry and gave sway to OEM[1] companies, each specialized in particular minicomputer applications.

Following the example of the DEC PDP computers, new bus standards were defined by different manufacturers. The possibility of selecting and configuring the most appropriate hardware for real-time applications was instrumental in the success of bus systems.

To appreciate the innovation brought by open bus systems, we have to take a look at the conventional structure for computers. A computer system is normally built around a **central processing unit (CPU)** to which are connected **peripheral unit**s performing different functions: keyboard, video interface, disk driver, input/output (I/O) cards (Figure 6.3). In this configuration, the peripheral units may communicate directly only with the CPU and only one peripheral unit at the time may be active exchanging data.

The CPU-centred configuration is inherently inefficient because all data has to pass through the CPU, even when the CPU does not need it. If the CPU operates much faster than the connected peripherals, the additional load for peripheral control is not much of a problem. This was the case for the first 30 years of computer history, but today the situation has changed as peripherals have gained much in speed. Data transfer between

---

1   Original equipment manufacturers, companies that build complete system applications, e.g. test units, using basic components delivered by other manufacturers.

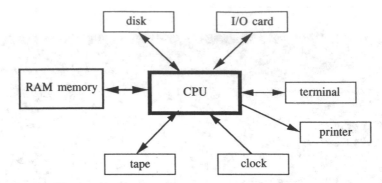

Figure 6.3   The conventional organization of computer systems.

disk and central memory, a very frequent operation, requires that the CPU copies data one byte at the time from one unit to the other. If the peripherals produce or consume data at a speed comparable to that of the CPU, they will generate a large fraction of the CPU load. Obviously it would be more convenient if the peripheral units could communicate directly with each other.

It is more effective to design a computer system where the peripheral units are more independent and have added computing capacity. The peripheral units are connected together with a bus by which each unit can communicate with all the others (Figure 6.4). On the bus, only one data exchange operation can take place at any given time, following appropriate coordination rules. At the centre of attention is the protocol, that is, the way data is exchanged among the connected boards. Here, details on how the data is processed by the boards, including the CPU, is not relevant.

The first buses were designed around specific processors and their lines were basically extensions of the processors' pins. The CPU controlled directly data exchange operations and timing. Newer buses are instead designed to be processor independent: their data and address formats and their exchange protocols do not depend on a particular processor architecture. In this way, even different CPUs can operate on the same bus if they follow the bus protocol.

In defining a new bus standard, there is much compromising on the physical dimension of the board. The dimension of the card implicitly defines the amount of logic circuitry

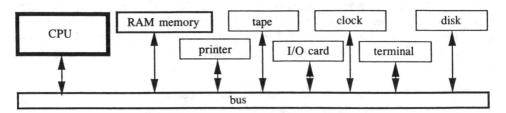

Figure 6.4   Principle of bus organization.

which will fit on it. Thanks to integrated circuits able to execute more and more functions with smaller components, the size of function boards for industrial applications has progressively been reduced down to typical board dimensions of about 15×30 cm^2. Larger boards may contain more functions, but they will also be more expensive. Smaller boards facilitate the selection of the right mix for a specific application and will be cheaper, but will use up slot space faster; in addition, they will load the bus with more frequent data transfer requests. As CPU power is becoming more affordable, it is economically feasible to build intelligent cards each with its own dedicated processor and local memory. The bus structure is a rational and modular scheme for different CPUs and peripherals to work together.

Buses support modularity which in turn gives greater flexibility for new applications. This is a key feature, especially when considering how often partial changes are required during the development of a new design. A bus allows resource sharing: several independent processors might, for instance, use the same memory or other units such as disks or printers.

6.2 BUS STRUCTURE AND OPERATION

6.2.1 General

A system bus is both a *physical* and a *logical* concept. Physically, a bus consists of about 50 to 100 conductors carrying electric signals in parallel (i.e. related signals are present on the conductors at the same time) between different boards containing electronic circuits. The bus wires may be soldered to the boards, but the most common method is to join together connectors into which the boards are inserted.

The logical concept of a bus is reflected in the rules and formats for data exchange, synchronization, handshaking and timing. Any board physically and logically compatible with a bus can operate on it and work together with other compatible boards.

A bus carries information in all directions. All boards receive the same data and must therefore autonomously recognize when they are addressed. Typical data exchange rates are of the order of a few tens of Mbyte s^{-1}. Because of delays for the propagation of electrical signals along the bus wires, the extension of a bus is limited to a length of a few decimetres. The bus and connected boards are usually mounted in a metal cage (the **backplane**) containing a set of connectors and, in some cases, a power supply and a cooling fan.

The most important parameters describing a bus system are:

- Mechanical and electrical data: board dimension, type of connector, power consumption, cooling requirements.
- Whether the bus is processor-oriented or processor independent.
- Address width or addressable memory space.
- Data width.
- Data transfer type synchronous/asynchronous.

- Data and address multiplexing.
- Clock frequency.
- Data transfer rate, typical and maximum.
- Number of interrupts and interrupt handling.
- Number of master units (that can control the data transfer).
- Additional features such as auto-configuration.

It is not possible to compare different bus systems only in the light of these parameters. A bus is a *system* solution and its analysis makes sense only if its many aspects are considered together. For instance, bus data tranfer rates are only 'half truths', especially when they are named by the bus vendors. It must be considered under which assumptions the given figures are valid. A high data-transfer rate could be a 'burst rate' reached using special test programs under particular conditions, while fetching single data items could in reality be much slower.

Data transfer rates are measured in multiples of bytes — usually Mbytes — per second. Take care when the rate is indicated in **words**. A word does not have a fixed length in terms of bits; it is often understood to be the same as the width of the bus data path. The data transfer rate is also known as the **bandwidth** of the data bus. However, this definition is not entirely consistent with other meanings of 'bandwidth' as, for example, in communication (see also Section 10.1.4).

6.2.2 Mechanical Construction

The trend toward standardization in bus architectures has imposed solutions for aspects of the basic hardware, for instance, board sizes and connector types. The most common format is the 19″ rack (Section 12.1.2), which has space for up to 21 boards in parallel (Figures 6.5 and 6.6).

The most popular type of board size is known as the Eurocard format (in the beginning it was used extensively in Europe), it is now normed in the United States as the IEEE 1101 standard. The smallest board 'single Eurocard' has dimensions of 100 × 160 mm (height × width) (Figure 6.7). The boards increase in height in 133.35 mm steps and in width in 60 mm steps. For instance, the 'double Eurocard' board has dimensions 233.35 × 160 mm (Figure 6.8). The 19″ and the Eurocard formats are used for different bus systems, for example, VMEbus, Multibus II and NuBus.

Two types of connectors are used for the connection of boards to buses: the edge connector and the plug-in connector. The **edge connector** (Figure 6.9) was used in earlier bus standards and is popular today particularly for personal computer applications due to its simplicity and low cost. However, edge connectors have a risk of poor electrical contact which may lead to operation errors.

Better performance is offered by the plug-in connector, which guarantees higher electrical and mechanical stability than the edge connector. A very common type of plug-in connector is the DIN 41612 (Figure 6.10). This connector is available in different versions with varying numbers of pins (Model A has 32 pins, B 64 pins and C 96 pins). Model

Figure 6.5 VMEbus rack. (Courtesy of Schroff, Germany.)

Figure 6.6 Backplane for VMEbus with 20 slots. This backplane supports boards in both single (100 × 160 mm) and double Eurocard format (233 × 220 mm). (Courtesy of Schroff, Germany.)

Figure 6.7   Single Eurocard format board (100 × 160 mm); an example of a CPU board for industrial systems. (Courtesy of Siemens, Germany.)

Figure 6.8   Double Eurocard format board (233 × 220 mm); Multibus II system with three 80186-based CPU boards, one exposed, one 80186-based Ethernet interface board and one 80386-based SCSI interface. (Courtesy of Intel.)

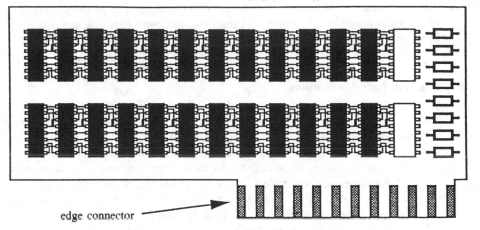

edge connector

Figure 6.9 Edge connector.

Figure 6.10 A 96-pin connector of DIN 41612-C. The connector is installed on a VMEbus backplane with nine slots. The small board to be inserted is a termination network. (Courtesy of Schroff, Germany.)

C allows enough signals to be carried so that apart from address, data and signalling more sophisticated applications can be made, and sophisticated, high performance buses such as VMEbus and Multibus II use this type of connector.

6.2.3 Bus Interface Electronics

In order to gain a good understanding of how a bus operates, it is important to take a look at the electronic components in the interface and how they interact. Each bus wire connects together circuits that must have matching signal and impedance characteristics. For the electrical connection of the boards to the bus, three different types of interface are used: TTL totem pole, open collector and tristate logic. The different electrical characteristics of the interfaces lead also to different modes of operation for the bus.

The **totem pole interface** is the normal output in TTL (transistor–transistor logic) integrated circuits (Figure 6.11). Here, either transistor Q1 or Q2 (but not both) is closed and conducting. The output voltage is equal to either the power supply level minus 1.1 V of voltage drop for Q1 and D1 or the ground level plus 0.4 V of voltage drop for Q2. Because of voltage drops in the output transistors, in TTL operations low level is defined as between 0 and 0.8 V and high level as anything above 2 V.

If a load is connected to the totem pole output, current will circulate in Q1 or Q2 depending on whether the load is connected to the power source or to ground and on the logical output state. In case of a load connected to the ground, Q1 must be able to *source* the current required by the load when the output is a logical '1' (high). Conversely, if the load is connected to power supply, Q2 must be able to *sink* the load current when the output is in low state. It is not possible to either source or sink any current in the other states.

Figure 6.11 also shows the typical TTL input. In order to input a logical '0', the input pin must be held at a voltage level of 0–0.5 V with respect to ground, but to do this requires that some current must circulate out of the input and to the ground. A TTL input is therefore equivalent to a load connected to the power source; the TTL output transistor Q2 must

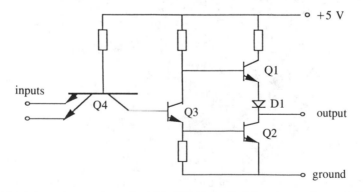

Figure 6.11   Internal structure of a TTL NAND gate. Transistors Q1 and Q2 and the diode D1 form the totem pole output.

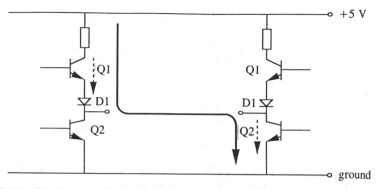
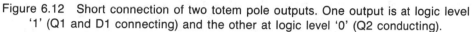

Figure 6.12 Short connection of two totem pole outputs. One output is at logic level
'1' (Q1 and D1 connecting) and the other at logic level '0' (Q2 conducting).

be able to sink this current from the cascaded inputs. For this reason the TTL logic is
also called **current sinking logic**. A typical sink current value is 1.6 mA at a voltage
level of +5 V. The TTL output circuits are usually designed to drive up to ten other TTL
gates, that is, the output transistor Q2 must be able to sink at least 16 mA.

Two totem pole outputs should never be connected together directly, otherwise if one
output is high and the other low at the same time, the power source is short circuited to
ground via the two gates (Figure 6.12). The gates may be damaged and the power spikes
can induce noise leading to erratic operations in the electronic logic.

A different type of TTL output is the **open collector** (Figure 6.13). This output is
obtained by cascading the TTL output with a transistor with its emitter connected at ground
level, the collector is the new output. When the transistor is closed, the output is at ground
potential; otherwise the output state is indefinite, insulated from ground via a high
impedance. To ensure that the output voltage is at power supply level when the transistor
is in open state, a **pull-up resistor** is connected to the output.

Several open collector outputs can be connected together in the **wired OR configuration** (Figure 6.14). The line level is high only if all transistors are open. When one or
more outputs are at a low level, the line is at low level. The wired OR configuration is

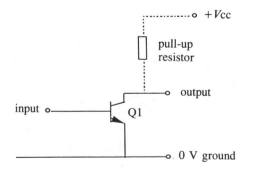

Figure 6.13 Open collector output. The output voltage is either 0 V or undefined.

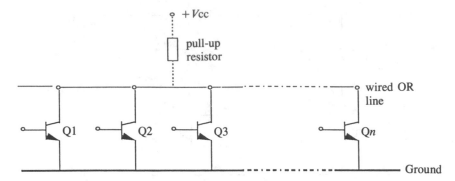

Figure 6.14 Example of wired OR connection. If at least one transistor is closed and conducting, the line is set at ground potential.

used when the boards have to communicate something general that has to be understood by all the others, e.g. that they are ready to operate. Every board may pull the line low. All boards also constantly monitor the line. It is enough for one board to pull the line low for the others to be able to detect it. Due to its simple operating principle, open collector logic was widely used in earlier bus systems.

Modern bus systems are based on **tristate logic** (Figure 6.15). Two states are logic high and low, the third is a high-impedance output. In the 'third state' the interface circuitry does not influence the bus lines, but the state of the lines is reported to the internal circuitry of the boards if needed.

Tristate logic is used mainly for data and address exchange. One board drives the bus with the usual high/low levels and all other units keep their output in the high impedance state, while at the same time monitoring the bus signals. Tristate logic outputs can be connected together without any risk. The basic state is the high impedance input, and only with explicit assertion will the tristate interface take one of the output states. Such assertion is given only when the board is enabled to control the bus lines.

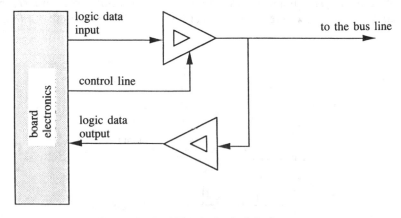

Figure 6.15 Tristate logic interface.

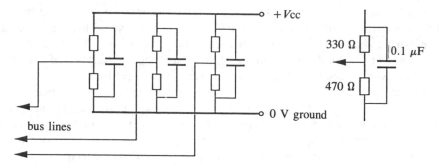

Figure 6.16   Bus termination network.

Current buses combine tristate logic interfaces to exchange data and addresses and wired OR interfaces to indicate readiness to send or accept data, coordinate operations and send interrupt requests.

At the high frequencies at which a bus operates, each conductor behaves as a transmission line with distributed impedance. As we have already seen in the description of electrical conductors (Section 4.7.2), the characteristic impedance of all connected units at both ends of each line must match, otherwise the signals might be reflected back at the end of the conductors. If the level of the reflected signals is high, they could interfere with normal operations.

To avoid signal reflections, a **termination network**, with impedance equal to the line impedance, is connected at both ends of a bus conductor (Figure 6.16). The termination network dampens out spurious oscillations caused by fast transients. Termination networks can be either built directly on the bus backplane or installed on boards which are inserted in connectors at both bus ends (Figure 6.10).

6.2.4 Bus Electrical Interface

The conductors of a bus can be divided into the following groups:

- Address lines.
- Data lines.
- Control lines: handshake, interrupt, clock.
- Power lines.
- Spare lines.

The **address lines** carry the source or destination address of the data currently present on the bus. Typical address size has increased since the first buses, when it used to be 16–20 bits, to today's typical 32 bits. Each connected board has a unique address range and must autonomously recognize when the address on the bus is within its range.

Data lines carry data to and from the addressed locations. Similar to address lines, their number has increased from 8 in earlier buses to 16 or 32 in more modern applications.

In some bus systems, the lines for addresses and data are entirely or partially shared: first the address is put on the lines and some instant later the data. This method is called **multiplexing** and allows a bus to be constructed using fewer lines. The price for multiplexing, however, is paid with lower transfer speeds. The transmission of address and data in succession takes more time than if it were done in parallel.

In some buses a distinction is made between memory boards and I/O boards. Some addresses could correspond to locations on two boards and, in order to avoid ambiguity, additional bus lines must indicate whether an address is meant for a memory bank or an I/O card. In **memory mapped I/O** the address range is unique and is divided between the boards; I/O ports are considered as memory locations with their specific address. With memory mapped I/O there is no need for additional destination qualifier lines.

Normally it makes no difference which slots the boards are inserted into; for the bus conductors it is immaterial where the signals are generated. When it is important to keep track of the physical slots where the boards are inserted, part of the address is hard-coded in the connector. This method is called **geographical addressing**.

Control lines are used to carry service signals, for instance, the type of operation — read or write, indication of readiness to send or receive data, or interrupts to request special handling by one of the processing units. Normally, one line carries a clock signal at a defined frequency to be used as general synchronization reference.

Many boards do not have autonomous power supply and draw their power from the bus. **Power lines** are then used to carry voltages typically at ± 5 V and ± 12 V levels. Power and ground are normally connected to several pins in order to distribute the current load of multiple boards among several conductors.

Spare lines are not connected at all, they are reserved for additional functions in future revisions of the bus standard. They might also be explicitly left free for the user's own applications.

Not all types of lines are present on all buses. Address and data sizes vary, the number of handshake lines depends on the data exchange protocol, and interrupts are supported differently from one bus to the other. Note that some of the bus lines carry their original meaning when they are at a high voltage level (**active high**) and some when they are at ground level (**active low**). To avoid confusion, a line is said to be either **asserted** or **negated**, independently of whether the asserted state is at a high or low logical level. Active low lines are indicated in bus data sheets with an asterisk (*) following the name or with a line above the name, for instance NACK* or $\overline{\text{NACK}}$. Active low lines often operate according to the wired OR principle; they are used when all boards must indicate readiness for an operation at the same time. If a board is not ready for operation, it just pulls the common line low so that the other boards may detect it.

The most important parameters to describe a bus are the address and data size and the data transfer rate. The latter is usually given in Mbyte s^{-1} and shows the maximum data transfer rate, that is, the physical limit of a bus operating under extreme conditions. The actual data transfer rate, or **throughput** of a system, may be much lower, as it depends on many other factors.

6.2.5 Bus Operations

The rules for the coordination of data exchange and other operations in a bus are as important as the meaning of the conductors in order to ensure compatibility and to make data exchange possible among the connected boards. All bus operations have to be exactly defined in relation to the used lines and the requiring timing. The set of rules for the coordination of all operations on a bus is called **protocol**. In order to be compatible with a bus and thus with the devices connected to it, a board must follow the bus protocol.

The board that at any moment sets the address and indicates what operation is to be performed is indicated as **bus master**. The **slave** is the unit addressed by the master and must react to its commands. For instance, in a data transfer operation between CPU and memory, the CPU acts as master, sets the address and indicates whether the operation is read or write. The memory (slave) recognizes its address and reacts by reading or putting the requested data on the bus. In some buses only one board can act as master, in other systems the master right can be passed among boards. If there is no clearly identified master and several boards tried to access the bus at the same time, the combination of conflicting signals might lock the bus (**bus contention**).

Buses can be **synchronous** or **asynchronous**. Synchronous buses operate following the timing of a **reference clock signal**. This signal is a rectangular wave at 10—20 MHz generated by one of the boards. A reference clock period is called a **bus cycle**. In synchronous buses, all operations must be carried out in a precise number of bus cycles. Asynchronous buses operate according to the principle of cause and effect. An operation may be carried out only after the previous operation is terminated. In asynchronous buses there is no synchronization clock, so that several handshake lines are necessary to carry different acknowledgments and indicate readiness for operation. In synchronous buses, readiness is implied by the clock: at a certain bus cycle all boards have to be ready for operation.

There are trade-offs between the two techniques. Synchronous buses are somewhat less complex, but are defined on the basis of the technology available at a certain time and offer comparatively little room for improvement. Asynchronous buses allow for greater flexibility due to the fact that boards operating at different speeds can be connected together on the same bus. If new components allow higher speeds, a board which uses them can be immediately integrated in an asynchronous bus. The more complicated circuitry required by asynchronous buses is usually not perceived as a problem since large scale integration allows the costs for interface logic to be kept low.

6.2.6 Synchronous Data Transfer

The most common bus operation is data exchange between two boards. This operation also exemplifies clearly the difference between synchronous and asynchronous operation. In data exchange, the board acting as master sets the bus address and asserts one or more

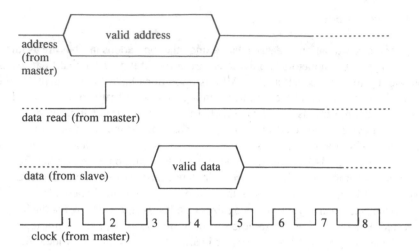

Figure 6.17 Synchronous data read operation. At each **clock pulse** determined signals must be on the bus. In this example, the **address** must be on the bus during clock cycles 2–24; the **data read** command at the rising edge of 3 and 4; and the requested **data** during the rising edges of 4 and 5.

control lines to indicate that the address is valid and whether the operation is read or write.

In a synchronous **read** operation the addressed device (slave) must put the data on the bus for a fixed number of bus cycles (Figure 6.17). In the example shown, all connected boards 'know' that the address is valid for 4 cycles after the beginning of the operation and that data on the bus is valid only during cycle No. 4. The synchronous **write** operation is very similar, the only difference is that the master is also responsible for putting the data on the bus.

The synchronous timing is always referred to the polarity change of the clock pulse. The actual protocols specify when all the signals must be set. For example, if the clock rate is 10 MHz, the interval between successive positive (or negative) edges is 100 ns. The interface logic on a board must be able to react within that time, still allowing for the electrical signals to settle on the conductors, so that the reaction time is of the order of 50 ns.

It may happen that the addressed board is not able to decode the address and process the requested data in time. In such a case, the board asserts a WAIT line to indicate to the other boards that the bus operation is delayed. When the slow board is ready to continue, it negates WAIT and normal operations are resumed. This action is called insertion of **wait cycles**.

6.2.7 Asynchronous Data Transfer

In asynchronous buses there is no clock line. Instead, several handshaking lines indicate when the sender and the receiver are ready for the information transfer, when address

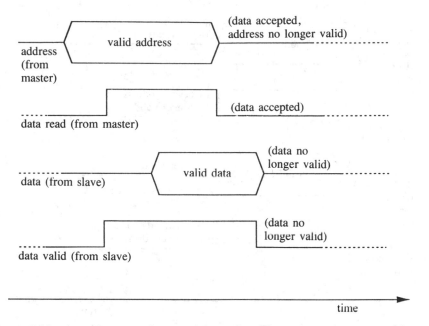

Figure 6.18   Asynchronous data read operation. The master places an **address** on the bus and confirms its validity with the assertion of a line (in the example: **data read**). The line indicates also that the operation is **read**. The slave board answers by writing the requested data on the bus and confirming with the **data valid** line. The master negates **data read** to confirm reception of data and the slave answers by cancelling the data and negating **data valid** as an acknowledgment.

and data are valid, and when the destination device has received the data. Asynchronous transfer builds on the cause–effect principle: a new operation may take place only after the previous operation is terminated.

In a data **read** operation (Figure 6.18) the master unit begins by putting the destination address on the bus and shortly thereafter confirms the address by asserting the ADDRESS VALID line. An additional line indicates that the operation is a data READ. With ADDRESS VALID asserted, all boards decode the address. The addressed board puts the requested data on the bus and confirms the operation by asserting the DATA READY line. The master unit acknowledges data reception with the DATA ACKnowledge line. Now the slave may negate the DATA READY line and reset the bus drivers; in reply the master negates its DATA ACKnowledge line. The data read operation is completed.

In some types of asynchronous buses the lines for address validation and for operation control are organized in a different way. For example, MEMORY WRITE asserted indicates at the same time that the address is also valid. With this organization, one line is needed for each one of the possible operations.

Similar to read is the data **write** operation (Figure 6.19). In this case the master unit puts on the bus the destination address and the data. After the right bus signal levels have been reached, the necessary lines are asserted to indicate that the operation is WRITE,

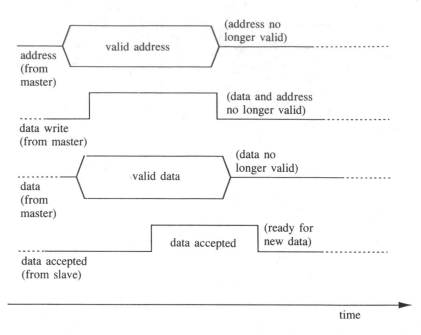

Figure 6.19   The asynchronous data write operation is similar to the data read (Figure 6.18). The master places an **address** and the **data** on the bus and confirms them by asserting the line **data write**. When the slave board has accepted the data, it confirms it by asserting the **data accepted** line. The master cancels the address and the data and negates **data write**. The slave answers by negating **data accepted**.

and address and data are VALID. At this time all boards read and decode the address; the addressed board also reads the data and when the storage operation is terminated, it signals this by asserting the DATA ACKnowledge line. The master board can then release the data ready line to indicate that data and addresses are no longer valid. The slave unit releases its data acknowledge line and the operation is complete.

6.2.8 Block Transfer and Direct Memory Access (DMA)

Data transfers between central memory and peripherals such as disk drives and video screen interface occur frequently in computer systems. Such transfers typically take place in blocks of hundreds or even thousands of bytes at a time. Other common memory transfers include copying block-organized data, for instance, program code, data vectors and character strings. It is relevant for this type of data that their content be located at consecutive memory addresses.

In **block transfer** the length of the block is passed along with the start address for the transfer. The data is then written out in succession by the addressed board following the bus clock when the transfer is synchronous, or after each handshake acknowledgment in asynchronous operations. In block transfer, there is no need to pass and acknowledge an address for each byte transfer. Instead, the read and write addresses for each byte transfer are automatically incremented at source and destination.

The block transfer technique requires some additional logic in the boards for the automatic generation of new addresses. Thanks to an improvement in transfer speed and system throughput, block transfer is becoming a standard feature in new bus systems.

Direct memory access (DMA) denotes a method for accessing the memory directly without having to pass via the CPU. In essence, DMA is an automatic address generator realized with a few chips and connected to the bus. Direct memory access generates the appropriate signals so that, for example, an I/O board can move data to the memory without explicit control by the CPU. This method differs from block transfer in that the address-generating circuitry is external to the boards and to the bus. With DMA the memory ignores whether the addresses are generated by the CPU or by other circuitry. As an example, DMA is used in IBM PC computers as a cheap way to get direct block transfer capacity.

6.2.9 Interrupt Handling

An interrupt is a signal that a peripheral board sends to the central processor (or to another board) in order to request attention. In response to an interrupt, the processor stops what it is currently doing and executes a service routine. When the execution of the service routine is terminated, the original process may resume its previous operations.

An interrupt indicates that an event asynchronous (i.e. not related in time) to the current program flow has taken place. It might be that new data is present on a communication channel, a button is pressed or an operation is completed and the interrupting device requires processor attention. By their very nature, interrupts can take place at any time.

In bus systems a board requests an interrupt by asserting an interrupt request line. When the processor is ready to execute the service routine, it signals it on interrupt acknowledge line. The generating board then negates its interrupt request.

When several boards request interrupts it is necessary to differentiate between them according to some precedence, as only one board at a time may receive attention. Most buses have several interrupt request lines and every line has a defined priority. Interrupts are serviced according to the priority of the line on which the request is made. The board servicing the interrupt writes the interrupt number on the bus so that the originating board can recognize it. This requires that only one line be used for interrupts from one board.

Another approach is the **shared interrupts**, where each interrupt line is shared by several boards. As each interrupt line may be asserted by more than one board, the service board must find where an interrupt originated by polling all the boards connected to the same line. The first board that acknowledges the request is then serviced. Note that with this method potential conflicts cannot be avoided when, for example, several boards generate the same interrupt at the same time.

A different method for sharing interrupt lines is known as **daisy-chain connection** (Figure 6.20). Normally, each board will keep the connection closed and, when a board requests attention, its logic circuitry prevents the acknowledge signal being carried further.

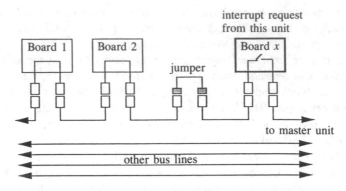

Figure 6.20   Daisy-chain connection.

On receiving an acknowledgment from the interrupt handler, the board writes an identification code on the bus. Two pins are needed on the bus connector for each daisy-chain line. When no board is inserted, the pins related to daisy-chain conductors must be short circuited with a jumper.

With daisy-chaining, where a few interrupt lines are shared by several boards, the boards installed closer to the servicing unit have an advantage because they control whether or not an acknowledgment signal is carried further. A disadvantage of daisy-chaining is that it takes time to transfer the acknowledgment signal along the chain.

Interrupt handling becomes complicated when several CPUs are installed in a system and when not only the source but also the destination of an interrupt has to be determined. Because of this, in the most recent bus designs interrupts are considered as messages and sent from the source to the destination board like any other data transfer. The software for handling interrupts is described in Section 7.6.5.

6.2.10 Bus Arbitration

On a bus, only one unit at a time may act as master. When several boards are able to become master, some method is needed to define which one should currently be master and when another has to take over. The selection of the bus master is known as **bus arbitration**.

There are two basic approaches to bus arbitration. In one solution, one of the units is designated as **arbiter** (a kind of 'super master'). A board wishing to get control of the bus indicates this to the arbiter via an interrupt on a BUS REQUEST line. The arbiter checks the relative priorities of the current master and of the requesting unit, and decides when master control has to be transferred.

In a different approach there is no arbiter as such. A board which intends to become master makes a request via a **bus allocation line** common to all boards (more lines might be used to distinguish between different levels of priorities). The current master identifies the requesting unit and compares the relative priorities. If the requester has higher priority, bus control is transferred. All units able to become master contain the bus arbitration logic.

Two selection principles may be followed to choose the master unit: **round-robin** and **priority allocation**. With round-robin, bus control is passed in an ordered fashion among all the modules requesting it (this is also known as 'fairness' mode). No difference is made among priority levels and master control is rotated among the boards according to a predefined order.

With priority allocation, control is passed to the module with the highest priority. The priorities may be changed in time in order to prevent modules with high priority and high activity to indefinitely hold control of the bus.

In some systems, bus arbitration is handled with dedicated lines and takes place concurrently with data transfers, in others data transfer and arbitration cannot be done at the same time. If there is a high arbitration activity, the way it is handled influences total system performance.

The principles for bus allocation are similar to those used for resource protection and in process scheduling algorithms. Resource protection is a general problem, independent of whether the resource is hardware or software. These principles are described in more detail in Sections 7.2 and 7.3.

6.2.11 Construction of a Bus System

Setting up a bus system does not need to be more complicated than just inserting all the boards in their slots. Some attention, however, is required in order to configure a system so that it can operate at maximum efficiency.

The boards should be ordered according to the importance of the interrupts they generate and, in multiprocessor systems, to their relative importance as bus master. We have seen that with daisy-chaining the boards closest to the central processor or to the bus arbiter are the first to receive attention. In that set up, the boards will be inserted on the bus in order of their priority. In some cases, the interrupt and bus grant lines are selected on the board with jumpers or DIP-switches. Other jumpers may be used to configure other operational parameters such as the default name for a drive or the type of input/output signals. If one or more slots between daisy-chained boards are free, they have to be short circuited with help of jumpers on the bus.

Depending on the operating system used, the connected boards and some operational parameters might have to be defined in the software in some 'system' or 'start-up' file. System configuration consists in several steps which can be successfully carried out only after the exact functional requirements of each board have been defined. Other important aspects are the insertion of termination networks at both ends of a bus and of pull-up resistors on open collector, wired OR lines.

Finally, before all the boards are inserted in their cage, it should be considered whether:

- The power supply is adequate for all boards.
- The ventilation is sufficient to carry away the generated heat.
- All the screenings and isolations have been properly done.

These suggestions are very general and not oriented to a particular type of bus. For all technical details, one should refer to the bus documentation and follow the indications reported there. Do not forget the close interaction between the hardware (the bus) and the software (the operating system and application programs). If the functional goals are clear (what the system is supposed to do), it is easier to tailor the bus and the operating system parameters accordingly. Do not rely too much on sophisticated self-configuration procedures to make up for poor system planning.

6.3 THE VMEBUS (IEEE 1014)

6.3.1 General Features

VME means VERSA module Eurocard. VERSA is the name of an earlier bus designed at Motorola and Eurocard is the standard board format that we have seen in Section 6.2.2. The VMEbus was designed by a group of companies led by Motorola; it is now defined as a standard in IEEE 1014. The VMEbus has several features which make it a powerful

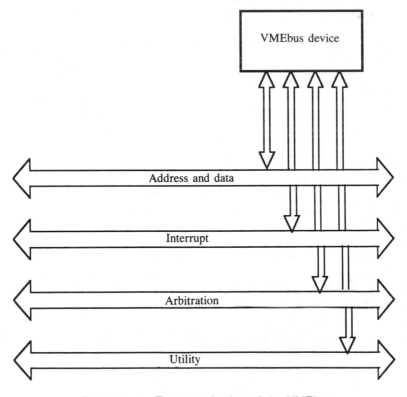

Figure 6.21   The organization of the VMEbus.

and flexible system bus for industrial applications. Its main features are:

- Address length 16/24/32 bits
- Data word length 16/32 bits
- Data transfer rate: theoretical maximum: 57 Mbyte s^{-1}, practical 30–40 Mbyte s^{-1}
- Seven interrupt signals, daisy-chained
- Support for multiprocessor systems, four priority levels for bus allocation
- Data block transfer, maximum block length 256 words

The VMEbus is available in single-height and double-height backplane formats, with one and two DIN connectors, respectively (in VMEbus the connectors are called J1 and J2). The single-height format supports 16-bit data transfer and 24-bit addresses (16 Mbyte address space), and the double-height format supports 32 data bit and 32 address bit or 4 Gbyte. Single- and double-height cards can be mixed in the same system and data transfers may take place alternatively with 16 and 32 bits, depending on which boards are currently addressed.

VMEbus has a maximum length of 500 mm, so that the maximum signal propagation delay allows up to 21 boards to be inserted on the bus. Particularly important in a VMEbus system is the card in the first slot, which must perform some system-wide functions, e.g. master arbitration in multiprocessing operations. The VMEbus consists of four sub-bases for data, arbitration, priority interrupt and general utility (Figure 6.21). The VMEbus connector has the specifications given in Table 6.1.

6.3.2 Data Transfer

The read/write operation is carried out asynchronously and with no need for multiplexing. Three address ranges are defined: short (16 bit), standard (24 bit) and extended (32 bit); the latter is possible only via the additional pins of the connector J2 of the double-height backplane. Data words of 8, 16, 24 and 32-bit length can be transferred on the bus; the data length is dynamic, which means that it can be changed at any time.

In data transfer, the master unit sets the address lines A01 to A31, the address modifier code lines AM0–AM5 and the control signal IACK*. Validation of the data is done with the data strobes DS0* and DS1*. Confirmation of address validity takes place with the falling edge of the address strobe AS*. The AS* line has the additional meaning of 'bus busy': as long as it is asserted, no other unit is allowed to initiate a bus operation or place a request to get control of the bus.

The line WRITE* indicates whether the operation is a read or write. The slave unit decodes the address and puts the related data on the bus (or reads it from the bus) and confirms completion of the operation by asserting the data transfer acknowledge line DTACK*. The master confirms data reception and acknowledgment by negating DS0* and DS1*. The slave then negates DTACK* and the data transfer cycle is terminated.

Data transfer on the VMEbus is oriented to the support of a multiprocessor environment. VMEbus includes six address modifier lines (AM0–AM5) which are set in parallel

Table 6.1(a)   VMEbus connector *P1/J1* pin assignment
(connector type is DIN 41612–G)

Pin number	Row A	Row B	Row C
1	D00	BBSY*	D08
2	D01	BCLR*	D09
3	D02	ACFAIL*	D10
4	D03	BG0IN*	D11
5	D04	BG0OUT*	D12
6	D05	BG1IN*	D13
7	D06	BG1OUT*	D14
8	D07	BG2IN*	D15
9	GND	BG2OUT*	GND
10	SYSCLK	BG3IN*	SYSFAIL*
11	GND	BG3OUT*	BERR*
12	DS1*	BR0*	SYSRESET*
13	DS0*	BR1*	LWORD*
14	WRITE*	BR2*	AM5
15	GND	BR3*	A23
16	DTACK*	AM0	A22
17	GND	AM1	A21
18	AS*	AM2	A20
19	GND	AM3	A19
20	IACK*	GND	A18
21	IACKIN*	SERCLK	A17
22	IACKOUT*	SERDAT*	A16
23	AM4	GND	A15
24	A07	IRQ7*	A14
25	A06	IRQ6*	A13
26	A05	IRQ5*	A12
27	A04	IRQ4*	A11
28	A03	IRQ3*	A10
29	A02	IRQ2*	A09
30	A01	IRQ1*	A08
31	− 12 VDC	+5 VSTDBY	+12 VDC
32	+5 VDC	+5 VDC	+5 VDC

to the address. The bus uses these lines to indicate whether the data transfer takes place in 'supervisor' (= privileged) or non-privileged mode, and whether the addressed location corresponds to data or programs. These indications do not have any effect on the actual data transfer, but can be used by the operating system and application software as an additional protection feature, for instance, to lock a protected data section from unauthorized access via the bus. Some of the address modifier codes are free to be defined by the user.

Table 6.1(b)   VMEbus connector *P2/J2* pin assignment

Pin number	Row A	Row B	Row C
1	User Defined	+5 VDC	User Defined
2	User Defined	GND	User Defined
3	User Defined	RESERVED	User Defined
4	User Defined	A24	User Defined
5	User Defined	A25	User Defined
6	User Defined	A26	User Defined
7	User Defined	A27	User Defined
8	User Defined	A28	User Defined
9	User Defined	A29	User Defined
10	User Defined	A30	User Defined
11	User Defined	A31	User Defined
12	User Defined	GND	User Defined
13	User Defined	+5 VDC	User Defined
14	User Defined	D16	User Defined
15	User Defined	D17	User Defined
16	User Defined	D18	User Defined
17	User Defined	D19	User Defined
18	User Defined	D20	User Defined
19	User Defined	D21	User Defined
20	User Defined	D22	User Defined
21	User Defined	D23	User Defined
22	User Defined	GND	User Defined
23	User Defined	D24	User Defined
24	User Defined	D25	User Defined
25	User Defined	D26	User Defined
26	User Defined	D27	User Defined
27	User Defined	D28	User Defined
28	User Defined	D29	User Defined
29	User Defined	D30	User Defined
30	User Defined	D31	User Defined
31	User Defined	GND	User Defined
32	User Defined	+5 VDC	User Defined

VMEbus supports data transfer in blocks of up to 256 words (**block** or **burst transfer mode**). In block transfer, master and slave unit automatically increment the addresses and transfer the new data word; the address placed on the bus is not changed during the process. Handshaking in block transfer is done with the lines DTACK* to acknowledge data reception and DS0*, DS1* to assert the validity of a new data item. A block transfer operation cannot be interrupted. Any unit wishing to get bus control has to wait until the transfer is completed and the line AS* is released.

A serious problem might arise in multiprocessor systems when a processor interrupts another that is operating on a memory location and then modifies the same location. This is an aspect of the resource protection problem, the implications of which will be illustrated in detail in Section 7.3.

The VMEbus defines a **read—modify—write** cycle in order to prevent data from being written over by a unit while it is used by another unit. The read—modify—write cycle is similar to the function of the **test__and__set** bit that prevents modification of data being accessed by several programs in a multitasking environment (see Section 7.3). In the read—modify—write cycle the operations of read and write by the master occur in immediate succession and no other unit is allowed to access the data item being processed. The line AS* is held asserted all the time to prevent another unit from getting control of the bus.

6.3.3 Priority Interrupt

The VMEbus supports seven interrupt request lines IRQ1*—IRQ7*. Each line is of wired OR type and can therefore be used by several boards concurrently. The interrupts have different priority levels of which IRQ7* has the highest priority. The VMEbus protocol does not specify how the interrupts are to be serviced.

The interrupt handler begins by asking the unit that generated the interrupt to identify itself with a status code. The interrupt handler places the binary code of the interrupt currently being serviced on the address lines A01—A03 and asserts the address strobe AS*, the data strobes DS0* and DS1*, and the interrupt acknowledge line IACK*. The IACK* line is daisy-chained through different boards; the board intending to acknowledge the interrupt will not pass the IACK* signal any further.

The board that generated the interrupt recognizes the number of the interrupt on the address lines and on reception of IACK* writes a STATUS/ID code on the bus data path, confirming it with the data acknowledge line DACK*. The board requesting the interrupt must hold the STATUS/ID word on the bus until the interrupt handler negates the data strobes DS0* and DS1*. The interrupt request procedure terminates in the same way as a normal data transfer, with the assertion and release of handshaking signals.

The self-identification of the modules based on the STATUS/ID code saves much overhead to the interrupt handler, which does not have to poll all devices to find the interrupt source. The STATUS/ID word may be up to 32 bits long (in the case of the extended data path with two connectors), so that considerable information may be carried in order to qualify the interrupt request.

Interrupts may also be serviced by a handler which is not the current bus master. In this case, the interrupt handler must first get control of the bus and become master using the normal arbitration procedure.

6.3.4 Bus Master Arbitration

The mechanism of bus arbitration resembles quite closely the operation of interrupt request and granting. Each master can request control of the bus via one of four bus request lines BR0*–BR3*, where BR3* has the highest and BR0* the lowest priority. These lines are of wired OR type so that several requests can be made concurrently.

The board installed in slot No. 1 acts as arbiter. When it receives a request, the arbiter first monitors the bus busy line BBSY*. As soon as the line is released, the arbiter may issue a bus grant along one of the lines BGX0*–BGX3*, related to the bus request lines BR0*–BR3*. The bus grant lines are daisy-chained so that each board decides whether to pass along the grant signals or to hold it. The board accepting the bus grant replies by asserting the bus busy line BBSY*. The arbiter then negates the bus grant line BGX* and the new master may proceed with its operation.

The principle for selecting the next board to become master when several requests are made at the same time is left to the arbiter. It is not explicitly part of the VMEbus specifications. The VMEbus master allocation scheme is not exempt from drawbacks. The boards closest to the arbiter might get a higher share of bus access. If many boards are daisy-chained, passing a grant signal takes time and contributes to slowing down the bus. It is important to consider these factors in choosing the disposition of the boards on the bus and the master selection principle with related parameters in the operating system.

6.3.5 Utility Signals

The VMEbus includes some utility lines to indicate wrong or erratic conditions. If a slave unit detects an error during a bus transfer cycle, it indicates this by asserting BERR*. The ACFAIL* line is set by a module monitoring the a.c. power supply and signals when power is about to be lost. Some generic system failure may be indicated by SYSFAIL*; it is up to the user to define in which situations the signal has to be used, while SYSRESET* indicates that a system reset is in progress or was initiated. On the VMEbus racks there is a push button to manually assert SYSRESET*. All these utility signals are of open collector wired OR type and can be asserted by any module.

A 16 MHz reference frequency is generated by the board inserted in slot No. 1. This signal is provided for convenience only and does not have any synchronization function for the bus operation.

6.3.6 VMEbus Extensions

Although the VMEbus has a high data throughput rate, at times a higher capacity may be needed. To avoid data transfer delays, different types of bus extensions operating in parallel and independently from the original VMEbus have been defined. The physical connection to VMEbus remains the same; the bus extensions use the free pins on the second connector J2. Bus extensions also provide a safety back-up in case of failure of the primary VMEbus.

Both the **VMXbus** (VME extension) and **VSBbus** (VME subsystem bus) use the 64 user-defined pins of connector J2 in the double-height module. Both buses support features such as interrupt handling and multiprocessing.

The **VMSbus** (VME serial bus) is a synchronous, serial bus with maximum data transfer rate about 3 Mbit s^{-1}. Bus access operates according to the collision detection method (see Section 10.5.2). Any module can initiate a transmission at any time when it detects that the line is free. If two units begin transmission at the same time, they detect that their output is garbled. They then halt the transmission and wait for a random time before making a new transmission attempt. Data transfers on VMSbus may be up to 8 bytes long. VMSbus uses two pins on connector J1 of VMEbus. One of the pins is used for the data signal (SER DATA*) of the open-collector type and the other for a reference clock signal (SER CLK). Data output on SER DATA* is synchronized with SER CLK, which is generated by the board in slot No. 1.

The **VXIbus** (VME extended instrumentation bus) is an extension of VMEbus to support fast instrument boards producing high quantities of data. The sizes of the VXIbus boards may be up to 'three-Eurocard' size with three connectors (including the original VMEbus connectors). The specifications of VXIbus not only encompass the electrical interface to the bus, but also other specific functional requirements. Among other things, a board must be able to identify itself on request and send replies using standard codes.

The VMEbus has been designed to efficiently support multitasking and multiprocessor environments. The defined features are the distinction of different data types on the bus, block data transfer, the read—modify—write cycle and a bus master arbitration mechanism. VMEbus still offers expansion possibilities that make it interesting for users wishing to customize it by adding special features.

6.4 MULTIBUS II (ANSI/IEEE 1296)

6.4.1 General Features

Multibus II is a synchronous, high performance bus system initially developed by Intel and later normed by IEEE in the standard ANSI/IEEE 1296. Multibus II has nothing in common with the other bus standard bearing the same name, Multibus I (see Section 6.5.5), which was also initially developed by Intel.

Multibus II is 'one step ahead' compared with other current industrial bus systems. The standard is not limited to lines and protocols, but deals also with the data that is transferred and other system aspects. The added functionality has its price, however: the bus requires additional intelligence in the connected boards. Multibus II boards use the 'double Eurocard' format 233 × 160 mm, with two 96-pin DIN 41612-C connectors, called P1 and P2. A maximum of 20 boards can be connected on the same backplane; the board in slot No. 0 is the **central service module (CSM)** and has special coordination functions.

Connector P1 is used for the bus signals while P2 is free for another communication path independent of Multibus and left to the user to define. Multibus II signals can be divided into five groups: address/data, central control, system control, arbitration and power

Table 6.2   Multibus II connector *P1* pin assignment
(connector type is DIN 41612–G)

Pin number	Row A	Row B	Row C
1	GND	PROT*	GND
2	+5 VDG	DCLOW*	+5 VDG
3	+12 VDG	+5 VDG Battery	+12 VDG
4	GND	SDA	BCLK*
5	TIMOUT*	SDB	GND
6	LACHn*	GND	CCLK*
7	AD0*	AD1*	GND
8	AD2*	GND	AD3*
9	AD4*	AD5*	AD6*
10	AD7*	+5 VDG	PAR0*
11	AD8*	AD9*	AD10*
12	AD11*	+5 VDG	AD12*
13	AD13*	AD14*	AD15*
14	PAR1*	GND	AD16*
15	AD17*	AD18*	AD19*
16	AD20*	GND	AD21*
17	AD22*	AD23*	PAR2*
18	AD24*	GND	AD25*
19	AD26*	AD27*	AD28*
20	AD29*	GND	AD30*
21	AD31*	RESERVED	PAR3*
22	+5 VDG	+5 VDG	RESERVED
23	BREQ*	RST*	BUSERR*
24	ARB5*	+5 VDG	ARB4*
25	ARB3*	RSTNC*	ARB2*
26	ARB1*	GND	ARB0*
27	SC9*	SC8*	SC7*
28	SC6*	GND	SC5*
29	SC4*	SC3*	SC2*
30	−12 VDG	+5 VDG Battery	−12 VDG
31	+5 VDG	SC1*	+5 VDG
32	GND	SC0*	GND

(Table 6.2). The system control lines indicate the content of the bus: request or reply operation, address or data, read or write command, etc. Bus access is requested through a common arbitration line and granting can follow the round-robin or a priority allocation method (Sections 6.2 and 7.2). The central control lines include the 10 MHz reference clock and an additional clock at 20 MHz. There are also lines to indicate bus failure, system start and power failure recovery.

Multibus II has two conductors, SDA and SDB, reserved for a 2 Mbit s^{-1} serial bus. The concept for this is similar to the VSBbus as subsystem bus for VMEbus.

6.4.2 Basic Operations

Multibus II introduces at bus level some ideas from communication theory, basically that communication is not only related to the electrical interface with its signal levels and timings, but the meaning of the transmitted data also plays a fundamental role (we will explore this concept in detail in Chapter 10). The basic idea of Multibus II is to coordinate the exchange of messages between independent boards.

Multibus II does not support configurations where the CPU is installed on one board and the RAM on another, and all data and program instructions must be fetched one by one. Rather, each board must contain all the necessary resources including CPU and RAM which allows part of the work to be performed locally. A board communicates with the others only to access their specific resources and functions. In this way less external memory access is needed and a significant part of data exchange is avoided altogether, which also optimizes bus throughput.

The Multibus II multiplexed address and data path is 32-bits wide and operates at a clock frequency of 10 MHz. A 4-byte word can be transferred at each clock pulse, corresponding to a maximum throughput of 40 Mbit s^{-1} (due to the transfer of overhead information, the effective maximum transfer rate is 32 Mbit s^{-1}). Four parity lines (another concept mutuated from communication technology) are used to verify the correctness of the received data.

The major functional difference of Multibus II compared with other bus systems is the message passing architecture. In Multibus II terminology the addressable memory range, the message formats and the transfer protocols for specific functions are collectively known as 'space'. Normal data transfer operations are part of the 'memory space'. There is also a 'system space' divided in 'message space' and 'interconnect space'. The message space deals with bus system arbitration and protocols; the interconnect space refers to the data and procedures for board identification and self-testing, and for system autoconfiguration.

Communication among boards takes place by exchanging **message packets** up to 32 bytes long. The messages are of two types: solicited and unsolicited (the latter can be only 28 bytes long), and contain the addresses of the source and destination boards. As the name indicates, solicited packets are sent only after an explicit request by the receiving unit; they are used primarily for data transfer. Unsolicited packets may be requests for data, signalling messages or interrupts.

6.4.3 Data Transfer

Data is passed from one board to another in messages with a maximum length of 16 Mbytes. Each data transfer is preceded by a negotiation among the boards exchanging the data.

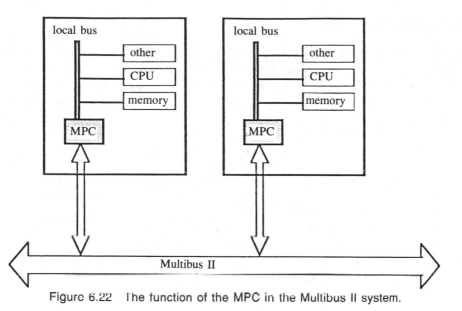

Figure 6.22   The function of the MPC in the Multibus II system.

The original message is divided in packets up to 32 bytes long, which are transmitted sequentially on the bus.

The message and packet-oriented operations of Multibus II require a notable amount of processing, which cannot be handled directly by the CPUs on the boards. Each Multibus II board must therefore have a dedicated bus interface chip, called the **message passing coprocessor (MPC)**. The MPC is connected to all resources on the board and is the only gate to the bus (Figure 6.22). The MPC takes care of all details of bus arbitration, divides the original messages into packets and reconstructs them at destination in an entirely transparent way for the board CPU. The MPC enhances system performance, as the board CPUs are relieved from bus control tasks and do not have to wait for bus arbitration or other operations to be terminated before performing their original functions.

6.4.4 Interrupt Handling

Multibus II has no interrupt lines; instead, interrupts are handled as unsolicited messages to be transferred with high priority. During normal operation, an interrupt message gets the next available bus cycle, which means that the latency time to access the bus is 1 μs. An interrupt message may be up to 28 bytes long, so that the interrupt source board may pass relevant data to the handler together with the interrupt itself. This avoids the need for an arbitration mechanism for requesting and gaining attention, and then actually servicing the interrupt. Again, this requires that the boards have enough local logic in order to autonomously preprocess the interrupt information and build intelligible packets. The higher functionality of Multibus II is obtained at the price of higher board complexity.

6.4.5 Interconnect Space

Each Multibus II board must be able to reply with self-identification data on request. This is done with a record (called **Interconnect Space Header Record**) which is hard-coded on each board and contains specific information such as board name, the identification code of the manufacturer and other board-specific data, for instance, a self-configuration record and diagnostic information.

The interconnect space is the basic concept behind the automatic configuration of a Multibus II system. At power start-up each board performs a self-test and writes the result in the configuration record. The central service module (CSM) then polls all the boards, reads their specific data and checks whether the tests yield positive results. In case a board resulted defective, it is marked as non-accessible, even if it remains physically installed on the bus. The system software can then 'ignore' defective boards.

Multibus II is somewhere between a 'down to the chip level' bus system and a more sophisticated communication network. The high level information requested by the boards means higher functionality as well as higher prices. Despite this, Multibus II indicates the direction for the development of future bus systems.

6.5 OTHER BUS STANDARDS

6.5.1 The IBM PC Bus

At the beginning of the 1980s, everybody was waiting for IBM to show a direction in personal computer design, whatever it might be, in order to follow it. The company finally introduced its personal computer (IBM PC) in 1981. Although technically the IBM PC was no better than other machines already present on the market, it immediately enjoyed large success thanks to the size and strength of the company producing it. Today, the IBM PC and its copies ('clones') are the most widespread personal computers in the world.

The IBM PC and XT are based on the Intel 8086/8088 CPU. On the motherboard there is space for up to 640 kbyte RAM memory. The CPU is connected to the bus and drives it directly; on the motherboard there are eight edge connectors into which peripheral cards may be inserted. Some interfaces (floppy and hard disk, screen and external printer) are built on boards which occupy a few slots, with about five slots remaining free for the user.

The IBM PC bus is strongly oriented to the control signals of the 8086 CPU. It has 62 parallel lines, the data path is 8 bits wide and the address path width 20 bits (which allows the direct addressing of 1 Mbyte). The data lines are bidirectional, while the address lines are always set by the processor or by the DMA circuitry.

The IBM PC bus does not support multiple bus masters. However, DMA operation is possible; it is controlled by a dedicated chip. This chip is a complete microcontroller able to manage the transfer of up to 64 kbytes of data between two boards by generating the increasing memory addresses and the required bus timing signals. From the point of view of the bus devices, an address is always present on the bus lines together with the required handshaking signals.

The IBM PC bus has six interrupt lines each with different priorities (the lowest-numbered interrupt has highest priority). The TTL-type interrupt signals are a known source of problems for the IBM PC bus. Two cards generating the same interrupt are incompatible and cannot be used together on the bus. A reconfiguration of the controller interrupts is not always possible.

In 1984 the personal computer IBM AT was introduced. This was based on the Intel 80286 CPU with word length of 16 bits and address width of 24 bits (16 Mbyte addressable). The IBM AT bus uses the same connector as the PC, together with an additional 36-pin connector. Other interrupt and handshaking lines have also been added. Boards designed for the IBM PC bus can also be used in the AT bus, but they will not offer the extended addressing and data transfer capabilities.

Despite what its importance might lead us to believe, the IBM PC bus does not have an official standard. The bus is described in the 'IBM-PC Technical Manual', but there has never been an official commitment by IBM to abide by it. The IBM AT bus is also known as the **Industry Standard Architecture (ISA)**. This name was defined by a group of companies, other than IBM, to indicate what they meant by 'IBM standard'.

More recently, the more powerful processors Intel 80386 and 80486 have replaced the older types 8086 and 80286 in a new generation of personal computers. When reference is made to an '80386 computer' or an '80486 computer', then, if the bus type is not explicitly mentioned, ISA is intended.

Because of its low flexibility and low speed, an IBM PC-based system is hardly of interest for industrial applications. However, the PC should not be written off entirely. Its main asset is the low cost of both central unit and peripherals together with a wide choice of interface cards, not to mention the huge amount of available software. For data collection applications with no special speed requirements, the IBM PC may be the most cost-effective solution.

The advanced performance of the 80386 and 80486 processors makes them of much more interest for demanding applications. The increased performance of interface cards and the availability of new control software will push computers that are based on these processors more and more towards the factory floor for applications such as machinery control, data collection and communication.

6.5.2 Micro Channel

Micro Channel architecture (MCA) was introduced by IBM in 1987; it is the basis of the personal computers of the PS/2 generation. Micro Channel is not compatible with the IBM PC and AT buses.

The basic Micro Channel connector has 68 pins on two rows and supports a data width of 16 bits with a 24-bit address range (16 Mbyte). An additional, 31-pin connector allows extension to full 32-bit words and 32-bit addresses (4 Gbyte). Data transfer is asynchronous, although some operations are so strictly timed that in practice they are carried out synchronously. Maximum data transfer rate is about 20 Mbyte s^{-1}. Even if its lacks compatibility, Micro Channel has grown out of the architecture of the IBM AT bus. In Micro Channel there are provisions for bus arbitrage and multimaster operation, but basically Micro Channel is oriented to a single processor. Operations with more than two bus masters are deemed to be inefficient (Shiell, 1987). Some lines of the basic connector carry audio signals that may be shared by the boards, and there is even a 20-pin extension for video signals.

Micro Channel supports the new concepts of automatic power-on tests, board self-identification with predefined codes and software-driven bus configuration. Jumpers and microswitches on the boards are explicitly forbidden and the whole configuration process, including the selection of new operating parameters for the boards, takes place under software control. The user configures the bus using only the keyboard and the screen.

Unlike the IBM PC series with an architecture that everybody was — unofficially — allowed to imitate, Micro Channel and the PS/2 computers are patented and IBM does not permit copies to be built by third parties. External companies may produce extension boards for the Micro Channel only with an official licence from IBM. The strict control by IBM over the bus has at least an organizational advantage. The hard-coded information on the boards will not be chosen at random and problems with incompatible boards should therefore not arise.

Micro Channel is a high performance bus that can be used successfully in advanced process control applications. Still open is the question of how many companies will produce interface boards and other hardware for industrial applications.

6.5.3 EISA

The EISA (Extended Industry Standard Architecture) standard was defined by a company consortium which included Compaq, Intel, Hewlett-Packard, Olivetti and many others. The stated goal of EISA is to have a well specified, high performance bus standard for advanced PC systems which is independent of the control of only one company (read: IBM). With the introduction of the PS/2 system and its Micro Channel bus, IBM declared in practice an end to open competition. As Micro Channel was subject to high licensing fees and to strong IBM control, the other manufacturers decided to go their own way in order to overcome both fees and control.

The EISA standard is based on the existing architecture of the ISA IBM AT computer. EISA is downward compatible to ISA; this means that an ISA board can operate on an EISA bus (but the opposite is not true). Physically, the EISA connector is similar to the ISA, with the pins for the new EISA functions and extended data and address paths placed between the pins of the ISA connector.

EISA is a synchronous bus with a 32-bit word length (word lengths of 8 and 16 bits for the older PC and AT boards are supported) and a 32-bit address length (4 Gbyte). Maximum data exchange rate is 33 Mbyte s^{-1} in burst data transfers.

The hybrid nature of EISA appears clearly from the interrupt handling scheme. EISA uses the same interrupt lines as ISA, but defines them as open collector, shareable lines.

The consequence is that EISA boards can share interrupt lines but ISA cards cannot because of their TTL-type interface. Consistently with modern bus specifications, EISA supports multimaster functions, board identification with hard-coded information and bus autoconfiguration.

Together with Micro Channel, EISA represents a new high performance standard for personal computers. The concept behind EISA with board identification and bus auto-configuration reflects the new functional requirements also found in other advanced systems. EISA is apt to be used in industrial control applications, but so far its use has been restricted to personal computers for business applications. A major advantage of EISA over Micro Channel is its wider manufacturer base, which will probably lead to a better selection of peripheral cards than for Micro Channel.

6.5.4 NuBus (IEEE 1196)

NuBus was initially developed at the Massachusetts Institute of Technology in 1978. It is now used as a backplane for the Apple MacIntosh computer series and is also available in the Eurocard format for industrial applications. NuBus is a synchronous bus with 32-bit multiplexed address and data. Its simple design is underlined by the low number of pins, 51 in total. The bus clock frequency is 10 MHz and its data transfer rate 37.5 Mbyte s^{-1}. NuBus supports multimaster operations.

NuBus has support for bus autoconfiguration with similar features to other advanced bus systems. For instance, in the MacIntosh version, each card must have a ROM containing MacIntosh-specific information such as a resource list, the required software drivers and even the board-specific display icon for the user interface (Chapter 11).

6.5.5 Multibus I (IEEE 796)

Multibus I is the name of an industrial bus originally developed at Intel but later opened to other manufacturers and normed by IEEE. Multibus I has two edge connectors on the board, one with 86 lines and the other with 60 lines. Not all the lines are defined and some are left as spare or for user-specific applications.

Multibus I supports asynchronous data transfers of 8 and 16-bit words with a maximum address width of 24 bits (16 Mbyte addressable). The bus has eight interrupt lines and supports multimaster operations. Multibus I is widely used in industrial applications and compatible boards are available from many different sources. Attention should be paid to the fact that some companies market Multibus I under different names, although the product is the same.

6.5.6 STD (IEEE 961)

The STD bus is one of the first general purpose buses for industrial applications. With its 8-bit word length, it is oriented to the 8-bit microprocessors of the first generation such as Intel 8080 or the Zilog Z80. The STD bus has a total of 56 conductors and supports synchronous data transfer with an address range of 16 bits (64 kbyte). The bus has 22 control lines for data transfer control, interrupts, etc.

The control lines of the STD bus present a peculiarity. Some of them are used for support of specific processors with different types of signals. Some STD cards can therefore operate only with a processor of one type but not with a different one; this is indicated in the board name, as in STD-Z80 compatible, STD-8080 compatible and so on. STD cards with no specific processor name should operate universally, but it always pays to double check the fine print in the documentation.

The STD bus lies at the low end of the industrial bus range. It has a limited data and address range, and by today's standards it is definitely slow. However, not unlike the IBM PC bus, it is popular for control applications where a simple microprocessor can do the job and there is no need to invest in more expensive devices that do not add anything to the required functionality.

6.5.7 S-100 (IEEE 696)

Some first generation personal computers at the end of the 1970s were based on the S-100 bus. Its data width is 16 bits and the address range, which initially was 16 bits, has been brought to 24 bits (16 Mbyte addressable). Data transfer on S-100 is asynchronous. There is support for multimaster operation: the bus allows for up to 16 master boards to be installed; the arbitration procedure is handled by a board acting as a permanent bus master.

Although its specifications were quite rigorous from the beginning, the S-100 has enjoyed more popularity as a backplane for hobby and business computers than in industrial applications. The selection of S-100 cards is also quite limited.

6.6 COMPARISON OF BUS SYSTEMS

Bus systems embody, in different fashions, the theoretical principles we have examined at the beginning of this chapter. It seems that different buses live similar lives. They are first designed and proposed by one or maybe a few cooperating industries. If the customers react positively to their market introduction and the number of manufacturers for bus-related components increases, there comes a time when an established and independent organization assesses and produces specific bus description documents ('standards'). The role of standardization committees is not to invent, but to structure and define without ambiguity existing bus know-how.

A notable exception to the usual bus life is the 'design-by-committee' Futurebus. The initial intention (from 1979!) was to define and standardize a complete 32-bit bus before

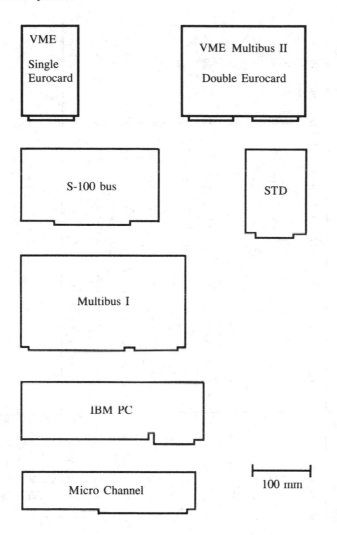

Figure 6.23 Comparison of form factors in various bus systems.

anybody started to manufacture it. After more than a decade, Futurebus still exists only on paper. A new version of Futurebus is under definition and its theoretical specifications outperform all other existing bus systems, with a data width of 256 bits and a maximum transfer rate of 3.2 Gbyte s^{-1} It remains to be seen whether the foreseen performance will also find a practical and working implementation.

In Table 6.3 different bus systems are compared and the board form factors are shown in Figure 6.23. The data transfer rates indicated in Table 6.3 are maximum performances;

Table 6.3 Comparison of different bus systems

Bus type	STD	S-100	Multibus I	IBM PC	IBM AT	VMEbus	NuBus	MultiBus II	EISA	Micro Channel
Standard (governing body)	IEEE 961	IEEE 696	IEEE 796	IBM	IBM	IEEE 1014	IEEE 1196	IEEE 1296	Different companies	IBM
Data width (bits)	8	16	8, 16	8	8, 16	8, 16, 24, 32	32	16, 32	8, 16, 32	16, 32
Address width (bits)	8	8, 16	8, 16	8	8, 16	16, 24, 32	32	32	32	16, 24, 32
Timing asynchronous/ synchronous	S	S	A	S	S	A	S	S	S	A/S
Multiplexing address/ data	N	N	N	N	N	N	Y	Y	N	N
Maximum throughput (Mbyte s^{-1})*	1	12	10	1	>1	40	37.5	32	33	20
Interrupt lines	2	9	8	6	11	7	—[1]	—[2]	11	11
Shareable interrupts	Y	Y	Y	N	N	Y	N	—[2]	Y[3]	Y
Multiprocessing	Poor	Good	Good	Poor	Poor	Very good	Very good	Very good	Good	Good
Autoconfiguration	N	N	N	N	N		Y	Y	Y	Y
Board dimensions approx (mm)	114×165	130×254	171×305	335×106	338×114	1×/2×EURO	2×EURO and others	2×EURO	338×114	325×65
Connector type	edge	edge	edge	edge	edge	1×/2×DIN	1×DIN	2×DIN, 1 used	edge	edge
Number of signals	56	100	86 and 60	62	62 and 36	96 and 32	51	96	197	136 and 62

*For maximum data width, address ranges: 16 bit = 64 kbyte;
20 bit = 1 Mbyte;
24 bit = 16 Mbyte;
32 bit = 4 Gbyte.

1×EURO — single Eurocard format 100 × 160 mm
2×EURO — double Eurocard format 233 × 220 mm

1×DIN — connector of DIN 41612 type

[1] NuBus has a separate interrupt line for each slot.
[2] Interrupts are handled and transferred as data.
[3] EISA cards can share interrupts; ISA ('IBM AT-compatible') cards installed on the same bus cannot.

the actual data throughput in real applications is a function of several factors and can only be assessed in practice. The type of timing synchronous/asynchronous is given as a general reference only; some buses follow a mixed synchronous/asynchronous timing and no clear-cut classification can be made. Where different data and address ranges are indicated, the largest can usually be reached only by using additional connectors and bus extensions.

6.7 SUMMARY

Bus systems are fundamental in control system applications, most of all because they allow for modularity. This allows the most appropriate hardware for a specific application to be selected and assembled together. The evolution in the structure of bus systems is due to increasingly sophisticated requests by customers and to the decline in the cost of computing power.

Integrating a bus system that will work satisfactorily may be quite easy but also turn out to be a major effort. In practice, every board may have to be configured with the help of jumpers or switches and, more often than not, some system software parameters have to be adjusted. Therefore, getting an application to run usually requires several trial-and-error steps.

The current trend in bus architectures is to hard-code basic identification information on each board. At start-up time, the system software polls all connected boards and configures itself accordingly without direct programmer intervention. This is so in the cases of Multibus II, EISA, Micro Channel and NuBus. The trade-off lies in the higher complexity of the additional control logic for each board; the most capable systems are, not surprisingly, on the high-cost side of the bus spectrum. Moreover, smaller boards in a sense mean more flexibility since it is easier to select the right mix for an application; with larger boards there is always a risk that many of the provided functions are not used.

The most important industrial bus systems have been described in this chapter. It is important to point out again that there is no 'best' system and that different buses only have different degrees of functionality. The real challenge lies in the selection of the right bus for a specific application, taking into consideration technical as well as economic constraints.

Other aspects of buses that are widely accepted today are address and data field lengths of 32 bits and standard norms such as Eurocard for form factor and board dimension, and DIN 41612 for the electrical connectors. Important bus standards for industrial applications are VMEbus and Multibus II. The standards EISA, Micro Channel and NuBus might come to play a significant role if appropriate hardware for industrial environments becomes available. Less ambitious bus standards such as S-100, STD and Multibus I are still widely in use in the industry because of their simplicity and large existing know-how base.

FURTHER READING

Lawrence and Mauch (1987) offer an excellent introduction to board and bus system applications. It strongly emphasizes system thinking and the relationship between hardware and software.

Tanenbaum (1990) is dedicated to the structure of digital design all the way from gates to complex computers. Relatively little space is given to bus operations, but the book is recommended for a broader view on the operations of bus-based systems. An overview of different bus systems is given in White (1989), with current trends in bus design described by Borrill (1989).

Peterson (1989) gives a comprehensive description of the VMEbus operations, balancing theory and practical examples. For further information on Multibus II, see, for instance, Finger (1987) and Hyde (1988).

Basic information about bus systems is found in the original documentation from the manufacturers and the standards organizations. The IEEE, for example, distributes the specification of all its normed systems. As mentioned, the various IBM buses and the EISA standard are not under the control of a public interest organization and are therefore not published for free use.

Many articles appear in the specialized press on the operations and performance of bus systems, with particular attention to the most popular types. There are also many books available. For instance, the operation of the IBM PC and its bus is described in Sargent and Shoemaker (1984); Ciarcia (1987) and Shiell (1987) have reported on the IBM Micro Channel. A comparison of the Micro Channel and NuBus-based PCs is given in Cornejo and Lee (1987); while Bailey (1989) deals with hardware for industrial NuBus applications. A general introduction to the EISA bus standard is given in Brett Glass (1989).

Finally, many manufacturers also produce and distribute qualitative information to support their bus products, which are often based on standard designs.

7

Operating Systems and Real-time Programming

Aim: To understand the issue of resource management by operating systems and get acquainted with the methods used in real-time programming, with respect both to theoretical constructs and practical solutions

The control of industrial processes is, in general, a complex task that has to be carried out by several computers linked together and with different specializations. The way computers are programmed depends mostly on the required response speed. Computers at the highest hierarchical level usually run programs for database management and statistical analysis. These programs are run, for example, once per day and there are no strict requirements on the response times. The programs are developed with the usual techniques used for administrative systems which are amply described in programming textbooks. The case of computers at the lowest level is different; these are directly in control of the physical processes. Here, the timing requirements are usually so strict that special programming methods and techniques must be used. These methods are the subject of this chapter.

Hardware is as important as the software for building efficient real-time computer systems. Hardware capacity must be available and the software has to exploit it. In a sense, hardware and software are logically equivalent; many solutions can be realized with hard-wired circuits as well as with program instructions. But there are situations where the software seems to fight against all the possibilities the hardware has to offer.

This chapter begins where the common language description handbooks end. Its purpose is to present what real-time programming looks like in practice. Ideally, a programmer should not leave the protected environment of a programming language from which — according to language handbooks — everything can be done. In the real world, programmers work with direct calls to the operating system, write resident code and mix languages. These are all horror actions according to textbooks for structured programming, but nevertheless belong to the hard facts of life. In this chapter we will not take a stand for or against certain programming techniques, but rather will inform on how some problems can be solved in practice: theory as a framework for practical action.

Section 7.1 deals with the basics about programs and processes, and Section 7.2

introduces some basic operating systems concepts, especially about multitasking and memory management. Sections 7.3 and 7.4 treat the problems of resource protection and mutual exclusion and how they are solved. Section 7.5 deals with interprocess communication and Section 7.6 with practical methods for real-time programming. A presentation of some languages and operating systems for real-time programming in Section 7.7 completes the chapter. Notice that the material is formal-theoretical in the first five sections and practice-oriented in the last two.

7.1 GENERAL CONCEPTS ABOUT PROGRAMS

7.1.1 Programs and Processes

A program describes the constant and variable data objects and the operations to be performed on them. A program is just pure information; as such, it can be recorded on any medium able to store information, for instance, paper or a floppy disk.

Programs may be analysed and written at several abstraction levels by using appropriate formalisms to describe the variables and the operations to perform at each level. At the bottom, the description is straightforward: the variables are stored in memory cells labelled with their location/address. At higher levels, the variables become abstract names and the operations are organized in functions and procedures. The programmer working at higher abstraction levels does not need to bother about which cells variables are stored in or about the aspect of the machine code generated by the compiler.

Sequential programming is the most common way of writing programs. The term **sequential** indicates that the program instructions are given in a fixed sequence, one instruction after the other. The purpose of a sequential program is to transform input data given in a certain form into output data of a different form according to a specified **algorithm** (i.e. solution method, Figure 7.1). In a sequential program, the only entities are the data and the code to act upon them. No time constraints are given; the result of a run depends only on the input data and the properties of the algorithm. The algorithm of a sequential program can, in principle, be coded in any programming language.

A sequential program acts like a **filter** on the input data. The filter abstraction is carried further in some operating systems (e.g. MS-DOS and UNIX) with device independence in program input and output. In such systems, the input/output of a program can be a file, a terminal screen or another program.

In real-time programming, the abstraction of resource independence cannot be made; on the contrary, one has to be constantly aware of the environment in which the program

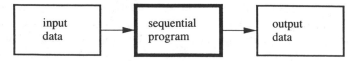

Figure 7.1   Data processing via a sequential program.

is operating, be it a microwave oven controller or a robot arm positioner. In real-time systems external signals usually require immediate attention by the processor. In fact, one of the most important features of real-time systems is their reaction time to input signals.

The special requirements of real-time programming and in particular the necessity to react quickly to external requests, are not approached adequately with the normal techniques for sequential programming. The forced serial disposition of instruction blocks that should be executed in parallel leads to an unnatural involution of the resulting code and introduces strong ties between functions which should remain separate. We have already seen in Chapter 2 what problems may arise when two different program modules are bound together.

In most cases it is not possible to build real-time systems using the normal methods for sequential programming. In real-time systems, different **program modules** or **tasks** have to be active at the same time, that is, operate in parallel, where each task is assigned to a specific function. This is known as **concurrent programming** to stress the cooperation among the different program modules.

The basic operating entity in real-time systems are the **processes**. There is a very important distinction between programs and processes. Programs are sets of information on how to operate on and transform the input data, while processes are programs in execution. A process consists of **code** (the program instructions), a **data area** where the process variables are stored and, depending on the actual implementation, a free work area (**heap**) and a **stack** (Figure 7.2). A program written in the same high level language

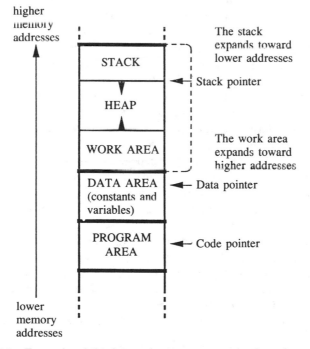

Figure 7.2 Example of the internal memory organization of a process.

and then compiled and executed on different machines will lead to different processes each with its own code, data, heap and stack areas.

Each process is at all instants in a well defined state, unequivocally described by the contents of the CPU registers, the locations of its code, data and stack areas, and a pointer to the next instruction for execution in the code area. This basic information about a running process is called its **canonical state** or **context**. The existence of a context is a general fact, whereas what registers, states and pointers are actually part of the context depends on the processor used.

The steps needed to transform a program in a process consist of storage on a computer-readable medium such as magnetic tape or floppy disk, compilation, linking, loading and execution. These steps are amply described in books on operating systems and will not be dealt with here.

The description used for algorithm definition in the top-down programming method also provides a complete documentation of the system. The importance of good documentation cannot be overestimated; it is enough to consider that **maintenance** and **corrections** are the major expenses related to the entire lifetime of a program. Maintenance is much easier if good documentation is available; it might, on the contrary, turn out to be impossible if the documentation is poor or insufficient.

7.1.2 Concurrent Programming, Multiprogramming and Multitasking

In both concurrent programming and real-time programming, the possibility of executing several tasks simultaneously on the same machine may be needed. These tasks share the resources of the system but otherwise are more or less independent from each other.

Concurrent programming is a macroscopic effect that can be realized either by using several processors where each task is run on an entirely dedicated processor, or by letting more tasks run on a single processor. The latter is the most common case, even though falling hardware prices make multiprocessors more and more economically feasible. Several processors may, of course, be active at the same time in a bus system.

In technical literature the term concurrent programming is sometimes used interchangeably with **multiprogramming**. Concurrent programming is the abstract study of programs with potential for concurrency, independently from the implementation details of the machine they run on. Concurrent programming is oriented to the execution on virtual processors without concern for the implementation details. Multiprogramming is the technique for letting several programs run on a single central processing unit.

A common term in real-time programming is **multitasking**. A **task** is a small program or module and multitasking is the technique of letting them run concurrently. In a sense, multitasking is the practical side of concurrent programming, where the actual aspects of the target machine are taken into consideration.

Concurrent programming is more difficult than sequential programming because the human capacity for following the development of interdependent processes and for examining their mutual interactions is limited.[1] Real-time programming is based on

1   Think about school studies in history: first comes national history and then Western or Eastern history in some chronological order. With this perspective, it is not natural to compare, for instance, what happened in different countries in 1848.

concurrent programming and refers also to techniques to increase the efficiency and execution speed of programs: interrupt management, exception handling and the direct use of operating system resources. Real-time programs also require special testing methods.

7.2 THE MANAGEMENT OF SYSTEM RESOURCES

7.2.1 The Function of the Operating System

An operating system is a very complex piece of software for administering the hardware and software resources of a computer system. An operating system offers a **virtual (logical) environment**, consisting of **CPU time** and **memory space**, in which the processes can be executed. With **virtual environment**, a conceptual environment is intended with specific features and that may or may not exist in physical hardware.

Multiprocessing is the basic conceptual tool for the design of multiuser as well as real-time operating systems; it deals primarily with resource allocation and protection. However, the goals of multiuser and real-time operating systems are not the same. A multiuser operating system, also known as a **time-sharing** system, allocates expensive resources to several users, checks that the users do not influence each other and divides the operating costs between them. In real-time programming the purpose of multitasking is to keep distinct operations separate from each other and to distribute the workload among different modules. In real time systems, the only 'user' is the system to be controlled.

In time-sharing systems, much attention is dedicated to the protection and separation of users by way of passwords, access control, etc. Real-time programming is less restrictive in this respect as the system designer(s) know what each module does. In situations where each CPU millisecond counts, no time can be wasted for access control overhead; file systems and protection mechanisms are not important parts of real-time operating systems. Time-sharing systems are also supposed to be 'fair' in some sense, trying not to put any user at a special disadvantage even under heavy machine load conditions. The same does not hold for priority-based real-time systems, where the processes are strongly differentiated.

We will now focus on multiprogramming on a single processor because this is the most common and most important case of digital control applications. Many of the ideas can be transferred to the multiprocessor case.

In multiprocessing, the basic entities are the processes or tasks and their contexts. The context of a process in execution can be 'frozen' at any time by saving the content of the CPU registers; while the initial process is suspended, the CPU can run other processes. To achieve multitasking on a single processor, the execution of each task is divided into several short intervals (Figure 7.3). The processor begins executing part of the first task, continues with part of the second, of the third, and so on. A time interval is assigned to each task, so that, for example, for 10 ms the processor is dedicated to the first task, then switches to the second, the third, etc. The macroscopic effect of the CPU time division among the processes is the parallel execution of n processes, each on a fully dedicated CPU of capacity $1/n$ (that is, n times slower) compared to the original one.

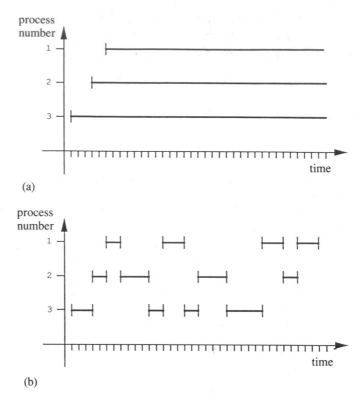

Figure 7.3   The principle of multitasking. (a) Macroscopic effect;
(b) CPU time division.

The execution of several tasks on different CPUs or on the same CPU are two different realizations of the same logical principle: in the first case the processes are distributed spatially, in the second they are distributed in time. Apart from overhead due to scheduling and intertask communication, if n processes run on k processors, each process is ideally assigned to a processor of capacity k/n of the original one.

A basic multitasking system consists of a procedure, implemented in hardware or software or a combination of both, to save the context of a process on the stack or at defined memory locations, and restore the context of another process to continue its execution where it was halted. A system program called a **scheduler** selects the next process to execute from among the loaded processes. The process switch operations are time critical and must be realized with maximum efficiency.

In processors not designed for multiprogramming, the process exchange module must save all registers and other context parameters on the stack and then save the pointers to the stack in a protected data area. Processors designed to support multiprogramming have compact instructions to save and recall the content of all the registers.

When the context of a process is saved, it is not necessary to also save the process variables. These are located in the process memory area which has to be protected by the operating system against changes by other processes. The same does not hold for the CPU registers, which are shared by all processes and whose content is changed all the time.

To be able to halt CPU execution at regular intervals in order for a different process to be executed, a timing device external to the CPU is needed. A **system timer** sends interrupt signals (**ticks**) to the processor at defined intervals; typical rates are one tick every 1 ms on very fast processors down to one tick every 50 ms on slower machines. At each tick, the CPU briefly suspends its operations to check whether the current process has to be interrupted and a new one loaded. The action that forces a running task to halt its execution in order to allow another task to run is called **preemption**.

The tick interrupt is not the only way to stop a process and transfer execution to another. A process can stop on its own either because it has reached the end or because it is idle waiting for an event, such as an I/O operation with a physical device which would take several ticks to complete.

7.2.2 Process States

A process executed in a multitasking environment can be in different states. These states are commonly shown with the help of a diagram (Figure 7.4); they are defined as follows:

- **Removed** The program is present on disk, ready to be loaded to internal RAM memory.
- **Waiting** The process is waiting for some external event (I/O data transfer, input from keyboard, an external interrupt) or internal (explicit signalling by another process) to be moved to the 'Ready' state.
- **Ready** The process can be executed whenever the CPU is available.
- **Running (Executing)** The process that is currently being executed.

Figure 7.4 shows which changes from one state to another are possible. The defined operations are:

1. From 'Removed' to 'Ready'. The process is loaded from disk to RAM memory, with relocation of all the relative addresses and assignment of the work areas (code, data, heap, stack) with the related pointers.

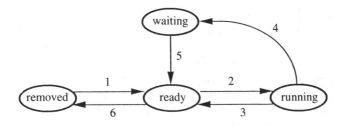

Figure 7.4 The states of a process.

2. From 'Ready' to 'Running'. The process is selected by the scheduler to run and is assigned CPU control via the process switch module.
3. The opposite change, from 'Running' to 'Ready', is controlled by the same process switch module when it is time to let another process run.
4. From 'Running' to 'Waiting'. The process enters an idle state to wait for an external event, often an I/O operation with units much slower than the CPU, or the process must wait for a determined period of time, because of an explicit instruction.
5. From 'Waiting' to 'Ready'. When the awaited event has occurred or the required time has elapsed, the process is not immediately executed but is put instead in 'Ready' state. The scheduler will later determine when the process can be executed again.
6. When the **end** instruction of a program is reached, the operating system may eliminate a process from central memory.

7.2.3 Strategies for Process Selection

There are several possible strategies for selecting from among the processes waiting in the queue, the one to run next. Several conflicting factors have to be considered: some processes need more execution time than others, must react quickly, are more important, etc. The decision of which process may continue execution at a given time is taken by the **scheduler**, a service module started every time a running process releases control of execution. The scheduler can follow different strategies, of which the most common are round-robin rotation and priority allocation. The strategies are similar to those used for bus arbitration (see Section 6.2.10).

The most simple selection strategy is **round-robin**: the processes are selected one after the other for execution, following a fixed order and for the same time interval. The main advantage of the round-robin method is its simplicity; on the other hand, there are notable drawbacks when processes with different requirements are allocated equal CPU resources.

A more complicated principle for process selection is based on the assignment of **priorities**. At each process change, the scheduler assigns execution to the process with highest priority. The priorities are defined by the programmer and in many cases can also be changed from within a process during execution.

Straight priority allocation leads easily to unfair situations. The process with highest priority would be always selected for execution (unless it is in waiting state) and be the only one to run. To avoid this situation, the scheduler decreases the priority of the running process at a constant rate. Eventually, the priority of the running process will be lower than that of some waiting process, which is then selected for execution. In this way, it is ensured that all processes are eventually executed. After some time, the priorities of the waiting processes are set back to their nominal values. This method is called **dynamic priority allocation**. It ensures that even processes with lower priority will be executed and that processes with high initial priority do not hold indefinite control of the CPU.

The consequence of different initial priority allocations is that processes with higher priorities will be executed more often than others. Processes which are called often and/or must be activated quickly have higher priorities; less important processes for which a longer response time is acceptable have lower priorities.

In real-time systems, however, the forced preemption of running processes may be undesirable. A different strategy is then employed: each process may start other processes and change the priorities of waiting processes. The responsibility for seeing that the play among the processes is carried out in the desired fashion lies with the programmer.

The minimal time interval assigned to each process before it is interrupted is called **time slice**; it has the length of a few ticks. The length of the time slice influences the performance of the system. If the time slices are short ($\sim 10{-}20$ ms), the system is quick to react to external events such as interrupts or terminal input, but the process scheduling overhead gets an important share of the total CPU time. With longer time slices, the processes execute more effectively with less overhead, but the reaction time gets appreciably slower.

The length of the time slice influences strongly the global performance of a system only when the running processes have similar priorities. Real-time operating systems do not only depend on the time slice. The operating systems are designed to react immediately to situations when changes in the process priorities may have occurred, such as the arrival of an external interrupt or the completion of a disk read operation. For each such event, the relative priorities of the waiting processes are computed anew and, if necessary, a process switch is performed.

The scheduling of processes based on priorities works correctly only when the various tasks have different priorities. It does not help to give maximum priority to all processes, as this certainly does not increase the execution speed of the CPU. Each process would still have to wait until all other processes have been executed before it can run again. A system where all tasks have the same priority works in a round-robin fashion. The best results in the operation of a real-time system come when the relative priorities are correctly defined and balanced.

7.2.4 Internal Memory Management

After the CPU, the other most important resource to manage in real-time systems is the central memory RAM. For this purpose, the methods used in real-time systems are generally simpler than the ones used in multiuser computing centres. In large operating systems with many users, most of the programs and data are stored in secondary memory (hard disk) and are loaded to RAM only when they are needed. This is acceptable for time-sharing and batch jobs when execution time is not very important, but not for real-time systems where all tasks should always be located in RAM ready for execution. However, disk memory support could still be necessary in real-time systems because the central memory is not always large enough to fit all programs and their data. A strategy is needed in order to allocate the available space as efficiently as possible.

A basic memory management technique is **segmentation**. A process is divided in some

Original program   Internal memory

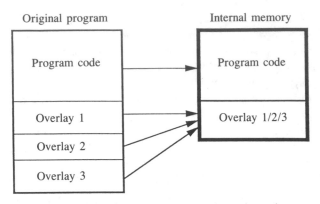

Figure 7.5   Use of program segments and overlays.

parts, called **segments** or **overlays**, which can be separately loaded in central memory (Figure 7.5). The programmer must explicitly define the program segments. Each program segment will execute and then call the next to continue. This method is straightforward but not particularly efficient because of its dependence on external disk memory.

Other memory management schemes are transparent to the processes, in other words, they do not have to be taken into consideration when writing the program. The most straight-forward method for memory management is division of the available memory in **partitions**, each assigned to a process (Figure 7.6). The size of the partitions is defined at the time of system initialization; it may be different for different partitions. When a process is loaded, it is put in the smallest partition large enough to contain it. If no partition is available, the process may have to be broken and loaded in several partitions. Many different algorithms are available to allocate the memory to different processes.

With partitions, once the memory space has been allocated at machine start-up, it cannot be reclaimed and reused. To be able to utilize all available memory over and over again, on middle and large-sized computers the **virtual memory management technique**

Internal memory

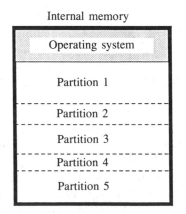

Figure 7.6   Memory division in partitions.

is commonly used. Virtual memory is based on the assumption that the total size of processes and data may be larger than the RAM space at disposal. A mass memory unit (e.g. a disk) that allows fast data exchange with central memory is used. The mass memory unit is large enough to hold the total memory space required by all processes. With virtual memory, a process may address a space larger than the one at disposal in central memory.

The main reason for which virtual memory is used is economic. The central memory is still much more expensive per unit of stored information than secondary mass memory. The central memory is also volatile and draws electric power, which both costs and heats up the system electronics. In case of a system crash, it is possible to restore operations almost to the point where the crash occurred if a constant copy of the processes is stored on disk. If a crash or a power failure occurs when the whole system is only loaded in RAM, then the whole process state is wiped out.

In real-time systems, virtual memory is of interest only when it is fast and efficient. To ensure fast reaction, the most important processes can be permanently stored in central memory partitions and less important ones be called at need. Another important considera- tion related to the use of secondary memory in real-time applications is whether it can be used in the operating environment. For instance, disk units (hard disks and floppies) cannot be used in environments with vibrations and shocks.

One of the major differences between multiuser and real-time operating systems lies in file management. The most important issues in multiuser systems are **directory structure** and **file protection**. The management and protection of directories, with the related controls and verifications at each access imposes an overhead seldom acceptable in real-time systems. A simplified use of mass memory, mainly for storing logs and reports and common ownership of all the tasks, does not warrant the need for a complex file system.

There is a strong analogy between CPU control allocation, resource protection and bus master arbitration in a multiprocessor bus system (Section 6.2.10). In all cases we deal with a limited resource (CPU time, memory, the bus) which has to be divided among several requesting units in a safe, efficient and fair manner. The criteria for assigning the resource, be it a simple round-robin scheme or a more complicated priority-based allocation, must avoid deadlocks and lockouts, assign the resource to all units requesting it and ensure maximum throughput for the whole system.

The most sophisticated operating systems allow the **tuning** of CPU and memory management parameters to achieve optimal performance. Process priorities and time slice length should be chosen and combined in order to increase the general performance, or throughput, of a system.

7.3 MUTUAL EXCLUSION, CRITICAL REGIONS AND DEADLOCK

7.3.1 Resource Protection

In multiprogramming there are often situations where the processes compete for resources. The consequences of the competition may lead to erratic behaviour and even to the complete halt of a system. Resources do not need to be expensive pieces of hardware, such as printers or magnetic tape units, they can be variables in central memory as well. The classic examples for software resource protection are seat reservations on airplanes and banking with balance control when several operations are executed — almost concurrently — on the same account. Before a flight, the airline seats exist only in the memory of the computer reservation system. Obviously, a seat cannot be allocated to two different customers if they happen to book the same flight at the same time. The seat information is therefore a type of resource to protect, in this case a resource existing only in software.

If different processes operate on common variables and read and modify them without a defined precedence order, their interaction could lead to undesirable results.

Let us consider two processes. Both processes access the same variable, first read its value and then modify it. If one process is interrupted just after the read operation and before it could change its value, the other process may modify the variable while the first is waiting. The first process then resumes execution without knowing that the variable has changed and then proceeds to work on an old value. After all, in a multitasking environment a process does not know when it is interrupted and when it starts again.

The problem is that the variable is accessed by both processes without restrictions. It does not help to check the formal correctness of different programs if the effects of the possible interactions among processes are not taken into account. A situation where the result depends on the relative random order of process execution is called a **race condition**.

This problem, known as **resource protection**, is central to the whole theory of multi-programming. The variable accessed by both processes has to be considered as a resource to be protected from their concurrent action. In order to avoid race conditions, access to the resource by the processes should not be free and indiscriminate but instead follow determined precedence rules.

The problem of resource protection has been studied for a long time and different solution strategies have been devised. These strategies vary with the type, technology and, most of all, access speed of the resource to be protected.

Slow units, which tend to be used for quite a long time by the process (e.g. printer or magnetic tape unit) are usually allocated exclusively to the requesting process on the basis of a precedence queue. Alternately, a resource is permanently allocated to a single process (called a **spooler**, from simultaneous peripheral operations on line) that accepts as input from other processes the names of the files or other data objects, organizes them according to defined precedence criteria and sends them one at a time to the requested unit. Spoolers are commonly used in multiuser operating systems.

Other methods are used for the protection of resources with very short access time and are continuously referred to by different processes, for instance, variables in central memory, records in a file or I/O interfaces on a data bus. This section is mainly devoted to such methods and will show different approaches together with their consequences.

The principal rule for resource protection is that a process should *never* change the state of a shared resource while another process has access to it. Or, more generally: a process should *never* access a resource currently used by another process, independently of whether or not it is going to change its state. The second rule is more restrictive but simplifies practical control operations because it is not necessary to keep track of what operations each process is going to perform on the resource.

The major difficulty in resource protection arises from the fact that in multitasking systems all processes can be interrupted at any time to allow other processes to be executed. The exact instants when the interruptions take place are not under the control of the programmer and cannot be known in advance.

A first, elementary, method to guarantee resource protection is to disable interrupts while a resource is accessed. This effect is achieved by blocking the reaction of the processor to the interrupt signals. As process switching is initiated via an interrupt, disabling the interrupt prevents process switching as well. A process is then guaranteed to work without interruptions when it is in a 'protected' area.

On the other hand, interrupts should normally be enabled to ensure quick reaction to special conditions requiring immediate attention. In a control system, part of the program modules are controlled by interrupts and disabling them can inhibit the processor from reacting to fully legitimate requests. In some systems, interrupts are not saved after they have occurred so they may go unnoticed if they arise when handling is disabled. Furthermore, interrupt disabling does not work in systems with several CPUs; a task running on a different processor might 'sneak' in the protected resource from behind.

Interrupt disabling should thus be used with extreme care and only when no other solution is feasible. It should also be limited to a few code instructions.

7.3.2 Mutual Exclusion

A different approach to the problem of resource protection is possible if we consider it to be a problem of mutual exclusion, that is, where access to a protected resource is done from only one process at a time. No process should then access a resource until the resource is explicitly released by the process that requested it first.

The goals of a correct execution of concurrent processes are that:

1. Only one process at a time has access to a protected resource.
2. The processes remain mutually independent. Stopping one process should not hinder the other process(es) from continuing their execution.

The above statements relate to two correctness properties, safety and liveness. **Safety** means that access limits have to be respected, so that a protected resource is not accessed by more than one process at a time. **Liveness** indicates that a program at some time will do what it is supposed to or, in other words, that it will not hang indefinitely. Safety can always be obtained by giving up some concurrency between tasks; the safest programs are in fact strictly sequential, where no parallel access to a resource from different parts of the program is possible.

To indicate concurrent processes the following notation, introduced by Edsger Dijkstra (1968), may be used:

```
cobegin
    x := 1;
    x := 2;
    x := 3;
coend;
write (x);
```

The execution of the instructions between the keywords **cobegin** and **coend** is parallel (Figure 7.7). The instruction **cobegin** does not impose conditions on the relative order of execution for the different processes and the instruction **coend** is reached only when all processes within the block are ended. If the execution were sequential, then the final value for x would be 3. With concurrent processes it is not possible to predict the final result with certainty; the tasks are executed, at least from an external viewpoint, at random. The final value of x in the example shown can equally well be 1, 2 or 3.

A practical and common method to manage access to resources is to use protection variables. A straightforward protection method is based on the use of one binary variable, $f1$. This variable is changed by both processes so that one of them has access to the protected resource when $f1 =$ **true** and the other when $f1 =$ **false**.

```
program protect_example (* resource protection *)

var     f1:     boolean;

begin
f1 := true;

cobegin

while true do (* repeat forever *)
    begin (* process A *)
    repeat until f1 = true;
    (* protected resource *)
    f1 := false;
    . . .
    end; (* process A *)

while true do (* repeat forever *)
    begin (* process B *)
    repeat until f1 = false;
    (* protected resource *)
    f1 := true;
    . . .
    end; (* process B *)

coend;
end. (* protect_example *)
```

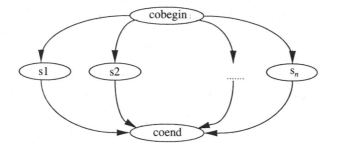

Figure 7.7 The precedence graph for cobegin.

This solution satisfies the mutual exclusion principle: the two processes control the value of $f1$ and enter the critical region only when $f1$ takes different values. A process in the critical region may be assumed to be the only one there. This solution, on the other hand, introduces new problems. The slowest process determines the total execution speed. It does not matter if A runs faster than B or vice versa, because every process must wait until the other has changed the value of $f1$ before it can run again. In addition, should a process stop execution for any reason, even the other would have to stop, at most after one loop. Finally, the test loops on $f1$ are a waste of CPU power.

These problems are a consequence of the introduction of control variable $f1$, that creates unnatural bonds among the processes in order to synchronize them. Modules which should, in principle, be independent are tied together by $f1$ so that it actually makes a sequential process out of the two modules. The same result would be obtained by eliminating $f1$ and executing both processes in succession in a single loop.

In a different solution, the protection variable $f1$ is reset after the test on its value and before entering the protected area:

```
repeat until f1 = true;
    f1 : = false;
    (* protected resource *)
    f1 : = true;
    . . .
```

In this case the processes are disjoint (and thus the liveness condition is satisfied), but the solution is not correct either. If the interrupt for process exchange halts process A after the test $f1 = $ **true** but before the assignment $f1 = $ **false** and process B does a similar check on $f1$, then both processes are able to access the protected resource against the safety property. The use of only one protection variable for a particular resource leads to the need for protecting the variable, which has now become a common resource itself.

Yet another solution may be based on two boolean variables, $f1$ and $f2$, where each flag protects its resource. This solution leads to a possible **deadlock** situation, where the processes block each other. A deadlock takes place if the value **false** is assigned to $f1$ and to $f2$ before the tests $f1$ =**true** and $f2$ =**true**. Both processes would have to wait forever for the release of $f1$ or $f2$ but, as it would depend on the other process, which is also locked, execution cannot go on.

Deadlock and simultaneous access to a protected resource are two symmetrical problems related to extreme situations. In one case each process waits for the others to move first, in the other many processes move together. The goal of correct process synchronization is to define procedures and coordination methods so that at least one process, but only one at a time, can enter a protected area. In the definition of such procedures, one should avoid defining the limits too narrowly, otherwise execution is no longer parallel and the processes will depend too much on each other.

A first solution to the mutual exclusion problem was proposed by the Dutch mathematician Dekker and is described in Dijkstra (1968). This solution is based on the use of three boolean variables $f1$, $f2$ and $f3$. The program control loop has the following aspect:

```
while true do
    begin
    f1 :=true;
    while f2=true do
        while f3=true do
            begin
            f1 :=false;
            repeat until f3=false;
            f1 :=true;
            end;
    (* protected resource *)
    f3 :=true;
    f1 :=false;
    . . .
    end.
```

The key of Dekker's algorithm is to pass not the right to enter a critical region, but the right to *try to enter* it. The processes are separated and stopping one process does not interfere with the execution of the others. Dekker's solution is very expensive because of repeated tests on the boolean variables and today it may be considered as a curiosity. Other similar solutions have been proposed.

In conclusion, it is better not to introduce new variables for synchronization of parallel processes because these bring new bounds and become themselves common resources. In solutions such as Dekker's, wait and test loops are executed in permanence with a consequent waste of CPU resources. The problem depends on the fact that the new variables for access protection have become common resources themselves.

To circumvent the problem, some computers have an instruction called **test__and__set (TAS)** which combines the check of the state of a boolean variable and its setting in one operation which cannot be interrupted. The importance of the **test_and_set** variable is due to the fact that on its base other synchronization and resource protection procedures can be built. The fact that the test on a variable and its modification are unified operations is sufficient to ensure protection.

The **test_and_set** instruction is functionally equivalent to the **read−modify−write** cycle in the VMEbus system (see Section 6.3.2). In both cases, it is unnecessary to be able to count on the indivisibility of the two operations, read and write. The instruction **test_and_set** is usually not present in high level programming languages and often is not even part of the instruction set of many microprocessors, yet it can be implemented if it is possible to disable the interrupts for the time of its execution.

7.3.3 Critical Regions

The concept of critical regions has been proposed by Brinch Hansen (1973) to avoid the inconveniences related to the variables for resource protection. A critical region is a part of a program where a protected resource may be accessed.

The notation for critical regions is:

REGION v **DO** s

v indicates the protection region and *s* the program code to operate on it. The rules to access a critical region are:

1. When a process intends to enter a critical region, it will receive permission to do it in a finite time.
2. Only one process at a time may enter or stay in a critical region.
3. A process remains in a critical region for a finite amount of time.

It is the responsibility of the compiler to verify that the variables in *v* are referred only from within *s*. The run-time operating system should check that only one *s* module at a time is executed and that no interrupts take place during this execution.

Not unlike other multiprogramming concepts, critical regions have more theoretical than practical importance, as they are not implemented in the proposed form in any important programming language. Nevertheless, they are still a useful model for the analysis of a problem and its solutions.

7.3.4 Deadlock

Deadlock is the state when some or all processes in a system are halted waiting for something to happen. If this 'something' can only be initiated by another waiting process, then all processes wait endlessly in a deadlock condition (Figure 7.8).

A different case of deadlock is when one or more processes still run but fail to make any progress. This situation is called **starvation**; this is the case when running processes continuously test the value of a condition variable which is not going to be changed because

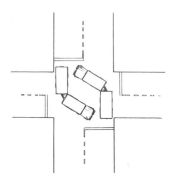

Figure 7.8 The deadlock.

the other processes are also busy testing. In other words, deadlocked processes are in the 'waiting' queue and starving processes are 'ready' or 'executing', but do not make any progress.

It has been shown that it is necessary that several conditions be true at the same time for deadlock to occur. If any of these conditions does not exist, deadlock cannot happen.

1. Mutual exclusion. There are system resources which can be used only by one process at a time.
2. Non-preempted allocation. A resource can be released only by the process that allocated it.
3. Successive allocation. A process can allocate the necessary resources one at a time.
4. Inverse-order allocation. The processes can allocate resources in a different order.

These four principles indirectly give the key for avoiding deadlock situations; it is sufficient that one of them is not true to make deadlock impossible.

The first principle cannot be changed as mutual exclusion is the principal condition to guarantee the ordered management of shared resources.

The second principle requires that the operating system recognize a deadlock situation and react accordingly, for instance, by forcing the release of a resource by a process. But recognition without ambiguity of a deadlock situation is very difficult and the forced release of some resources may lead to other types of practical problems. The forced release of a resource is interesting only with internal resources (variables stored in RAM memory) and in situations seldom leading to deadlock.

Following the third principle, the alternative to allocating one resource at a time is to assign all needed resources at once. This solution is not feasible, as many resources would remain unused for a longer time, or allocated for the whole execution of a process when their actual use may be more limited.

Violation of the fourth principle leads easily to deadlock conditions. If two processes need two resources **A** and **B** where one allocates them in order **A-B** and the second in order **B-A**, it is sufficient that the first process allocates **A**, is interrupted and control passed to the second process, which allocates resource **B**, for deadlock to be verified. Each process is now waiting endlessly for the other to release its resource.

On the other hand, the fourth principle suggests the only practical way to avoid deadlock conditions. If a precise precedence order in which the resources can be requested is defined and all processes follow this order, deadlock can be avoided. This method is quite easy to apply in real-time systems where all processes are in the hands of the same programmer(s). It is then advisable to define at the beginning of the project development phase an allocation order for all hardware and software system resources to be shared by more processes. Each process will then reserve the resources it needs as long as they are available, waiting when some resource is busy. After having been used, the resources will be released in an order inverse to their allocation. This is a reasonable protection against deadlock.

7.4 PROCESS SYNCHRONIZATION: SEMAPHORES AND EVENTS

Mutual exclusion imposes some conditions on access to a resource by two or more different processes. This problem can be considered from a different viewpoint: a process can proceed beyond a certain point only after another process has reached some other point of its execution. If the points in the processes are located before and after the protected resource, then mutual exclusion is achieved.

The introduction of a time precedence order in the execution of several processes is called **synchronization**. Process synchronization is the most natural function in an operating system and is used in practice for the implementation of resource protection: the access to a resource is ordered in time with the help of a synchronization mechanism.

7.4.1 Semaphores

We have seen that the introduction of extra variables for resource protection is not free from problems, as the protection variables become common resources themselves. The root of the problem is that the operations of check and change of the value of a variable are separated and can be interrupted at any time. Moreover, continuous tests on the values of the variables waste CPU time. The **semaphore** was proposed by Dijkstra (1968) as a basic synchronization principle to overcome the problems related with protection variables. The semaphore is probably the most common method for process synchronization.

A semaphore is an integer variable which can only be 0 or take positive values. Its initial value is defined in its first declaration in a program and is usually equal to 0 or 1. Semaphores which can take only values 0 and 1 are called binary semaphores.

Two operations are defined on semaphores: **signal** and **wait**. The **signal** operation increases the value of the semaphore by 1; it does not have any other effect on the executing process. The **wait** operation leads to different results, depending on the current value of the semaphore. If this value is greater than 0, it is decreased by 1 and the process calling the **wait** function can proceed. If the semaphore has value 0, execution of the process calling **wait** is halted until the value is increased again by another process with a **signal** operation. Only then is it possible for **wait** to decrease the value of the semaphore and proceed with execution.

It is very important that the operations of test and decrement of the **wait** function are executed in one step only. The operating system is not allowed to break the execution of **wait** after the test on the value and before the decrement operation. The semaphore **wait** has the same operational significance as the function **test__and__set**.

The names of the functions **signal** and **wait** have mnemonic meaning: **signal** is associated with a 'go' to a process and **wait** is self-explanatory: if the semaphore has value 0, the process must wait for a **signal**. If several processes are waiting for the same **signal**, only one of them may continue execution when **signal** is given. Depending on the implementation, the processes may wait in an ordered 'First In, First Out' queue or also be selected at random to proceed. The semaphore alone does not imply a given wait and execution order.

Note that the functions **wait** and **signal** have different names in the literature. Among others, the names **secure/release** and **Down/Up** have been proposed. Also, the implementation details of the semaphore vary with different authors although the basic function is the same.

With the introduction of semaphores, the resource protection problem of the preceding section gets a straightforward solution, as shown in the following example:

```
program sem__example (* resource protection *)

var    P1:     semaphore

begin
P1 := 1;
cobegin

while true do (* repeat forever *)
    begin (* process A *)
    wait(P1);
    (* protected resource *)
    signal(P1);
    . . .
    end; (* process A *)

while true do (* repeat forever *)
    begin (* process B *)
    wait(P1);
    (* protected resource *)
    signal(P1);
    . . .
    end; (* process B *)

coend;
end. (* sem__example *)
```

One semaphore variable is sufficient for access coordination, compared with the different flags of the preceding examples. The processes executing **wait** are put in the waiting queue and do not have to consume CPU time to test the state of the semaphore. The operating system releases a process when **signal** is executed.

With the use of semaphores, the two processes can access the common resource in an ordered manner. No unnatural bonds are introduced: if one process runs faster than the other one, it will just access the resource more often in a given time interval. A process is forced to wait for the other one only when the latter is in the protected area. Liveness is also guaranteed. If a process should for any reason stop running, provided this happens outside the protected area, the other is not hindered from continuing its execution.

7.4.2 Synchronization

The semaphore can help in synchronizing related activities. For instance, if a process has to operate on data only after this has been read from an external port, the code can have the following aspect:

Process read data	Process change data
while true **do**	**while** true **do**
begin	**begin**
(* get new data *)	**wait**(data__available);
signal(data__available);	(* process new data *)
end;	**end**;

This solution has the advantage that, if the data processing algorithm is not ready with execution and new data is available, the presence of the data is indicated with a semaphore value higher than 0. The processing routine can then catch up later with the lost data.

Synchronization errors due to incorrect use of semaphores may be difficult to trace. A process not executing a **wait** instruction can enter a protected region together with another process, leading to unforeseeable results. Of course, it cannot be said that such an error will show up during testing, and it may never even happen during the whole lifetime of a system. It is easier to find the opposite error: a missing **signal** operation should, at a certain point, lead at least one process to halt, which is promptly detected.

A compiler does not usually have the possibility of checking whether semaphores are used correctly, i.e. if **wait** operations are matched in other points by **signals**, and if the semaphores in programs is arbitrary in the same way as other instructions, and depends on the algorithm logic. The burden to verify the correctness of the code lies, then, with the programmer. The use of structured programming methods helps considerably in this task.

7.4.3 Events

Semaphores are an effective method for the solution of synchronization and resource protection problems, but they are not entirely free from drawbacks themselves. One frequently encountered case is when several processes access common data. Each process has to modify the data only if it satisfies some condition, which may vary for the different processes. All processes have the following structure:

> **begin**
> **wait until** condition;
> **modify** data;
> **end**

Access to common data must be protected. One possible solution makes use of one semaphore to control access to the common variables and of another semaphore to indicate whether the common data has changed. All processes refer to the semaphore *mutex* (**mut**ual **ex**clusion) and to *change* (change in the data) in a similar way:

```
var     mutex, change:    semaphore;
        waiting :         integer;

begin
wait(mutex);

while not condition do
    begin
    waiting := waiting + 1;
    signal(mutex);
    wait(change);
    wait(mutex);
    end;

(* operations on the common variables *)

while(waiting > 0) do
    begin
    waiting := waiting - 1;
    signal(change);
    end;

signal(mutex);
end;
```

The semaphore *mutex* protects the common resource; *change* is used to wait until some change has taken place in the common variables so that it is worth repeating the test on the condition values.

According to the definition of semaphores, a variation in *change* lets only one of the waiting processes proceed with execution. It is then necessary to keep track (with the variable *waiting*) of the number of waiting processes and send at the right instant the required number of **signal**(*change*) to free them.

This solution is somewhat unnatural and difficult to analyse. An alternative solution, based on continuous tests by the processes, is not any more satisfying because of the waste of CPU time.

To solve this particular kind of problem, a new synchronization variable has been defined, **event** with the associated operations **await** and **cause**. Following an **await**(*event*) operation, a process is put in waiting queue until the value of *event* changes. The change is controlled via the function **cause**. On *event* change all waiting processes are released, and not only one as in the case of semaphores.

An event can be implemented with a binary variable as well as with a counter, but even if its definition differs, the basic principles remain the same. Similar to the semaphore functions **wait** and **signal**, the names **await** and **cause** have also not been universally accepted and other names have been proposed.

An event variable alone does not protect a critical section from the concurrent access of several processes. Semaphore variables associated with the resource to be protected are still needed. The solution of the problem of the last example with the help of event variables has the following aspect:

```
var     mutex:    semaphore;
        change:   event;
begin
while not condition do await(change);
wait(mutex);
(* operations on common variables *)
signal(mutex);
end;
```

This solution requires that the event *change* is connected to the semaphore *mutex*; that is, **signal**(*mutex*) also operates as **cause**(*change*). It is not sufficient to write the two instructions in succession because process execution *could* be interrupted when it is exactly between them, with unforeseeable consequences. At each change in the event value, all processes test *condition* and only the processes for which *condition* is verified are allowed to proceed. Access to the common resource is protected with the semaphore *mutex*. Here, only one process is allowed to proceed.

Even though it might not sound so, this solution is simpler than the one based only on semaphores. It is also more efficient because the processes do not waste time in testing the value of the semaphores but perform the tests only when it is worth doing it, i.e. after some variables have changed. It requires, however, that the operating system allows for execution of **signal** and **cause** as a single, undivided operation.

The use of semaphore and event variables alone does not guarantee absence of deadlocks. If, in the above example, the functions **await**(*cause*) and **wait**(*mutex*) were exchanged in one of the routines, then the principle of ordered allocation of resources would have been violated and deadlock could happen. Semaphores and events are convenient high level substitutes for the single operation **test_and_set** and help avoid continuous test loops but, if they are misused, they can lead to race conditions and deadlock.

7.4.4 An Example of Resource Protection: the Circular Buffer

A very important problem with an elegant solution based on event variables is the **circular** or **bounded buffer** (Figure 7.9). A circular buffer is a finite memory area used for data exchange between two or more processes. Applications of the circular buffer are found in communication problems, where the relative speeds of the transmitting process, the communication channel and the receiving process are different. The processes operating on the circular buffer are of two kinds: **producer** and **consumer**. The producer writes data into the buffer and the consumer reads data out of it.

Producer and consumer must work independently from each other. Both have bounds: the producer can operate only when it has sufficient space at its disposal to insert new data and the consumer may read data only if this is present in the buffer. The producer must stop when the buffer is full, the consumer when the buffer is empty. The circular buffer is a clear example of a common resource for which the normal protection rules hold: when a process writes or reads data, the other must wait. The buffer is protected with one semaphore.

To avoid continuous tests for 'data present' or 'free space', an event variable is defined: *buf_change*. *Nmax* is the dimension of the buffer and N the number of elements currently contained in it. The producer and the consumer programs have the following aspect:

var	mutex:	semaphore;
	buf_change:	event;
	N :	integer;

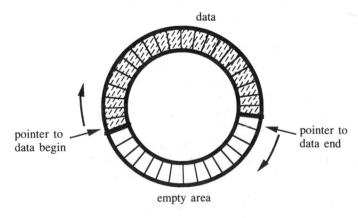

Figure 7.9   The circular buffer.

```
begin (* producer *)
while true do (* repeat forever *)
    begin
    wait(mutex);
    while(N = Nmax) do await(buf__change);
    (* insert a new data item *)
    N := N + 1;
    cause(buf__change);
    signal(mutex);
    end;
end; (* producer *)

begin (* consumer *)
while true do (* repeat forever *)
    begin
    wait(mutex);
    while(N = 0) do await(buf__change);
    (* take out one data item *)
    N := N - 1;
    cause(buf__change);
    signal(mutex);
    end;
end; (* consumer *)
```

The producer and consumer processes are halted until a change of the event variable *buf__change* shows that it is time to repeat the condition test, as the number of the elements in the buffer has changed.

7.5 PROCESS COMMUNICATION: MONITOR, MAILBOX, RENDEZVOUS

When several processes work in mutual cooperation, they need to exchange information. A multitasking operating system must provide an adequate method for this purpose. Data exchange should be transparent for the processes, in other words, it should not influence the data that have to be transferred. The data and the communication format should be defined within the processes and not depend on the particular communication method used.

7.5.1 Common Memory Areas

A first method for data exchange is the use of common memory areas where different processes have access to read and write. These memory areas are common resources to be protected, e.g. with semaphores, as we have already seen in the case of the bounded buffer. The main advantage of common memory areas is that access to them is direct and immediate apart from semaphore wait operations. The areas can be easily organized for structured data exchange: a process might write fields one by one and another process read whole records at a time.

On the other hand, common memory areas have to be located at known addresses in primary memory. This is not difficult to do in assembler but is much trickier in high level languages if the language does not allow direct access to absolute memory locations.

7.5.2 Monitors

The 'classic' method for resource protection and interprocess communication is the **monitor**. Monitors are a theoretical construct introduced to simplify the writing of programs that exchange messages with each other or which need to be synchronized. The monitor consists of a reserved data area and associated procedures with exclusive right to operate on the data. External procedures are not allowed to access the monitor data area directly, but have to call monitor procedures; in its turn, the monitor gives service to only one external procedure at a time. It is the responsibility of the operating system to check that execution of a monitor procedure is not interrupted by another procedure from the same monitor. In this way, the monitor can guarantee that execution of one called procedure is completed before another procedure has access to the same data area.

The general structure of a monitor follows:

```
Procedure monitor (* monitor example *)

var    struct1:    record;    (* protected area *)

procedure write_data (monitor_input); entry;
    begin
    . . .
    with struct1 do
    . . .
    end;

procedure read_data (monitor_output); entry;
    begin
    . . .
    with struct1 do
    . . .
    end;


    . . .
end. (* procedure monitor *)
```

The procedures *write__data* and *read__data* are called **entry** procedures to the monitor; they are the only ones which can be called from external processes. External processes do not have direct access to the protected area *struct1* and to the internal monitor procedures not explicitly defined as **entry**.

Monitors are implemented as critical regions with their reserved access routines. Internally monitors use semaphores to control access to their own data areas. The main advantage of monitors is that the details of data protection and task synchronization are left to the operating system.

Monitors have been proposed in some multitasking languages (e.g. Concurrent Pascal), but they are not implemented in any commonly used programming language. It is often possible to build one's own monitor structures in programming languages like C and Pascal, when a protection and synchronization mechanism is available. In such a case thorough testing of the monitor procedures is imperative before using them.

7.5.3 Mailboxes

A different communication method that allows data exchange and process synchronization at the same time is the **mailbox**. A mailbox is a data structure oriented to **messages** which can be deposited and collected (Figure 7.10). Different mailboxes may be defined within the same system to allow the exchange of different types of messages.

In many operating systems (e.g. VAX/VMS) mailboxes are considered to be logical files and the access procedures are similar to those that access physical store devices. The allowed operations on mailboxes are: creation, opening, message writing, message reading, closing, deleting.

Mailboxes do not have an independent structure. They are located in central memory or on disk, and exist only as long as the system is powered up and operating. If they are physically located on disk, mailboxes are classified as temporary files, to be deleted at system shutdown. Mailboxes do not have generic identifiers or names; they are labelled with logical identifiers (most often numbers) defined when they are created. All processes that use mailboxes address them with their logical identifiers.

To create a mailbox, the operating system defines pointers to a memory area for read/write operations and the associated variables for access protection. The memory space allocated for a mailbox depends on the way it is implemented in the system. The main implementation methods are: a **buffer area** whose size can be defined when the mailbox is created, or a **linked list** structure which does not put any limits on the number of messages that the mailbox can hold. Unlike a buffer with its single-character organization or a

Figure 7.10   Mailbox operation.

common memory area, where exact knowledge of the addresses is required, the mailbox is a complete data structure with a logical identifier.

In the most common implementations, a process intending to send a message writes it in the mailbox in a similar way to writing in a file:

put_mailbox (#1, message)

In the same fashion, a process intending to receive a message reads it from a mailbox with an instruction of the kind:

get_mailbox (#1, message)

Writing a message in a mailbox is a fairly easy operation: the content of the message is simply copied in the indicated mailbox. Possible programming errors here are the request to operate on a non-existent mailbox or lack of memory space to store the new message, indicating that either the mailbox is too small or the deposited messages have not yet been read out.

In reading from a mailbox, the message which was written first is passed out and copied into the receiving data structure. After being read, a message is deleted from the mailbox.

A read operation from an empty mailbox may lead to different results depending on the actual implementation. The output might consist in an empty string (i.e. of length 0) or the read operation might be blocked until a new message is available to be read. In the last case, to avoid the undesired halt of a process, a function to indicate the number of messages currently stored in the mailbox is usually available. A preventive test operation then helps to avoid reading from an empty mailbox.

A semaphore is equivalent to a mailbox where messages of length 0 are stored. The operations **signal** and **wait** are equivalent to mailbox **write** and **read**; the current value of the semaphore is equivalent to the number of deposited messages.

Message passing is a very important communication method in distributed systems, where several CPUs at distinct locations are connected together. In such configurations, there is always a risk that a message gets lost. An acknowledge message can confirm reception of a message, but there is no way to guarantee that the acknowledgment does not get lost in its turn, prompting the sender to initiate transmission again. This problem does not have a general solution, message passing strategies have instead to be defined on a case-by-case basis. For instance, each message could be labelled and numbered so that sender and receiver can determine if the passing order is respected. This method is used in some types of communication protocols (see Section 10.4).

7.5.4 The Rendezvous

The **rendezvous** (French for 'meeting') is a function for both synchronization and intertask communication implemented in the programming language ADA. The rendezvous is an asymmetrical operation: one task requests a rendezvous and the other declares that it is ready to accept it. The task requesting the rendezvous operation must know the name of the task to call, while the called task does not need to know who the caller is. The principle is the same as for a subroutine call, or a blind date.

The rendezvous works as follows. The task to be called has an interface toward other processes, called an **entry point**. The **entry point** is associated with a parameter list where single parameters are qualified as **in**, **out** and **in out**, depending on whether they are input or output for the routine. Inside the called task, one or more **accept** instructions show where the parameters of the external calling process must be passed.

The calling task and the task to be called execute independently. When one of the tasks reaches its **accept** or **entry** instruction, it has to wait. When the other task has also reached the corresponding instruction, the rendezvous takes place. The called task continues execution with the instructions in the **accept** block, while the other waits. When the called task reaches the end of the **accept** block, both tasks can freely continue execution. Unlike the semaphore, the rendezvous is a synchronous function: both tasks have to stop at the meeting point and, only after both have reached it, may execution continue.

Different tasks may refer to the same **entry** call. If more entries are called than can be accepted by the system, the tasks are put in an ordered wait queue where precedence is given to the task which waited longest ('First In First Out' order). The other tasks must wait until the called task again reaches its **accept** instruction.

The rendezvous combines data transfer (via the parameter list) with task synchronization. Even synchronization alone is possible, if the parameter list is omitted.

7.5.5 Comparison of the Methods for Synchronization and Communication

The main problems related to concurrent programming, mutual exclusion, synchronization and interprocess communication may seem to be distinct, but they are in effect equivalent. A synchronization method can be used to implement mutual exclusion and communication functions. Similarly, with a method of interprocess communication it is possible to realize synchronization and mutual exclusion functions.

The relation among the principles is of practical importance when a system offers only one method and the others have to be derived from it. Message passing and access to common memory areas is slower than the control and update of semaphore and event variables, and involves data processing overhead. For each function, the most straightforward implementation method should be chosen and strange constructs should be avoided as much as possible.

When it is possible to choose from among different synchronization and communication functions, the function most apt to solve the specific problem should be used; the resulting code will be clearer and probably even faster. It is very important to consider how efficiently the solutions are implemented in practice in the actual software environment.

7.6 METHODS FOR REAL-TIME PROGRAMMING

Real-time programs differ from sequential programs for the following reasons:

- The execution flow is not only determined by the processor but also by external events.
- Normal programs act on **data**; real-time programs act on **data** and on **signals**.
- A real-time program may explicitly refer to the time.
- There are timing constraints. Failure to compute a result within a specified time may be just as bad as computing a wrong result (the right data too late is wrong data). A typical predictable response time of 1 ms is generally required, in some cases even 0.1 ms may be necessary.
- The result of a real-time execution depends on the global state of a system and cannot be predicted beforehand.
- A run is not terminated when the input data has ended. A real-time process waits for new data to be available.

The particular aspects of real-time programming require the use of special techniques and methods, which are not necessary in sequential programming. These techniques are mainly related to control of program execution flow from the external environment and from time. The most important of them are **interrupt interception**, **exception handling** and the direct use of operating system functions.

7.6.1 The Programming Environment

Before examining the issues related to real-time programming, we have to consider the environment where the programs will run. A typical real-time environment is a mini-computer, a bus system, a PC or a board-based microcomputer system connected with the outside world via hardware interfaces.

The software for a real-time system might range from ROM-stored routines to complex operating systems allowing both program development and execution. In large systems, development and execution take place on the same machine. Smaller systems might not be able to support the development tools; the programs may have to be developed on more powerful machines and then downloaded to the target system. A similar case is given by **firmware**, software embedded in electronic appliances during their manufacture. Firmware is hard-coded in read-only memory (ROM); it is developed on a different machine from where it is run.

The first action for a programmer is to become familiar with the programming environment and the software tools available. The issues to be faced will range from data-type representation in hardware and software, leading to the discovery that some systems order bits in one direction and some in another, some collocate data straight in memory and others use 'backward storage', where the low level byte of a word gets a higher memory address than the high level byte. The number of such issues is very high and the attentive programmer knows how to separate general data and code structuring from the technicalities of the actual implementation machine.

It is essential to become acquainted early on with the functions provided by the actual environment and define alternatives. For example, the microprocessor Motorola 68000 has the function **test__and__set** in its instruction set, so that intertask communication can be implemented via shared memory areas. The VAX/VMS operating system offers mailboxes, and process synchronization can be implemented by a message-passing mechanism. As many multitasking and real-time systems are developed by programmer teams, clarity is required from an early stage on which techniques to use.

Of great importance is the structuring of hardware and software resources, that is, the assignment of bus addresses and interrupt priority levels for the interface devices. Hardware address definition depends little on software development, so that it can be handled at an early stage. Relative service priorities depend on the type of hardware and the functions to be performed. Their definition should not be postponed until coding time, otherwise conflicts between program modules and risk of deadlock are unavoidable consequences.

The software should be built as for an operating system: in a **modular** and **layered** fashion, as this considerably simplifies the construction of complex systems. The main objects of attention are the interfaces rather than the content of single modules. It is very important to define with precision interfaces or interaction points between the modules. These points are used for synchronization and communication between the processes.

For the tasks to exchange information in shared memory areas or with the help of messages, the exact area structure and message format must be defined in advance. This does not mean that changes in their definition might not take place after software development has started, only that the later they are done, the more expensive they will be in terms of code rewriting, testing, etc. On the other hand, it is to be expected that some changes will be made anyway in the course of software development, as insight into the problem increases with progress on the work.

7.6.2 Program Structure

Real-time programming is a special form of multiprogramming in which, besides the development of cooperating tasks, attention has to be dedicated to the timing issues of the system interacting with the external world. The principal feature of real-time programs is that they must always be running and never halt. If they are not currently running, they are idle and wait to be resumed via an interrupt or event. Error situations which could lead to the arrest and abort of a process must be recognized in time and corrected from within the process itself.

The major steps in the development of a real-time system are easily identified. Initially, the problem is analysed and described. The system functions have to be divided into elementary parts, and a **program module (task)** is associated with each of them. For instance, the tasks for the control of a robot arm could be organized as follows:

- Read path data from disk.
- Compute next position.
- Read actual position from sensors.
- Compute appropriate control signal for positioning.

- Execute control action.
- Verify that reference and actual positions are within the allowed range.
- Accept data from operator.
- Stop on emergency (asynchronous command, interrupt driven).

The tasks are sequential programs. They have the aspect of closed loops repeating indefinitely, continuously processing the data and the signals at their input. At some point in the code there is usually an instruction to make the loop wait for an external event or for a given time. The code is structured in such a way that the **end** instruction is never reached:

```
while true do (* repeat forever *)
    begin (* handling routine *)
    wait event at #2,28 (* external interrupt *)
    (* handling code *)
    . . .
    end; (* handling routine *)

end. (* never reached *)
```

Each module is developed indicating clearly the areas where protected resources are accessed. Entering and exiting those areas is coordinated by some method: semaphores, messages or monitor. In general, a program in a protected area must stay safe there until it leaves the area. Interruptions in the execution of the process should not influence the resources in the protected area. In this way, the chances that the module behaves correctly are increased.

Memory allocation by the processes will have to be considered. If all the modules do not fit in the memory together, they will have to be divided into segments to be loaded in at request. A common technique is overlaying, where a main module permanently resides in memory and reads from disk storage overlays with code and/or parameters for some specific operations as described in Section 7.2.4.

In real-time systems, the processes might have to access common subroutines. In one solution, the subroutines are linked together with the separate tasks after compiling, but this means that several copies of the same code are loaded in memory.

A different approach is to load in memory only one copy of the subroutines, but still access them from several programs. Such subroutines must be **reentrant**, that is, they can be interrupted and called several times without interference. Reentrant code operates only on the internal registers and on the stack; it does not address any fixed memory location. During execution, there will be one active stack for each process, so that a reentrant module shared by different processes can be interrupted at any time and restarted from a different position in its code, using a different stack. A reentrant procedure can thus be found in many different process contexts at the same time.

7.6.3 Priorities

Many of the support systems for multiprogramming have the possibility of assigning execution priorities to the different tasks. In many cases, priority assignment is dynamic, which means that the priorities may be modified by the processes as well as by the operating system. Other systems place restrictions on the definition and the later change of process priorities.

The most important modules, or the ones which should have fast execution response, get higher priority. It is necessary to pay attention to the conventions in the system used, whether highest priority is associated with a higher or lower numerical value. Priorities have a relative meaning and make sense only if they are different from one module to the other.

7.6.4 Intrinsic Functions of Operating Systems

A typical situation encountered in real-time programming is the need to use routines from the operating system because there is no equivalent instruction in the programming language used.

To overcome this problem, many programming languages offer an interface to the operating system so that its modules can be called directly from programs written in high level languages. There are different kinds of program interfaces to the operating system: direct calls, primitive functions and access via library modules.

Direct calls take place with a specific language instruction or a software interrupt, transferring the execution to a system routine. The necessary parameters are passed to the operating system via registers and the stack. When the high level language does not allow direct manipulation of the registers and of the stack, it is necessary to use an interface written in assembler. The parameters are passed from the main program to the assembler module and from this to the operating system. After the system call has been completed, the results are passed back to the assembler module and from this to the calling program.

Primitive functions (also called simply **primitives**) are used to communicate directly from a process to the operating system routines. Primitives are used in real-time processes to communicate with the operating system routines so as to influence the flow and order of process exchange operations, as with the start or halt of other processes, the change of process priorities or the disabling and enabling of interrupt handling.

Parameter passing and returning the results from operating system modules is in itself not a difficult operation, but attention is called for. The compiler, lacking access to the information related to the call procedure, cannot verify its correctness. An error in passing some variable is overseen at compiling time, it may go unnoticed by the linker as well and will lead to problems during execution.

In other cases, **library modules** from the operating system can be used to access the internal resources of operating systems. These modules are already precompiled and have only to be linked to the programs using them. It is necessary to verify with the system documentation which parameters are requested by the service modules, the aspect of the data interfaces and the methods for linking to programs written in a high level language.

7.6.5 Interrupt and Exception Handling

Real-time systems interact with the external environment via hardware interfaces. Access to the interfaces and to external data is made either on request (polling) or via interrupts.

In **polling**, the CPU asks all interfaces in succession whether they have new data to report. If this is the case, the program must fetch the data from the input channel and process it. In polling, attention must be paid to the device polling order and how often polling takes place.

With **interrupts**, request for attention comes from external devices when new data is available. Interrupts are asynchronous events with respect to the running process and require immediate attention. On reception of an interrupt signal, the processor stops, saves the context of the process currently executing, reads from a table the **address** of a **service routine** for the interrupt and jumps to it (Figure 7.11). The service routine is called **interrupt handler**.

When the CPU transfers control to an interrupt handler, it might save only the pointers to the code area of the running process. It is the duty of the interrupt handler to save, in temporary buffers or on stack, all registers it is going to use and restore them at the end. This is a critical operation, and it might be necessary to disable interrupt servicing under execution of the first instructions of the handler in order to avoid the handler itself being interrupted in turn.

In interrupt management a very important factor is the **response time**, which should obviously be as little as possible. The response time is the sum of the interrupt latency (how long it takes for the interrupt to get attention) and the time needed for a context switch, until the interrupt handler actually runs. The typical system load also plays a role. If the workload is so distributed that the CPU has to service many interrupts at the same time, new ones will have to be queued until the CPU is available.

Interrupt service routines should be as compact and short as possible. If a complex action is needed following an interrupt, it is better if the action is performed by a regular process. The interrupt service routine should do only the minimum necessary, for example, get an input value and then pass a message to the other routine, signalling that an interrupt has occurred and service is requested. It is always good practice to write reentrant code for system routines and for interrupt handlers. In this way, conflicts arc avoided in case a handler is interrupted and called again before it has terminated its execution in the first context.

A problem similar to interrupt servicing is reaction to **exceptions**, i.e. unusual conditions that result when the CPU cannot properly handle the execution of an instruction and that hinder the normal continuation of a process. Examples of exceptions are division by zero and addressing a non-existing memory location. Names for different kinds of exceptions are also **traps, faults** and **aborts**.

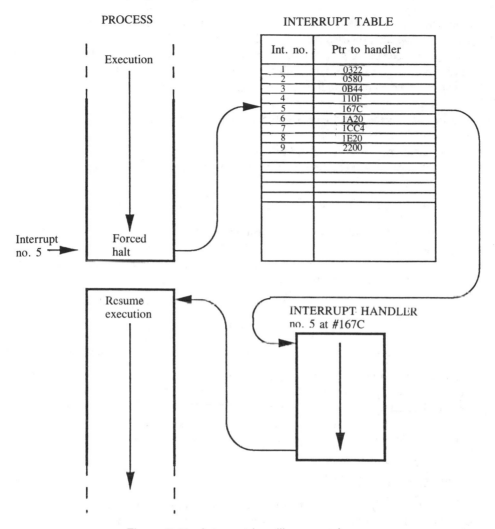

Figure 7.11 Interrupt handling procedure.

The common handling of exceptions by an operating system is the termination of process execution and indication of the error situation with messages written in clear text on the device used for the output messages. While acceptable in interactive multiuser sequential processing, in real-time systems the abrupt halt of a process must be avoided. Think about the possible consequences if a microprocessor-controlled fly-by-wire or car automatic braking system (ABS) should halt because of an unexpected **divide-by-zero** exception. In real-time systems all possible exceptions should be analysed beforehand and appropriately handled.

A very tricky aspect of exception handling is the verification that an exception does not arise again. Put another way, exception handling must address the cause and not the symptoms of the abnormal situation. If an exception is not handled correctly, it may arise again prompting the processor to jump to its specific handling module. For instance, the divide-by-zero exception handler must check and modify the operands and not just resume operations to the point before the fault took place. This would lead to an indefinite loop.

The effective memory address of any program module is known only at loading time. At system start-up, a module writes the memory addresses where the interrupt handlers are loaded in the interrupt service table. The interrupt routines are then accessed by referencing this table. Exception handling modules are written in a fashion similar to interrupt handlers. Their addresses are put in the interrupt address table at predefined locations. The possible exceptions and the pointer storage locations depend on the actual system.

7.6.6 Time-related Functions and Time Efficiency

Real-time processes may refer to time waiting for some interval or until a given time. These functions usually have the form:

wait(n)   (n = time in seconds or milliseconds)

and

wait until (time)   (time = hours, minutes, seconds, ms)

When one of these functions is executed, the operating system puts the process in a waiting queue. After the requested time has elapsed, the process is moved from the waiting queue to the ready queue.

The worst, but not uncommon, method to solve a 'time-waiting' problem is to introduce a closed loop to check the system time variable in the so-called busy−wait:

repeat (*do nothing*)
until (time = 12:00:00);

In general, these active waiting loops are nothing else but a waste of CPU time and should be avoided. But there are cases where reality looks different. In a system where an A/D conversion takes 20 μs and a process switching operating 10 μs, it is more economic to run as busy waiting for the 20 μs before new input data is fetched than to start the task exchange procedure implicit in a 'well-behaved' wait operation. Each case is judged on its own; obviously they require advanced system knowledge and the correct feel.

An important aspect of processes started periodically (such as filtering or regulation algorithms) is the **accumulated time error**. This depends on the fact that a process is not executed immediately after it is moved out of the waiting queue but has to wait for different time intervals in the queue of executable processes until its execution turn arrives (Figure 7.12). Requested and real execution time are not the same.

Accumulated time errors can take place if the running time for a new activity is computed as:

new execution time = end of old execution time + interval

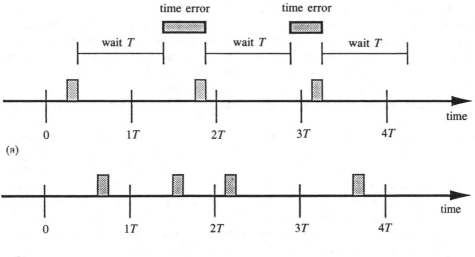

(a)

(b)

Figure 7.12 (a) The wrong way to execute periodic tasks (it leads to accumulated time errors); (b) the correct solution (it does not lead to accumulated time errors).

the latter is an example of an instruction like **wait 10 seconds** written at the end of a loop. The correct solution is obtained by using the equation:

new execution time = old **reference** execution time + interval

The principle appears from Figure 7.12, where the nominal times are drawn on the *x* axis. As absolute time is taken as reference, accumulated time errors are avoided.

Running time efficiency is one of the most important aspects in real-time systems. The processes must execute quickly and compromises between good and structured versus time-efficient code often have to be made. It is a fact of life that if short-cuts are needed to achieve some result, they will be taken anyway. If they can not be avoided, good documentation on what is done and why is imperative.

7.6.7 Testing and Debugging

The proof of the correctness of a program is a fundamental step in its development. It is imperative to check that the program performs its function without errors. In practice this means that there is little room for a formal theory of testing, which is compensated by experience and tradition.

Errors are elusive. Many errors tend to show up sporadically and cannot be reproduced easily. No proof can guarantee that a program is entirely error-free and no tests can ensure finding all the errors. The goal of a test procedure is thus to find as many errors as possible and guarantee that a program is *reasonably* safe. To quote Dijkstra, 'Testing can only prove the presence of errors, but not their absence'.

A comprehensive test requires adequate design and preparation; it is necessary to combine practical checks with analytical verifications. To start with, the test procedures and data are described in a special **test requirement document**. After the tests have been performed the results will also be recorded there. If several people participate in system development, it is advisable that the team assigned to test definition and execution be different from the one responsible for program development.

In testing real-time systems, further complication stems from the many possible interactions among tasks. The probability of introducing a new error when an old one is fixed is very high; the experience gained so far with large programs (of the order of 10^5 lines of code) suggests that this probability lies between 15 and 50 per cent.

The basic test methods are: **exhaustive** and **per sample**. In an exhaustive testing all possible combinations of input and output data are checked; this method can obviously be employed only when these combinations are small in number. Most often the sample method is used. A representative number of input/output combinations is selected. The sample data should also include extreme situations, as, for example, data outside the allowed range that the module under test has to recognize and process correctly.

To begin with, the program modules are tested separately. Testing should ensure that each code line is executed at least once. That is, if the program module contains branch instructions such as **if-then-else**, then the condition data during testing should lead first to the execution of one branch **then** and then of the other **else**.

In this testing phase, **debug programs** are very useful. They allow direct verification and manipulation of the CPU registers and of the memory locations while running the actual machine code. When debuggers are used, breakpoints are introduced in the code. At each breakpoint it is possible to check the state of the registers and of the variables, and compare them with the logic of the process.

Only after all the modules have been separately checked and the errors corrected, are the modules allowed to interact with each other. The most probable interactions will be checked on the base of the reference documents. Attention has to be paid to the functions that are necessary for system security.

The multiple interactions between program modules may lead to system errors even if the single modules operate correctly. In multitasking operations, conventional one-step-at-a-time debuggers cannot be used. In their place, so-called 'post-mortem' debuggers are used (the name is explicit about the circumstances of their use). At the time of a fatal crash, the memory content is frozen and printed out with the help of the 'post-mortem' debugger. The programmer is left with the — not easy — task of finding reasons for the crash from the memory dump. If the system has been designed to handle interrupt signals and features internal exception management, the correctness of the reactions must be checked. Error situations will be provoked on purpose to check the outcome of the processing.

The results of each separate test and of the global operations are reported in a document which forms the basis for taking the appropriate correction acts. Do not forget that errors are much more difficult and expensive to correct the later they are discovered. Investment in test procedure is not only investment in the quality of a system, but also in its general economy.

7.7 LANGUAGES AND OPERATING SYSTEMS FOR REAL-TIME PROGRAMMING

7.7.1 Real-time Requirements for Languages and Operating Systems

Real-time systems can be implemented with conventional, sequential programming languages running in a multitasking environment or with high level languages that directly support multitasking. The use of high level languages makes program development and maintenance easier, and the programs are more easily portable among different systems.

Real-time programming requires special features not commonly found in normal languages for sequential programming. A language for real-time programming, or the supporting operating system, should have the following capabilities:

- Definition of tasks to be executed in parallel
- Synchronization among tasks
- Data exchange among tasks
- Time-related functions
- Direct access to external hardware points
- Interrupt handling support
- Exception handling support

Few languages have all these features. Many languages have only a few of them, which may still may be sufficient for some applications. Some companies have developed special languages to support their own product lines. These languages do not claim universality and are rather oriented to a specific machine and its interfaces. Typically, they are based on existing languages (FORTRAN, BASIC) with extensions for real-time functions, as the names 'Process BASIC', 'Real-time FORTRAN' indicate. Some languages do not support real-time programming directly, but are designed to be easily extensible (e.g. Modula-2, C++).

There are languages that have been purposely developed for generic real-time applications and are not oriented towards a particular hardware system. This is what happened in Britain with CORAL66 and in West Germany with PEARL, where the respective governments supported these languages for military and economic reasons. With the exception of ADA, no language for real-time programming enjoys universal diffusion; all have remained restricted to their original environment.

During the 1970s, the idea of a single, general, multipurpose programming language enjoyed wide support. Eventually, this idea led to the development of ADA. The main concept was that the programming environment, i.e. the language, should be entirely separated from the details of the technical application. The programmer should not deal with machine-level details but reason only in terms of abstract structures and data types.

Experience has shown that this approach is not realistic. Universal, strongly-typed programming languages guarantee a certain safety for a program but at the same time they are less flexible. There is a trade-off: when a language is safe it is also bulky and difficult; the generated code also tends to be heavy and relatively inefficient. An open language like C builds on a few basic ideas but gives the experienced programmer more flexibility and power. No language is best, each language fits a given application and environment (and suits a programmer) better than others.

7.7.2 The Programming Language ADA

Towards the mid 1970s, the US Department of Defense (DoD) decided to have a single programming language for their control and real-time systems, as an alternative to the hundreds then used, in order to limit software development and maintenance costs. In 1979, the DoD finally adopted the language proposal made by the French company Honeywell Bull. The name of the language comes from Augusta Ada Byron, Countess of Lovelace (1815—52), who is considered to be the first programmer in history. She wrote the programs for the analytical engine (a mechanical computer which was never built) designed by the English inventor Charles Babbage.

ADA is defined as a complete program development environment with editor, debugging facility, library management system, etc. ADA is an American standard (ANSI/MIL-STD-1815A) and the language definition includes the control methods to check conformity to the standard. No dialects are allowed: to be validated, a compiler must show that it can run all tests correctly.

ADA has a structure similar to Pascal, though ADA is much richer in all respects, especially regarding the features of real-time systems. Processes are declared in ADA as **tasks**, defined as entities with parallel and independent execution, each task on a dedicated logical (virtual) processor. Tasks can be declared as interrupt and exception handlers.

A new concept introduced in ADA is the **package**, a module with its own definitions of data types, variables and subroutines, and where it is explicitly indicated which of the routines and variables are accessible from outside. Packages can be compiled separately and later linked together in the same executable program. This feature supports modular program development and the creation of application libraries.

Low level programming is not effectively supported in ADA, a sign of the belief that all kinds of operations should be dealt with at high level only. A peculiar feature of the language is that ADA has no internal input/output facilities. Instead, it uses application packages with predefined functions for hardware interface management and access to external data for each specific processor.

The main disadvantage of ADA is its complexity, which makes the language difficult to learn and to use. The existing compilers are still expensive products and require powerful CPUs. So far, ADA has not reached the popularity initially envisioned, and it is doubtful if it ever will.

7.7.3 BASIC

BASIC is the easiest of the high level programming languages to learn. It was developed in 1964 to support interactive program development with teletypes. Because of its simplicity, BASIC is often criticized by expert programmers and there is no doubt that this language is not a good choice for the construction of large structured systems. On the other hand, small applications can be developed in a much shorter time using BASIC than other languages. Moreover, BASIC is available on almost all microcomputers.

BASIC can be compiled but is most often interpreted, that is, each instruction is translated in machine code only at the moment of its execution. BASIC is particularly important for the development of small application tasks to be run within a larger system. BASIC should not be used for larger applications (of the order of more than 500/1000 lines of code). Nevertheless, BASIC is the optimal choice for non-professional programmers who need to solve specific problems quickly.

7.7.4 C

The programming language C, although it lacks almost all features that computer theoreticians deem to be necessary for a good programming language, has enjoyed wide success during the 1980s. C has become very popular for all applications requiring high efficiency, such as in real-time programming. For normal microprocessors used in control applications C compilers and development systems are available, often from different sources. There is a clear trend in the industry towards widespread use of C and the operating system UNIX (which is written in C) because applications written in C are portable with relatively little effort from one machine to another.

The philosophy of C is to structure a program in functions. C has weak type control and allows the programmer to do almost everything, down to register and bit manipulation. This freedom makes the language unsafe because the compiler does not have the possibility of checking whether suspect operations were intended or not. The small number of predefined functions and data types makes the programs easily portable among different systems. C supports both good (i.e. structured) and poor programming styles, leaving the responsibility for good development to the programmer. This is of particular importance when programs have to be maintained: a poorly written and sparsely commented C program is as cryptic as a program written in assembler. C is defined in an international standard (ISO 9899).

C++ is a much more powerful language than C, from which it is derived. C++ offers greatly improved data abstraction through the class concept, similar to the abstract data type with its distinction between data and operators. The C++ class is much easier to use in practice than similar concepts in other languages, as C++ supports object-oriented programming and the stepwise refinement of data types. The main strength of C++ is its ability to support the design and implementation of easy-to-use program libraries. Real-time programming is not directly implemented in the C++ language, but can be supported via specially developed program modules and class libraries. C++ is described in Stroustrup (1991).

7.7.5 FORTRAN

FORTRAN was the first high level programming language and probably contributed more than any other to the diffusion and practical use of computers. Released in 1957, it is still largely used especially for numerically oriented computations. In general, FORTRAN has limited possibilities for type definition, a quite complicated way of dealing with non-numeric data and in general lacks too many of the important features of real-time languages to be seriously considered for this purpose. Newer FORTRAN versions have imported features from other languages and support more advanced data structures. In this sense, the difference between FORTRAN and other languages is becoming smaller.

Due to its established use in scientific application, it is not seldom that data is processed on line with existing FORTRAN programs or that new programs for analysis and statistics are written in FORTRAN. In such cases, the real issue is to coordinate the data transfer between a real-time database where the data is collected and the application modules written in FORTRAN. Coordination of this type of data transfer under real-time conditions is the responsibility of the operating system. FORTRAN is not recommended for other applications, such as writing device drivers or other modules at operating system level, because other languages are better suited to this purpose.

7.7.6 Pascal and Modula-2

Pascal was introduced by the Swiss Nikolaus Wirth in 1971 as a didactical language to teach good programming technique. It quickly outgrew its initial scope and is now used for quite different applications. The fortune of Pascal, like BASIC, lies in the spread of micro and personal computers, for which it was widely available. Modula-2 was introduced a few years after Pascal to correct some of the original problems of the latter and to offer an environment more oriented to multitasking and real-time applications. Pascal and Modula-2 are very similar in style and structure, although Modula-2 has more functions and reserved words.

In Pascal and Modula-2 it is assumed that the programmer always remains within the original language environment, which, as we have seen, is unrealistic in practical real-time work. Flexibility in their use is much greater if some code for special applications (device drivers, interrupt handlers) is written in assembler. Both languages usually support linking to external Assembler modules and some popular packages for personal computers (e.g. Turbo-Pascal) even allow for on-line assembly programming. Pascal and Modula-2 are a good choice for program support. Their emphasis on structure makes them immediately readable, a crucial factor for later program maintenance. The program structure is quite evident from the written code, much more so than in languages such as BASIC or FORTRAN.

7.7.7 The Operating System UNIX[2]

UNIX is a multitasking, multiuser operating system and, in fact, one of the most widespread operating systems in the world. UNIX is written entirely in C apart from a few hardware-

2   UNIX is a registered trademark of AT&T Bell Laboratories.

oriented, machine language routines which have to be developed independently for each system. Adapting a UNIX system on a different machine requires in principle that only the machine language kernel is written anew. This means that UNIX can run on many different computer hardware systems with an identical operating system interface. Only the basic management routines reside in the kernel, all others run as user processes. A typical UNIX operating system consists of about 10 000—20 000 program lines in C and 1000—2000 in assembler. The kernel is a single memory-resident program about 100 kbyte to 1 Mbyte long, depending on the actual machine and the desired functionality.

Several authors have pointed out that the UNIX kernel lacks a well thought-out structure. The kernel is not modular and it is not possible to interrupt system calls, which means that response times become longer. In addition, because its design is based on a very comprehensive process description, process exchange takes a comparatively long time. Another elegant UNIX concept that leads to practical problems in real-time applications is piping in interprocess communication. The pipe is a kind of logical channel for data and messages. Writing and reading in pipes requires time-consuming process changes. Newer versions of UNIX (V.2, V.3) support semaphores, shared memory and interprocess signals.

Another common and founded critic to UNIX is its unfriendly user interface. In fact, the older and most cryptic commands are still in use, or have been changed with other commands with names and abbreviations at least as unnatural as the previous ones. More positive is that the UNIX commands are very short and can be concatenated on the same line, where the output of a command is understood as the input to the next. In this way complex operations can be concentrated to a few lines and do not require the writing of long control files. On some systems, user-interfaces with windows and menus are available just to 'translate' the selected actions into UNIX commands.

Due to its large diffusion in scientific and technical environments, it was all too obvious to try to adapt UNIX to real-time environments. But here there are several difficulties, in large part due to the fact that real-time in a strict sense was not part of the initial UNIX requirements. In fact, in 'hard' real-time, reaction times of the order of 1 ms are needed, in some cases even 100 μs. With UNIX, even the most powerful systems cannot guarantee a reaction time of less than 10—100 ms.

Nevertheless, the advantages of portability and of a common environment are so many, and not all applications are sto strict in their requirements. Many industries therefore support UNIX applications, for example, General Motors which is known to be an innovative force in industrial automation. In a broader sense, the real issue is not the reaction time for a specific application but the cost to support different operating systems also considering factors such as program maintenance, personnel training, duplicated work, etc. It seems now that UNIX and C have realized in practice what ADA was supposed to do in theory.

7.7.8 OS-9

OS-9, from the company Microware, was first released in 1977 for the Motorola 6809 microprocessor. Today it is a modern operating system primarily used on the 68000—68040

CPUs. OS-9 is a popular system, with several hundreds of thousands licences sold.

On a brief look, the operating system seems quite like UNIX, with similar commands, compatible system calls, tree structures, pipes and uniform handling of the I/O. But despite the similarities, OS-9 is a different product. OS-9 is hardware and real-time-oriented; the goal of program development is usually to transfer later the program to a target system and not to run it on the development system. OS-9 itself is also modular so that the software to be run on a target system need only include selected parts of the operating system. All code generated in OS-9 is position independent, reentrant and can be transferred to ROM. The uniform I/O handling makes it easy to, for instance, use input data from a disk file, a comfortable feature during program development.

For interprocess communication signals, events, pipes and data modules are available. When a signal is sent from one process to another, the addressed process is forced to execute a user-defined intercept routine in what is basically a forced subroutine call. A pipe is, as in UNIX, a sequential data stream from one process to another. The fastest way to transfer data is via data modules, i.e. shared memory areas in RAM to which all processes can refer.

OS-9 includes many possibilities for system tuning, with different ways to influence the priority behaviour. The system also supports hardware interrupts. OS-9 and its tools are written in the language C. Many C programs that are written for UNIX environments can be compiled and run on OS-9 systems without any changes.

7.7.9 VAX/VMS

VMS is the operating system for the DEC computers of the 32-bit processor VAX series. Its popularity in control applications is mostly due to the high quality of the hardware on which it runs and to the large number of available development tools. VMS can be used in both real-time and multiuser environments, with the related security controls of the latter.

VMS offers a wide range of functions, and a standard and clean interface for direct calls from programs. In this way all languages can, at least in principle, be integrated with all functions of the operating system. As real-time features, VMS offers mailboxes in the form of logical, record-oriented files, the possibility of having resident routines and interrupt handling. In VMS a process can control its own execution (priority, memory allocation) and can create and control the execution of other processes.

As with all large operating systems, VMS has performance problems when tight timing is concerned. For this reason, and because of the popularity of the VMS system, a special version of VMS tailored for real-time applications, called VAX/ELN, has been developed. VAX/ELN consists of two different products: a run time environment to execute the application programs on the target machine and a development package with compilers for different languages. System development takes place on a large unit with all the necessary resources for preparing a system that contains only the program modules required for a specific application. The final system is then downloaded to the target machine.

7.7.10 MS-DOS

MS-DOS, produced by Microsoft Inc., owns its popularity to the wide diffusion of the IBM personal computers (IBM PC) and 'compatibles' for which it is the native operating system. MS-DOS does not support multitasking and offers some, although limited, possibilities to write resident code to be started via interrupts or via a clock tick (at a frequency of 18.2 times per second).

MS-DOS applications are, however, very widespread, and there are several MS-DOS-based products with multitasking features. These products actually use MS-DOS only as a startup system in order to be loaded from a diskette, and then take independent control over the whole computer, from the screen down to the single device drivers at which point MS-DOS is forced out of the game.

In fact, one of the advantages of the IBM PC is that it allows programming from scratch of all the chips in the machine and thus the construction of a new operating system. On the other hand, not everybody is too keen to write an operating system anew just to be able to run a few routines in parallel.

The IBM PC and MS-DOS have to be taken as they are, not too advanced or sophisticated, but with an enormous product base which keeps down costs. If an application can be constructed with an MS-DOS computer, some peripheral cards and little software, and it fits its purpose, there is no reason not to use it. But high expectations of performance or reliability are easily disappointed. For more information on the IBM PC, see Section 6.5.1.

7.8 SUMMARY

An operating system offers processes a logical environment consisting in CPU time and memory space for their execution. Operating systems for multiuser and real-time applications are basically similar, but programming is done in practice in different ways. Real-time applications require reaction times of the order of 1 ms. In real-time programming functions are used to coordinate the work of different processes that are not necessary in conventional programs.

The central problem in multitasking and real-time programming is the access to protected resources. Access can be coordinated with synchronization or communication. The most basic, low level primitive is the **test-and-set** function. Semaphores and mailboxes are the most used synchronization and communication principles, and it depends on the particular operating system as to how they are defined and used in detail.

The theory of concurrent programming studies the related issues and proposes solutions to implement the necessary functions in practice. Curiously, many of the constructs proposed by the theory have not been implemented in any important programming language. Their utility lies more in the theoretical analysis of problems, even if actual solutions are later implemented in some other way. However, the constructs proposed by the theory do have much importance, and their correctness has been proved in advance. This is very important in real-time systems where program testing is particularly difficult. The application of already verified methods gives reasonable security for the correctness of the related applications.

It is very elusive to forecast anything for programming languages. Despite all negative comments, FORTRAN and COBOL are still alive and widely used, even if they are based on programming concepts more than three decades old. The fortunes of C and Pascal came almost by chance, as these languages were initially defined with much lower ambitions.

For microcomputer-based real-time programming systems, the best choices today are Pascal/Modula-2 and C, eventually integrated by assembler routines when the problems can not be solved effectively in the language used. Pascal and Modula-2 support a better programming style than C and are a good choice for work to be carried out by groups. BASIC can in many cases be the optimal choice for the development of applications on ready systems: it is easy and allows productive work to be carried out quickly. In all cases, the integration of language and operating system is a very important aspect to consider.

FURTHER READING

Young (1982) describes theoretical computer language issues such as data typing and constructs with attention to the development of ADA. It is very factual on language issues but does not deal much with practical real-time problems.

Glass (1983) offers a collection of different papers about real-time systems, with emphasis on the management of development work. It is a good complement to other tutorial books, even if some of the cases described seem somewhat out-of-date today.

Tanenbaum (1987) is a highly authoritative work on operating systems which presents both established and new ideas in a comprehensive way. The author has a 'do-it-yourself' approach to describe UNIX and even presents an equivalent operating system (MINIX) for personal computers. A solid introduction to operating systems on the base of the VAX architecture is to be found in Levy and Eckhouse (1980).

The problems and solutions for resource protection and concurrent programming are discussed in Dijkstra (1968), Ben-Ari (1982) and Brinch Hansen (1973). The deadlock problem and avoidance methods are discussed in Coffmann *et al.* (1971) and Coffmann and Denning (1973).

Handbooks and tutorials for programming languages number in the hundreds for *each* language. For operating systems, with the exception of MS-DOS and UNIX, there is much less choice. The best (only) reference literature might — alas — be the original system documentation. Good luck!

Current information and interesting programming hints are found in magazines such as *Byte, Computer Language, Dr Dobb's Journal, Communications of the ACM, Datamation*. Readers are urged to select one or two journals covering their area of interest and read them regularly.

8

Sequential Control

Aim: To describe sequencing networks and their implementation in software, particularly structured functional diagrams

In this chapter binary control will be discussed. In general, binary control is simpler than conventional feedback control because both the measurements and the control actions are of the on/off type. However, specific issues have to be considered. We have already seen a simple example of sequencing networks in Chapter 2. Models have been discussed in Chapter 3 and instrumentation was reviewed in Chapter 4.

In the process industry and in industrial automation there is a wealth of applications of switching circuits. Switching theory is of fundamental importance in many fields such as telecommunications and is the very principle on which digital computers are based.

Elementary swiching theory is recollected and the basic logical gates are presented in Section 8.1. The structuring of sequencing networks as ladder diagrams is discussed in Section 8.2. Sequencing has traditionally been realized with relay techniques. Until the beginning of the 1970s electromechanical relays and pneumatic couplings were dominating the market.

During the 1970s **programmable logical controllers (PLCs)** became more and more common, and today sequencing is normally implemented in software. Even though ladder diagrams are being phased out of many automation systems they still are used to describe and document sequencing control implemented in software, as described in Section 8.3.

The logical decisions and calculations may be simple in detail, but the decision chains in large plants are very complex. This naturally raises the demand for structuring the problem and its implementation. Sequencing networks operate asynchronously, i.e. the execution is not directly controlled by a clock. The chain of execution may branch for different conditions and concurrent operations are common. Section 8.4 is dedicated to Grafcet, an important notation to describe binary sequences, including concurrent processes; it is used both as a documentation tool and a programming language. Applications of function charts in industrial control problems is the subject of Section 8.5.

8.1 ELEMENTARY SWITCHING THEORY

8.1.1 Notations

In this section we will describe elementary switching theory that is relevant for process control applications.

There are several examples of binary elements, and several devices have been described in Sections 4.4 and 4.6. Both an electric switch or relay contact and a valve intended for use in logic circuits are binary and are designed to operate in the on/off mode. A transistor can also be used as a binary element operating only in an on/off mode, either conducting or not conducting current.

A binary variable is represented as a variable having values 0 or 1. For a switch contact, relay contact or a transistor (labelled X) the statement $X=0$ means that the element is open (does not conduct current) and $X=1$ means closed (conducts). For a push button or a limit switch, $X=0$ means that the switch is not being actuated and $X=1$ indicates actuation.

Often a binary variable is represented as a voltage level. In positive logic the higher level corresponds to logical 1 and the lower level to logical 0. In TTL (transistor-transistor logic) logical 0 is typically defined by levels between 0 and 0.8 V and logical 1 any voltage higher than 2 V. Similarly in pneumatic systems $X=0$ may mean that the line is exhausted to atmospheric pressure while $X=1$ means a pressurized line.

8.1.2 Basic Logical Gates

A brief recapitulation of Boolean algebra is made here. We denote negation of X with $\bar{X}$. Two normally open switch contacts connected in series constitute an AND gate which is defined by **Boolean multiplication** as

$$X = A \times B \tag{8.1}$$

We note that $X=1$ only if both A and B are equal to 1, otherwise $X=0$ (Figure 8.1). The multiplication sign is often omitted, just as in ordinary algebra. An AND gate can have more than two inputs, since any number of switches can be connected in series. Adding a third switch results in $X=ABC$. We use the ISO (International Organization for Standardization, Section 12.1.4) symbol for the gate.

A common operation is a logical AND between two bytes in a process, called **masking**.

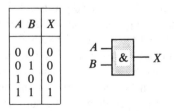

A B	X
0 0	0
0 1	0
1 0	0
1 1	1

Figure 8.1   The truth table and ISO symbol for an AND gate.

input register	11011000
mask	01101101
output	01001000

Figure 8.2   Masking two bytes with an AND operation.

The first byte is the input register reference while the other byte is defined by the user to mask out bits of interest. The AND operation is made bit by bit from the two bytes (Figure 8.2). In other words, only where the mask byte contains 'ones' the original bit of the reference byte is copied to the output.

If two switches A from B are connected in parallel, the operation is a **Boolean addition** and the function is of the OR type. Here, $X=1$ if either A **or** B is actuated and the logic is denoted (Figure 8.3) by:

$$X = A + B \tag{8.2}$$

As for the AND gate, more switches can be added (in parallel), giving $X=A+B+C\ldots$. The ≥ 1 designation inside the OR symbol means that gate output is 'high' if the number of 'high' input signals is equal to or greater than 1.

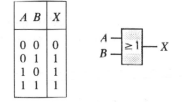

Figure 8.3   The truth table and ISO symbol for an OR gate.

A logical OR between two bytes also makes a bit by bit logical operation (Figure 8.4). The OR operation can be used in order to unconditionally set one or several bits to 1. There are some important theorems for one binary variable, such as

$$\begin{aligned} X + X &= X \\ XX &= X \\ X + \bar{X} &= 1 \\ X\bar{X} &= 0 \end{aligned} \tag{8.3}$$

input register	11011000
mask	01101101
output	11111101

Figure 8.4   Masking two bytes with an OR operation.

Likewise for two variables we can formulate and easily verify:

$$X + Y = Y + X$$
$$XY = YX$$
$$X + XY = X \qquad \text{(absorption law)}$$
$$X(X + Y) = X \qquad \text{(absorption law)} \tag{8.4}$$
$$(X + \bar{Y})\, Y = XY$$
$$X\bar{Y} + Y = X + Y$$
$$XY + \bar{Y} = X + \bar{Y}$$

The De Morgan theorems are useful in manipulating Boolean expressions:

$$\overline{(X + Y + Z + \ldots)} = \bar{X}\bar{Y}\bar{Z} \ldots \tag{8.5}$$
$$\overline{(XYZ\ldots)} = \bar{X} + \bar{Y} + \bar{Z} \ldots$$

The theorems give possibilities for simplifying complex binary expressions, thus saving components for the actual implementation. The behaviour of a switching circuit can be represented by **truth tables**, where the output value is given for all possible combinations of inputs.

8.1.3 Additional Gates

Two normally closed gates in series may define a **NOR gate**, i.e. the system conducts if *neither* the first *nor* the second switch is actuated. According to De Morgan's theorem this can be expressed as

$$X = \bar{A}\bar{B} = \overline{(A+B)} \tag{8.6}$$

This shows that the NOR gate can be constructed from the combination of a NOT and an OR gate (Figure 8.5). The circle at an input or output line of the symbol represents Boolean inversion.

A NOR gate is easily implemented electronically (Section 8.1.5) or pneumatically. Moreover, any Boolean function can be obtained from NOR gates only, which makes it a **universal gate**. For example, a **NOT gate** is a NOR gate with a single input. An OR gate can be obtained by connecting a NOR gate and a NOT gate in series. An **AND gate** is obtained by using two NOT gates and one NOR gate (Figure 8.6) and is written as:

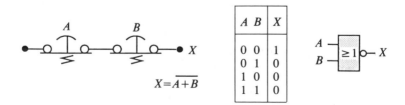

A B	X
0 0	1
0 1	0
1 0	0
1 1	0

$$X = \overline{A+B}$$

Figure 8.5   A NOR gate, its truth table and its ISO symbol.

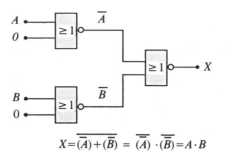

$$X=\overline{(\overline{A})+(\overline{B})} = (\overline{\overline{A}}) \cdot (\overline{\overline{B}})=A \cdot B$$

Figure 8.6 Three NOR gates acting as an AND gate. Note that this is not the minimal realization of an AND gate!

$$X = \overline{(\overline{A}) + (\overline{B})} = (\overline{\overline{A}})(\overline{\overline{B}}) = AB \tag{8.7}$$

A **NAND gate** is defined by:

$$X - \overline{(AB)} = \bar{A} + \bar{B} \tag{8.8}$$

The system does *not* conduct if both A and B are actuated, i.e. it conducts if either switch is not actuated. Like the NOR gate, the NAND gate is a universal gate (Figure 8.7). The NAND and NOR operations are called **complete operations**, because all others can be derived by using either of them. No other gate or operation has the same property.

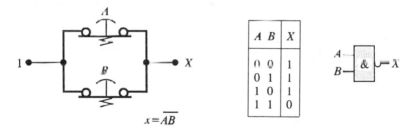

A B	X
0 0	1
0 1	1
1 0	1
1 1	0

$$x=\overline{AB}$$

Figure 8.7 A NAND gate, its truth table and its ISO symbol.

A circuit with two switches, each having double contacts (one normally open and the other normally closed), is shown in Figure 8.8. This is an **exclusive OR (XOR)** circuit, and the output is defined by:

$$X= A\bar{B} + \bar{A}B \tag{8.9}$$

The circuit conducts only if either $A=1$, or $B=1$, but $X=0$ if both A and B have the same sign (compare with the OR gate). For example, such a switch can be used to control the room light from two different switch locations A and B. In digital computers XOR circuits are extremely important for binary addition.

An exclusive OR (XOR) between two bytes will copy the 1 in the input register only where the mask contains 0. Where the mask contains 1 the bits of the first operand are

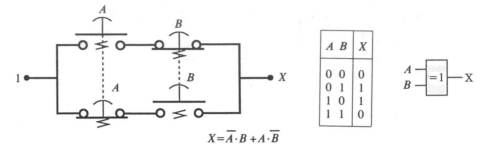

$$X=\overline{A}\cdot B + A\cdot\overline{B}$$

Figure 8.8   An exclusive OR gate, its truth table and its ISO symbol.

inverted. In other words, in the positions where the operands are equal the result is 0 and in the reverse where the operands are not equal, the result is a 1 (Figure 8.9). This is often used in order to check if and how a value of an input port has been changed between two readings.

input register	11011000
mask	01101101
output	10110101

Figure 8.9   Masking two bytes with an XOR operation.

Example 8.1 Simple combinatorial network

A simple example of a combinatorial circuit expressed in ISO symbols is shown in Figure 8.10.

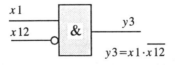

$$y3 = x1\cdot\overline{x12}$$

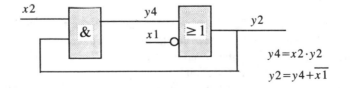

$$y4 = x2\cdot y2$$
$$y2 = y4 + \overline{x1}$$

Figure 8.10   Simple combinatorial circuit.

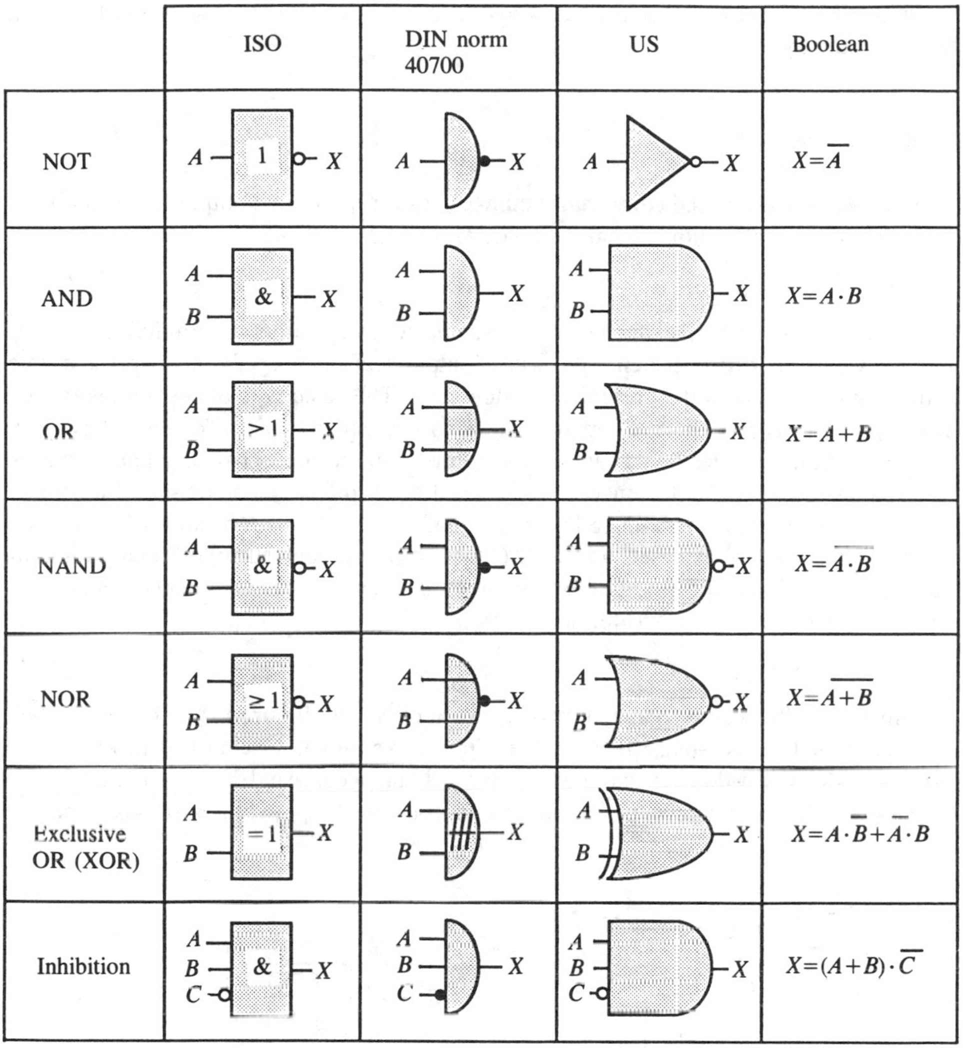

	ISO	DIN norm 40700	US	Boolean
NOT	A —[1]o— X	A —▷•— X	A —▷o— X	$X=\overline{A}$
AND	A —[&]— X (B)	A, B —▷— X	A, B —▷— X	$X=A\cdot B$
OR	A —[>1]— X (B)	A, B —▷— X	A, B —▷— X	$X=A+B$
NAND	A —[&]o— X (B)	A, B —▷•— X	A, B —▷o— X	$X=\overline{A\cdot B}$
NOR	A —[≥1]o— X (B)	A, B —▷•— X	A, B —▷o— X	$X=\overline{A+B}$
Exclusive OR (XOR)	A —[=1]— X (B)	A, B —▷— X	A, B —▷— X	$X=A\cdot\overline{B}+\overline{A}\cdot B$
Inhibition	A, B —[&]— X, C —o	A, B —▷— X, C —•	A, B —▷— X, C —o	$X=(A+B)\cdot\overline{C}$

Figure 8.11 Commonly used logical gate symbols.

Standards other than ISO are often used to symbolize switching elements. The ISO symbols are not universally accepted and in the USA there are at least three different sets of symbols. In Europe, the DIN standard is common. Three common standards are shown in Figure 8.11. In principle all switching networks can be tested by truth tables. Unfortunately, the number of Boolean functions grows rapidly with the number of variables n, since the number of combinations becomes 2^n. It is outside the scope of this text to discuss different simplifications of Boolean functions. A method known as Karnaugh map may be used if the number of variables is small. For systems with many variables (more than about 10) there are numerical methods to handle the switching network. The method by Quine–McCluskey may be the best known, and is described in standard textbooks on switching theory.

8.1.4 Flip-flops

Hitherto we have described **combinatorial networks**, i.e. the gate output X depends only on the *present* combination of input signals $\mathbf{U} = (A,B,\ldots)$, or

$$X(t) = f([\mathbf{U}(t)] \tag{8.10}$$

The gates have *no memory* so the network is a *static system*. In order to introduce a memory function we define **flip-flop elements**, whose output depends not only on the present state of the inputs but also on the previous flip-flop state. The basic type of flip-flop is the *SR* **flip-flop** (Set–Reset). The two inputs S and R can be either 1 or 0. They are, however, not permitted to both be 1 or 0 at the same time. The output is called y and normally also $\bar{y}$ is an output. If $S=1$ then $y=1$ ($\bar{y}=0$) and the flip-flop becomes **set**. If S returns to 0 then the gate remembers that S had been 1 and keeps $y=1$. If R becomes 1 (assuming that $S=0$) the flip-flop is **reset**, and $y=0$ ($\bar{y}=1$). Again R can return to 0 and y remains 0 until a new S signal appears. Let us call the states at consecutive moments y_n and y_{n+1}. Then the operation can be written as:

$$y_{n+1}=\bar{R}.(S + y_n) \tag{8.11}$$

An *SR* flip-flop can be realized by two logical elements (Figure 8.12). By adding two AND gates and a clock-pulse input to the flip-flop we obtain a **delay (*D*) flip-flop** or a **latch**. The delay flip-flop has two inputs, a data (u) and a clock pulse (*CP*) (Figure 8.13). Whenever a clock pulse appears the output y accepts the D input value that existed before

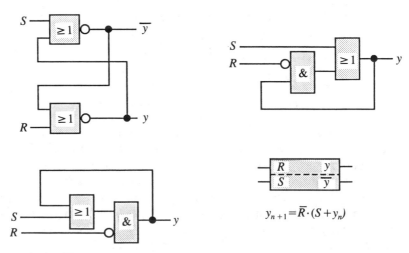

Figure 8.12   Three different realizations of a flip-flop gate and its ISO symbol.

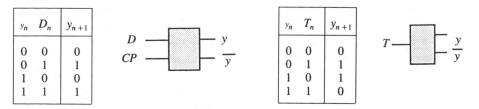

Figure 8.13 A delay (*D*) flip-flop and a trigger (*T*) flip-flop with their truth tables.

the appearance of the clock pulse. In other words, the *D* input is delayed by one clock pulse. Note that the new state y_{n+1} is always independent of the old state.

By introducing a feedback from the output to the flip-flop input and a time delay in the circuit we obtain a **trigger** or a **toggle (*T*) flip-flop**. The *T* flip-flop (Figure 8.13) is often used in counting and timing circuits as a 'frequency divider' or a 'scale-of-two' gate. It has only one input, *T*. Whenever an upgoing pulse appears in *T* the output *y* flips to the other state.

All three types of flip-flops can be realized in a **JK flip-flop**, with *J* being the set signal and *K* the reset signal. It frequently comes with a clock-pulse input. Depending on the input signals the *JK* flip-flop can realize a *SR* flip-flop, a latch or a trigger.

8.1.5 Realization of Switching

In Section 4.6.1 we considered the realization of switching functions for actuators. They are mostly made for switching *power* as opposed to signals. Here we focus on *signal switching* in an electronic environment. Electromechanical relays for signal switching were mentioned in Section 1.4. They will be commented on further in the next section. Electronic logic gates can be implemented with diodes that provide AND and OR gates. It is not suitable to cascade several gates in series, so they are not very attractive.

A common way to implement the gates, however, is by transistor logic, since the signal then can be amplified back to normal levels. As pointed out in Section 6.1 several gates can be mounted on a single chip to form an integrated circuit (IC) and can be packed into medium (MSI) or large scale integration (LSI). Figure 8.14 shows one type of three-input NOR gate implemented with a simple *n-p-n* bipolar junction transistor.

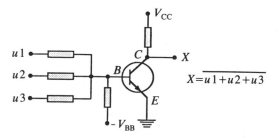

Figure 8.14 Transistor realization of a NOR gate.

As long as the three inputs $u1$, $u2$, $u3$ are *low* (0), the transistor base B is below zero voltage and the transistor does not conduct. Then the transistor output X, connected to the collector C is *high* (1), at somewhat less than the supply voltage V_{CC}. If any of the inputs goes high, the transistor conducts and a current flows from C to the emitter E producing a voltage drop across the resistor, sufficient to make the output X *low*.

A realization of a NAND gate was shown in Section 6.2.3, and as has been shown in Section 8.1.3 all other basic switching functions can be constructed from NOR gates alone. TTL logic has been dominating for a long time. Within the TTL family there are several types with different power consumption and speed.

Conventional TTL circuits have been largely replaced by LS-TTL (low-power Schottky TTL) elements. They contain so-called Schottky diodes to increase the switching speed and use considerably less power than the older TTL type. Many TTL circuits today are being succeeded by CMOS (complementary metal-oxide semiconductor) that are based on FETs (field-effect transistors) rather than bipolar transistors. A CMOS circuit has about three orders of magnitude lower power consumption than corresponding TTL gates. Also the CMOS are less sensitive to electrical noise and the level of the supply voltage. Furthermore, CMOS circuits are in principle slower and are easily damaged by static electricity. Another new generation of the CMOS circuits is the high speed (HC) CMOS logic.

Complex circuits can of course be manufactured as medium or large-scale integrated circuits. This is, however, not economically justifiable in very small quantities. By using so-called **programmable logic devices (PLD)** one can obtain semicustomized ICs quite inexpensively. PLDs are fuse-programmable chips that mostly belong to the LS-TTL family. The circuit contains a large array of programmable gates connected by microscopic fused links. These fuses can be selectively blown by using a special programming unit.

In the PLD family there are **programmable array logic (PAL), field-programmable logic arrays (FPLA)** and **programmable read-only memory (PROM)**. In the PAL system there is a programmable AND-gate array with the AND gates connected to a fixed OR-gate array. In the FPLA system both the AND and the OR gates are programmable. In both PAL and FPLA there are chips available with NOR, XOR or D flip-flops so that a complete sequential system can be included in one chip.

To make programming of PLDs less complex there are several software packages available for personal computers. They convert the Boolean equations to be implemented into data for feeding to the programming unit. Testing of the programmed chip is another attractive software feature.

8.2 LADDER DIAGRAMS

8.2.1 Basic Description

Many switches are realized by solid-state gates, but electromechanical relays are still in use in many applications. Statistics show that the share of electromechanical relays versus

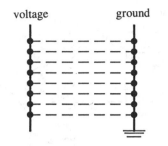

Figure 8.15 Framework of a ladder diagram.

the total number of gates in use is decreasing. This does not mean that their importance is dwindling; relays remain, in fact, a necessary interface between the control electronics and the powered devices.

Relay circuits are usually drawn in the form of **ladder diagrams**. Even if the relays are replaced by solid-state switches or programmable logic they are still quite popular for describing combinatorial circuits or sequencing networks. They also serve as a basis for writing programs for programmable controllers.

A ladder diagram reflects a conventional wiring diagram (Figure 8.15). A wiring diagram shows the physical arrangement of the various components (switches, relays, motors, etc.) and their interconnections, and is used by electricians to do the actual wiring of a control panel. Ladder diagrams are more schematic and show each branch of the control circuit on a separate horizontal row (the rungs of the ladder). They are meant to emphasize the function of each branch and the resulting sequence of operations. The base of the diagram shows two vertical lines, one connected to a voltage source and the other to ground.

Relay contacts are either normally open (n.o.) or nomally closed (n.c.), where *normally* refers to the state in which the coil is not energized (cf. Figure 4.17). Relays can implement elementary circuits such as AND and OR as well as *SR* flip-flops. The relay symbols are shown in Figure 8.16.

Figure 8.16 Relay symbols for n.o., n.c. contacts and relay coil.

Example 8.2 Combinatorial circuit

The combinatorial circuit of Figure 8.10 can be represented by a ladder diagram (Figure 8.17). All the conditions have to be satisfied simultaneously. The series connection is a logical AND and the parallel connection a logical OR. The lower case characters (x,y) denote the contacts belonging to a relay while the capital symbols (X,Y; the ring symbol) denote the coil.

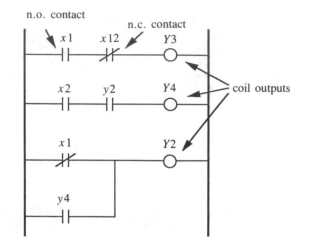

Figure 8.17   The combinatorial circuit in Figure 8.10 represented
by a ladder diagram.

The relay contacts usually have negligible resistance, whether they are limit switches, pressure or temperature switches. The output element (the ring) could be any resistive load (relay coil) or a lamp, motor or any other electrical device that can be actuated. Each rung of the ladder diagram must contain at least one output element, otherwise a short circuit would occur.

Example 8.3   A flip-flop as a ladder diagram

A flip-flop (Figure 8.12) can also be described by a ladder diagram (Figure 8.18). When a **set** signal is given, the S relay conducts a current that reaches the relay coil Y. Note that the R is not touched. Energizing the relay coil closes the relay contact y in line 2. The **set** push button can now be released and current continues to flow to coil Y through the contact y, i.e. the flip-flop remains *set*. Thus, the y contact provides the 'memory' of the flip-flop. In industrial terminology, the relay is denoted as a **self-holding** or **latched** relay. At the moment the **reset** push button is pressed, the circuit to Y is broken and the flip-flop returns to its former **reset** state.

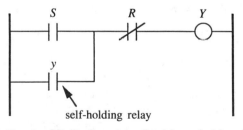

Figure 8.18   An SR flip-flop described by a ladder diagram.

8.2.2 Sequential Circuits

In Section 3.7 we described sequential systems where the outputs depended not only on the momentary values of the inputs (as in combinatorial networks) but also on those of earlier inputs. A sequence chart thus contains memory elements and states. Many sequence operations can be described by ladder diagrams and can be defined by a number of states, where each state is associated with a certain control action.

Only one state at a time can be active. Therefore some kind of acknowledgment signal is needed in order to change from one state to another. The acknowledgment signal is in fact the **reset (R)** signal in Figure 8.18. The sequence can be described as a series of *SR* flip-flops where each step is a rung on the ladder. When the acknowledgment signal is given, the next flip-flop is set. The structure of the sequence is shown as a ladder diagram in Figure 8.19. The execution jumps one step at a time and returns to Step 1 after the last step.

Step 1 can be initiated with a **start** button. When running in an infinite loop it can also be started from the last step, *X*. When *X* is active together with a new condition for the startup of Step 1, then the Step 1 coil is activated, and the self-holding relay keeps it set. When the condition for Step 2 is satisfied the relay Step 2 latches Circuit 2 and at the same time guarantees that Circuit 1 is broken. This is then continued in the same fashion. In order to ensure a repetitive sequence, the last step has to be connected to Step 1 again.

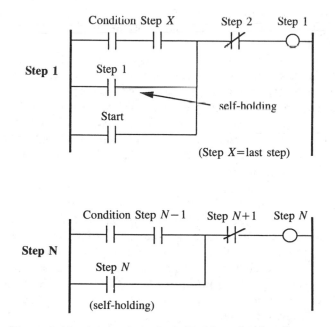

Figure 8.19   A sequence described by a ladder diagram.

This is an example of an **asynchronous** execution. In switching theory there are also **synchronous** charts, where the state changes are caused by a clock pulse. In industrial automation applications we mostly talk about asynchronous charts, since the state changes do not depend on clock pulses but on several conditions in different parts of the sequence. In other words, an asynchronous system is **event-based**, while a synchronous system is **time-based**. Moreover, we are dealing with design of asynchronous systems with **sustained input signals** rather than pulse inputs.

8.3 PROGRAMMABLE CONTROLLERS

8.3.1 Basic Structure

Programmable logical controllers (PLC) are microcomputers developed to handle Boolean operations. A PLC produces on/off voltage outputs and can actuate such elements as electric motors, solenoids (and thus pneumatic and hydraulic valves), fans, heaters and light switches. They are vital parts of industrial automation equipment found in all kinds of industries.

The basic operation of a PLC corresponds to a software-based equivalent of a relay panel. However, a PLC can also execute other operations, such as counting, delays and timers. Since a PLC can be programmed in easy-to-learn languages it is naturally much more flexible than any hardware relay system and a single PLC can replace hundreds of relays. PLCs are in fact more flexible than programmable logical devices but usually slower, so PLDs and PLCs often coexist in industrial installations to offer the best and most economic solutions.

The PLC was initially developed by a group of engineers from General Motors in 1968, where the initial specification was formulated: it had to be easily programmed and reprogrammed, preferably in-plant; easily maintained and repaired; smaller than its relay equivalent; and cost-competitive with the solid-state and relay panels then in use. This provoked a great interest from engineers of all disciplines in how the PLC could be used for industrial control. A microprocessor-based PLC was introduced in 1977 by Allan-Bradley Corporation in USA. It was based on an Intel 8080 microprocessor with circuitry to handle bit logic instructions at high speed.

The early PLCs were designed only for logic-based sequencing jobs (on/off signals). Today there are hundreds of different PLC models on the market. They differ by their memory size (from 256 bytes to several kilobytes) and I/O capacity (from a few lines to thousands). The difference also lies in the features they offer. The smallest PLCs serve just as relay replacers with added timer and counter capabilities. Many modern PLCs also accept proportional signals and they can perform simple arithmetic calculations and handle analog input and output signals and PID controllers (Chapter 9). This is the reason why the letter L was dropped from PLC, but the term PC may cause confusion with personal computers so we keep the L here.

PLCs can be programmed using both relay-type ladder diagrams (mostly in the United

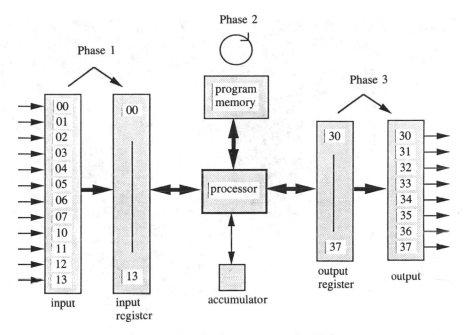

Figure 8.20   Basic structure of a PLC.

States) and logical gate symbols (mostly in Europe) but programs similar to BASIC are becoming more common.

Figure 8.20 shows the basic structure of a PLC. The inputs are read into the input memory register. This function is already included in the system software in the PLC. An input/output register is often not only a bit but a byte. Consequently one input instruction is capable of giving the status of 8 different input ports.

The instructions fetch the value from the input register and operate on only this or on several operands. The central processing unit (CPU) works towards a result register or accumulator (A). The result of an instruction is stored either in some intermediate register or directly in the output memory register that is written to the outputs. The output function is usually included in the system programs in a PLC. A typical commercial PLC is shown in Figure 8.21.

A PLC is specifically made to fit an industrial environment and for exposure to hostile conditions, such as heat, humidity, unreliable power, and mechanical shocks and vibrations. Also a PLC comes with input/output modules readily available for different voltage levels and can be easily interfaced to hundreds of input and output lines.

8.3.2 Basic Instructions and Execution

In order to make a PLC system useful in industrial automation, it has to work in real time. Consequently the controller has to act on external events very quickly, in other words,

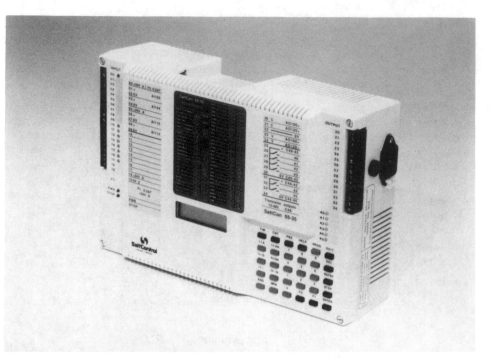

Figure 8.21   A small PLC. (Courtesy of SattControl, Sweden).

have a short response time. There are two principal ways to sense the external signals, by **polling** the input signals regularly or by using **interrupt** signals (see also Section 7.6.5). A polling method has the drawback that some external event may be missed if the processor is not sufficiently fast. On the other hand such a system is simple to program. A system with interrupts is more difficult to program but the risk of missing some external event is much smaller. The polling method is usually used in simpler automation systems while interrupts are used in more complex control systems.

The 'programming' of a PLC consists mainly of the definition of sequences. The input and output functions are already prepared. The instructions from a ladder diagram, a logical gate diagram or Boolean expressions are translated to machine code. At execution, the program memory is run through in a cyclic manner in an infinite loop. Every scan may take some 15–30 ms in a small PLC and the scanning time is approximately proportional to the memory size. In some PLCs the whole memory is always scanned even if the code is shorter. In other systems the execution stops at an **end** statement that concludes the code; thus the loop time can be made shorter for short programs.

The response time of the PLC of course depends on the processing time of the code. While the instructions and the output executions are executed the computer system can not read any new input signals. Usually this is not a big problem, since most signals in industrial automation are quite slow or last for a relatively long time.

The ladder diagram can be considered as if every rung of the ladder were executed at the same time. Thus it is not possible to visualize the ladder diagram being executed sequentially on a row-by-row basis. The execution has to be very fast compared to the timescale of the process under control.

A small number of basic machine instructions can solve most sequencing problems. There are four fundamental instructions:

ld, **ldi** A number from the computer input memory is loaded (**ld**) or inverted (**ldi**) before it is read into the accumulator (A).

and, **ani** An AND or AND inverse instruction executes an AND logical operation between A and an input channel, and stores the result in A.

or, **ori** An OR or OR inverse instruction executes an OR logical operation between A and an input channel, and stores the result in A.

out The instruction outputs A to the output memory register. The value remains in A, so the same value can be sent to several output relays.

Note that the logical operations may be performed on bits as well as on bytes.

Example 8.4 Translation from a ladder diagram to machine code

The translation from ladder diagram to machine code is illustrated by Figure 8.22. The gate y11 gives self-holding capability.

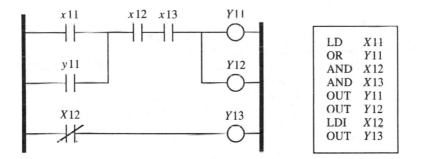

Figure 8.22   Translation of a ladder diagram to machine code.

A logical sequence or ladder diagram is often branched. Then there is a need to store intermediate signals for use later. This can be done with special help relays, but in a PLC it is better to use two instructions **orb** (OR block) or **anb** (AND block). They use a memory stack area (Last In First Out) in the PLC to temporarily store the output.

Example 8.5 Using the block instruction and stack memory

The ladder diagram (Figure 8.23) can be coded with the following machine code:

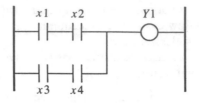

Figure 8.23   Example of the use of a stack memory.

ld $x1$ Channel 1 is read into the accumulator (A).
and $x2$ The result of the AND operation is stored in A.
ld $x3$ The content of A is stored on the stack. Channel 3 is read into A.
and $x4$ The result of lines 3 and 4 is stored in A.
orb   An OR operation between A and the stack. The result is stored in A. The last element of the stack is eliminated.
out Y1   Output of A on Channel 1.

Example 8.6 Using the block instructions and the stack memory

The logical gates in Figure 8.24 are translated to machine code by using block instructions. The corresponding machine code is:

ld $x1$ Load Channel 1.
and $x2$ The result is stored in A.
ld $x3$ The content of A is stored on the stack. Status of Channel 3 is loaded into A.
and $x4$ The result of lines 3 and 4 is stored in A.
ld $x5$ The content of A is stored on the stack. Status of channel 5 is loaded into A.
and $x6$ The result of lines 5 and 6 is stored in A.
orb   Operates on the last element in the stack (the result of lines 3 and 4) and the content of A. The result is stored in A. The last element of the stack is removed.

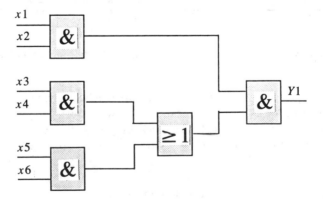

Figure 8.24   Example of a logical circuit.

anb Operates on the last element in the stack (the result of lines 1 and 2) and the content of A. The result is stored in A. The last element of the stack is removed.

out Y1

8.3.3 Additional PLC Instructions

For logical circuits there are also operations such as XOR, NAND and NOR, as described earlier. Modern PLC systems are also supplied with instructions for alphanumerical or text handling, and communication as well as composed functions such as timers, counters, memory and pulses.

A pulse instruction (PLS) is used to give a short pulse, e.g. to reset a counter. A PLC may also contain certain delay gates or time channels so that a signal in an output register may be delayed a certain time. Special counting channels make it possible to count a number of pulses.

Different signals can be shaped, such as different combinations of delays, oscillators, rectangular pulses, ramp functions, shift registers or flip-flops. As already mentioned, advanced PLCs also contain floating point calculations as well as prepared functions for signal filtering and feedback control algorithms.

8.3.4 Programming a PLC

Usually, a PLC is programmed via an external unit; this unit is not needed for the PLC operation and may be removed when the PLC is in operation. Programming units range from small hand-held portable units, sometimes called 'manual programmers' to personal computers.

A manual PLC programmer looks like a large pocket calculator, having a number of keys and a simple display. Each logic element of the ladder diagram is entered separately, one at a time, with series or parallel connections achieved by using keys for AND, OR and NOT.

A more sophisticated programmer consists of a personal computer with a graphic display which is becoming increasingly common. The display typically shows several ladder diagram lines at a time and can also indicate the power flow within each line during operation to make debugging and testing simpler. Other units are programmed using logical gates instead of a ladder diagram. The program is usually entered by moving a cursor along the screen (using arrow keys or a mouse). When the cursor reaches the location where the next element is to be added, confirmation is given via additional keys.

An increasing number of PLCs are programmed in English-statement type languages, because of the increasing use of PLCs for analog control. With the combination of a Boolean language and several other types of instructions, it is clear that the structuring of large programs soon becomes extremely difficult. Therefore the demand for high level languages increases significantly as the complexity of the PLC operations increase.

8.4 SPECIFYING INDUSTRIAL SEQUENCES WITH GRAFCET

The need for structuring a sequential process problem is not apparent in small systems but becomes crucial very quickly. As a control system becomes more complex the need for better functional descriptions increases. In other words, each block has to be able to include more and more complex functions. This means that logical expressions in terms of ladder diagrams or logical circuits are not sufficiently powerful to allow a structured description. In order for a more rational top-down analysis to be possible, the functional diagram **Grafcet** (GRAphe de Commande Etape-Transition) was developed by a French commission in the late 1970s and has been adopted as the French national standard. Since 1988 Grafcet has been specified in a European standard (IEC 848).

8.4.1 The Grafcet Diagram

Grafcet is a method that was developed for specifying industrial control sequences diagrammatically. A similar method originated in Germany, called FUP (FUnction Plan). The basic ideas behind the two methods are the same and the differences are of minor importance.

By way of an example, we will illustrate the use of Grafcet for a batch process. A tank is to be filled with a liquid. When it is full the liquid is heated until a certain temperature has been reached. After a specified time the tank is emptied, and the process starts all over again. The Grafcet diagram of the sequence is shown in Figure 8.25.

An indicator *Empty* signals that there is no liquid left. A *Start* signal together with the *Empty* indication initiates the sequence. In Step 2 the bottom valve is closed and the pump started. An indicator *Full* tells when to stop the pumping and causes a jump (called a **transition**) to Step 3, and the pump is switched off and a heater switched on. It remains on until the final temperature has been reached (*Temp*) and there is a jump to Step 4. The heater is switched off and a timer is started. When the waiting time has elapsed (*time_out*) there is a transition to Step 5 and the outlet valve can be opened. The sequence then returns to *Start*.

The figure illustrates that the Grafcet consists of a column of numbered **blocks**, each one representing a **step**. The vertical lines joining adjacent blocks represent **transitions**. Each transition is associated with a logical condition called **receptivity**, which is defined by a Boolean expression written next to a short horizontal line crossing the transition line. If the receptivity is logical 1 the transition is executed and the system passed to the next step.

A Grafcet diagram basically describes two things according to specific rules:

- Which order to execute actions.
- What to execute.

The function diagram is split up into these two parts. The part describing the order between the steps is called the **sequence part**. Graphically this is shown as the left part of Figure 8.25, including the five boxes. The sequence part does not describe the actions to execute. This is described by the **object part** of the diagram, that consists of the boxes to the right of the steps.

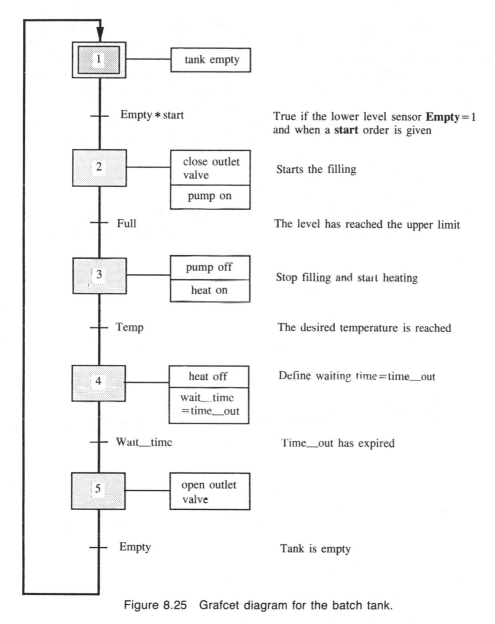

Figure 8.25 Grafcet diagram for the batch tank.

A *step* can be either active or passive, i.e. being executed or not. The *initial* step is the first step of the diagram. It is described by a box with a double frame. An **action** is a description of what is performed at a step. Every action has to be connected to a step and can be described either by a ladder, logical circuit or Boolean algebra. When a step becomes active its action is executed. However, a logical condition can be connected to the action so that it is not executed until both the step is active and the logical condition is fulfilled. This feature is useful as a safety precaution.

Several actions can be connected to one step. They can be of the type *outputs, timers* or *counters* but can also be controller algorithms, filtering calculations or routines for serial communication. A *transition* is an 'obstacle' between two steps and can only originate from an active step. Once a transition has taken place, the next step becomes active, and the previous one inactive. The transition consists of a logical condition that has to be true in order to make the transition between two steps possible.

By combining the three building blocks *initial step, steps* and *transitions*, it is possible to describe quite a large number of functions. The steps can be connected in:

- simple sequences
- alternative parallel branches
- simultaneous parallel branches

In a **simple sequence** there is only one transition after a step and after a transition there is only one step. No branching is made. In an **alternative parallel sequence** (Figure 8.26)

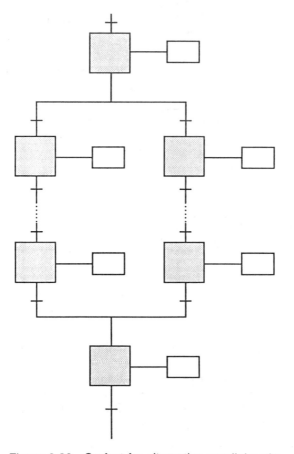

Figure 8.26   Grafcet for alternative parallel paths.

there are two or more transitions after one step. This means that the flow of the diagram can take alternative ways. Typically this is an *if-then-else* condition and is useful to describe, for example, alarm situations. It is very important to make sure that the transition condition, located immediately before the alternative branching, is consistent, in other words, the alternative branches are not allowed to start simultaneously. A branch of an alternative sequence always has to start and end with transition conditions.

In a **simultaneous parallel sequence** (Figure 8.27) there are two or more possible steps after a transition. Several steps can be active simultaneously. In other words, this is a concurrent (parallel) execution of several actions. The double horizontal lines define the parallel processes. When the transition condition is true, both the branches become active *simultaneously* and are executed *individually* and *concurrently*. A transition to the step below the lower parallel line cannot be executed until both the concurrent processes are completed.

The three types of sequence can be mixed, but has to be done in a correct way. For example, if two alternative branches are terminated with a parallel ending (two horizontal bars) then the sequence is locked, since the parallel end waits for both branches to finish,

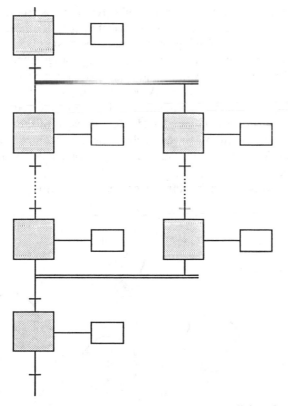

Figure 8.27   Grafcet for simultaneous parallel paths.

while the alternative start has started only one branch. Also, if simultaneous parallel branches are finished with an alternative ending (one horizontal bar) then there may be several active steps in the code, and it is not executed in a controlled manner.

8.4.2 Computer Implementation of Grafcet

In Grafcet there are several inherent real-time features that have to be observed during implementation. As we have seen in Chapter 7, the realization of real-time systems requires intensive effort with considerable investments in time and personnel. The implementation of the Grafcet function diagram into computer code is not part of the definition and of course varies in the different systems. Obviously any implementation makes use of real-time programming tools.

Grafcet compilers are available for many different industrial control computers. Typically, the block programming and compilation are performed on a PC. After compilation the code is transferred to the PLC for execution. The PC is then removed in the real-time execution phase. More advanced PLC systems have Grafcet compilers built into the system.

The obvious advantage of Grafcet and similar types of abstract descriptions is their independence of specific hardware and their orientation to the task to be performed. Unfortunately, it has to be said that high-level languages such as Grafcet do not yet enjoy the success they deserve. It seems odd that so many programmers start anew with programming in C or Assembler, while control tasks of the type we have seen are much more easily solved with a functional block description.

As in any complex system description, the diagram or the code has to be suitably structured. A Grafcet implementation should allow division of the system into smaller parts and the Grafcet diagram into several subgraphs. For example, each machine may have its own graph. Such hierarchical structuring is of fundamental importance in large systems (see further, Chapter 12).

Of course Grafcet is also useful for less complex tasks. It is quite easy for the non-

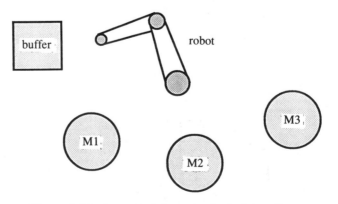

Figure 8.28   Layout of the manufacturing cell.

specialist to understand the function as compared to the function of a ladder diagram. By having a recognized standard for the description of automation systems, the chances for reutilization of computer code are increased.

The translation of Grafcet to computer code depends on the specific PLC. Still, even if there is no automatic translator from Grafcet to machine code the functional diagram is very useful, since it allows the user to structure the problem. Many machine manufacturers today use Grafcet to describe the intended use and function of the machinery. Of course it makes subsequent programming much simpler if Grafcet can be used all the way.

An implementation tool also may simulate the control code of a Grafcet diagram on the screen. During the simulation the actual active state is shown.

8.5 APPLICATION OF FUNCTION CHARTS IN INDUSTRIAL CONTROL

The use of Grafcet for sequential programming is demonstrated for a manufacturing cell in a flexible manufacturing system. The cell consists of three NC machines (e.g. a drill, lathe and mill), a robot for material handling and a buffer storage (Figure 8.28).

At the cell level we do not deal with the individual control loops of the machines or of the robot. They are handled by separate systems. The cell computer sends on/off commands to the machine and its main tasks are both to control the individual sequencing of each machine (and the robot) and to synchronize the operations between the machines and the robot. The control task is a mixture of sequencing control and real-time synchronization of the kind described in Chapter 7. We will demonstrate how Grafcet expresses the operations. The implementation of the function chart is then left to the compiler.

The product that is to be manufactured has to be handled in the three machines in a predefined order (like a transfer line). The robot delivers new parts to each machine and moves them between the machines.

8.5.1 Synchronization of Tasks

The synchronization of the different machines is done by a **scheduler** graph with the structure indicated in Figure 8.29. The scheduler communicates with each machine and with the robot, and determines when they can start or when the robot can be used. It works like a scheduler in a real-time operating system (Section 7.2), distributing the common resource — the robot — as efficiently as possible. The scheduler has to guarantee that the robot does not cause any deadlock. If the robot has picked up a finished part from a machine and has nowhere to place it, then the system will stop. Consequently the scheduler has to match demands from the machines with the available resources (robot and buffer capacity).

The scheduler graph is described by a number of parallel branches, one for each

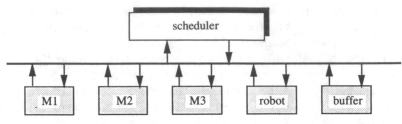

Figure 8.29   Logical structure of the machine cell.

machine, robot and buffer. Since all the devices are operating simultaneously, the scheduler has to handle all of them concurrently by sending and receiving synchronization signals of the type *start* and *ready*. When a machine gets a *start* command from the scheduler it performs a task defined by its Grafcet diagram. When the machine has terminated it sends a *ready* signal to the scheduler.

Figure 8.29 shows that no machine communicates directly with the robot. Instead all the communication signals go via the scheduler. The signals are seen as transition conditions in each Grafcet branch. By structuring the graph in this hierarchical way it is possible to add new machines to the cell without reprogramming any of the sequences of the other machines. The robot has to add the new operations to serve the new machine.

A good implementation of Grafcet supports a hierarchical structuring of the problem. The total operation can first be defined by a few complex operations, each consisting of many steps. Then it is possible to go on to more and more detailed operations.

8.5.2 Machine Sequencing Control

For each machine a sequencing control can be defined in a separate function chart. The Grafcet sequence for the drill is indicated in Figure 8.30. The *start drill* condition is a global variable and is a signal from the scheduler graph. If the drill has not acknowledged the start order within a given time *tim1* then an alarm is activated and the graph is directed into an alternative parallel branch. Likewise, if the drill has not reached the workpiece within a given time then another alarm is activated. The graph for the individual machine may be lengthy but still straightforward with a number of alternative routes due to alarms. When it is finished a variable *ready_drill* becomes true. This variable is noted in a logical condition in the scheduler graph. The scheduler is now activated to tell the robot to pick up the material.

The robot can perform a number of tasks, described as a structure of alternative parallel branches, such as:

- Pick up a part from one machine
- Deliver a part to the buffer
- Take one part from the buffer and fix it in a machine

The importance of structure is obvious. The sequencing operations of the cell could have

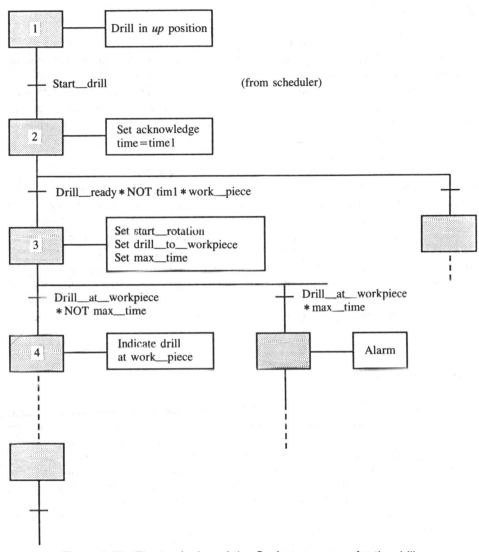

Figure 8.30 The beginning of the Grafcet sequence for the drill.

been written in machine code or as a ladder diagram. However, long codes in low level languages are not meant for people to read, understand, debug or maintain. For a high level language such as Grafcet, the code itself is good documentation.

Many Grafcet programs support some type of simulation feature and the system can be executed without the external signals connected to the real machines. The logic can be tested and the result may be displayed by more or less sophisticated animations.

8.6 SUMMARY

Binary control, or on/off control, is a vital part of many automation applications. It is based on switching theory. An important aspect from the theory is that with few fundamental gates (NOR, NAND) all combinatorial networks can be realized.

There are two classes of sequential systems. Combinatorial networks have no memory and the output is a logical function of the present input values. In a sequencing network there is a memory function so that a series of logical steps can be defined. In industrial automation the systems are normally asynchronous, which means that the sequences are triggered by external events and not by periodic clock pulses.

Switching elements can be realized in discrete components, integrated circuits, programmable array logic or in software. Programmable logical controllers together with programmable logical arrays are becoming the normal way to implement logical networks or sequencing control.

The description of a sequencing network can be done on a gate level with either logical circuits or ladder diagrams. However, there is an apparent need to structure complex sequencing systems, and high level languages and function charts for sequencing control have been developed. A good language implementation has to support hierarchical structuring of the code. The function diagram Grafcet has been used to demonstrate some high level language principles in sequencing systems.

FURTHER READING

Switching theory is introduced in Lee (1978) and Fletcher (1980). Deppert and Stoll (1988) show that pneumatic circuits may be an interesting alternative to electrical devices in logical circuit design.

Pessen (1989) contains a good overview of sensors, actuators and switching elements for both electric and pneumatic environments. Programmable logic devices are described in several articles in *BYTE* magazine (1987) and Warnock (1988) presents a lot of practical information on PLCs, their construction, use and applications. However, the manuals from the different PLC manufacturers provide full details of the facilities and programming methods for a given model.

There are two primary sources for Grafcet information, *GRAFCET* (1979) and *Manual 02.1987* (1987) available from Telemecanique. FUP is defined by the German standard DIN 40719 while the European Grafcet standard is described in IEC Standard 848.

Several articles on components, PLCs and market reviews appear regularly in the journals *Control Engineering, Instrument & Control Systems, Machine and Design* and *Product Engineering*.

9

Feedback Control

Aim: To familiarize the reader with the most common controllers and their digital implementations in process control systems

In this chapter we will discuss continuous and discrete time controllers. The purpose is to familiarize the reader with different control structures, so that their general performance and use can be appreciated. It is outside the scope of this text to go into the details of analysis or of the tuning of the controllers. Instead we emphasize implementation aspects and controller structures.

Controllers can be designed from either continuous or time-discrete process models. This is explained further in Section 9.1. In Section 9.2 we review how simple continuous controllers can be used in feedforward and feedback combinations to obtain good control performance. On/off controllers are quite common in the process industry and are briefly discussed in Section 9.3. The **proportional-integral-derivative (PID)** *controller* is the dominating controller type in process computer applications and its properties are discussed in more detail in Section 9.4. This has to be discretized in time for computer realization and the discrete version of the PID is derived in Section 9.5. Practical controller aspects are summarized in Section 9.6.

Simple controllers can be combined into more complex structures, as discussed in Section 9.7. Process types where the limitation of PID controllers is apparent are discussed in Section 9.8. A general discrete controller and its properties are described in Section 9.9. If a system dynamic description is available in state/space form, then state feedback can be applied. Its properties are demonstrated in Section 9.10.

9.1 CONTINUOUS VS TIME-DISCRETE CONTROLLERS

Many industrial processes are characterized by several inputs and outputs (Section 2.3.4). In most cases, however, the internal couplings are not significant and the processes can be controlled by many local controllers, one input/output pair at a time. This is the normal structure in **direct digital control (DDC) systems**. The PID or similar controller is the most common type, so it is considered worthwhile devoting much space on the computer implementations of the PID controller.

A computer makes it reasonable to also realize other control structures, for example, non-linear or self-tuning controllers. Once the controller parameters are known, the implementation of the control algorithms is usually quite straightforward. However, every implementation has to be supplied with a 'safety umbrella' of routines to test the performance of the chosen control method.

9.1.1 Sampled Signals

When a feedback control strategy is implemented digitally, the continuous signal from the sensor is sampled, a control action is calculated and the controller output is sent via a DAC to the final control element (Chapters 4 and 5). The control signal $u(t)$ usually remains constant during the sampling interval. Sometimes the digital output signal is converted to a sequence of pulses representing the change in the actuator. Control valves driven by pulsed stepping motors (Section 4.5.3) are often used with digital controllers.

The execution of control algorithms is normally clock driven so that the controller must be started periodically. This is different from the asynchronous execution of logical circuits described in Chapter 8. One controller at a time is computed, and consequently it is not suitable to require every controller to execute at exactly the same time. Controllers are often realized in dedicated computers close to the physical process.

9.1.2 Continuous vs Discrete Control Design

A controller implemented in a digital control system is by definition digital and time discrete. Traditionally, however, most dynamic systems are described by ordinary differential equations, derived from physical laws such as mass and energy conservation (Chapter 3). There are two ways to synthesize a controller from the process description. In the first one, a continuous controller is derived from either a continuous state/space or a transfer function description of the process. The continuous controller is then discretized in time to fit the computer implementation. The other way is to discretize the plant dynamic model (Section 3.4) and make the complete design with discrete time models and methods.

Generally speaking, a continuous controller synthesis followed by discretization of the controller usually demands a shorter sampling interval, which implies a higher load for the computer. Since most PID controller design is made in this way, however, we will begin here with continuous PID controllers and then discretize them in time. The controller output is assumed to be stepwise constant, which is the normal output from DACs (Sections 5.2 and 5.3).

Controllers derived from a time-discrete description of the process look similar, but have different coefficients. This means that computer implementation is also similar, and the software can be prepared for a general controller. The parameters are not defined in the software structure but are supplied separately.

9.2 CONTINUOUS CONTROLLERS

In Chapter 3 different classes of dynamic systems were identified, such as external or internal, time-continuous or time-discrete descriptions. Different ways to achieve the dynamic models were indicated. The system representations are the basis for different controller synthesis procedures.

The transfer function $G(s)$ of a *linear* dynamic system is defined as the ratio between the Laplace transforms of the output and input (Section 3.3.4)

$$\frac{Y(s)}{U(s)} = G(s) \tag{9.1}$$

We will consider systems with only one input and one output so the system has only one transfer function. This transfer function $G(s)$ of the plant is considered to be *fixed*, i.e. its parameters (Equation 3.4.3) can not be changed. They are given by the construction or design of the process itself. It should be kept in mind that it is usually not trivial to find $G(s)$. However, many control schemes do not require a detailed model of the process.

9.2.1 Simple Controllers

The controller is usually a dynamic system such as the plant, and consequently it can be defined by a transfer function. In the simplest case, the input to the controller is the output error, or the difference between the reference value (**setpoint**) $u_c(t)$ and the measurement $y(t)$:

$$e(t) = u_c(t) - y(t) \tag{9.2}$$

The Laplace transform of the error is:

$$E(s) = U_c(s) - Y(s) \tag{9.3}$$

The transfer function $G_{REG}(s)$ of the controller (where REG means regulator) is defined as the ratio between the error input and the controller output $U(s)$,

$$U(s) = G_{REG}(s) E(s) = G_{REG}(s) [U_c(s) - Y(s)] \tag{9.4}$$

This is the most common form of simple feedback systems and is represented by the block diagram in Figure 9.1. The controller has two inputs, the measurement value and the setpoint (reference) value and one output, the controller signal. In this simple case, however, the controller uses only the difference between the two inputs.

From a mathematical point of view, the transfer function $G_{REG}(s)$ is treated in exactly the same way as the process transfer functions $G(s)$. The fundamental difference is that

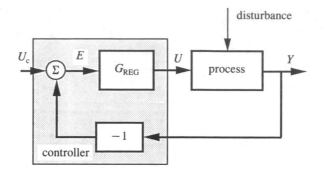

Figure 9.1   The simplest controller structure.

the coefficients of the controller transfer function $G_{REG}(s)$ are not fixed, but can be tuned. They are available for the control engineer to adjust so that the total system (the closed loop system) behaves in the desired way. The closed loop system in Figure 9.1 has the transfer function:

$$G_c(s) = \frac{Y(s)}{U_c(s)} = \frac{G_{REG}(s)\ G(s)}{1 + G_{REG}(s)\ G(s)} \tag{9.5}$$

It is reasonable to think that the more parameters a complex controller $G_{REG}(s)$ has, the more degrees of freedom it gives, and that with more parameters the behaviour of the closed loop transfer function can be changed more arbitrarily. Below we will illustrate how complex the controller needs to be to achieve the desired results.

9.2.2 Feedforward from the Reference Value

The simple control structure shown in Figure 9.1 only reacts on the error $e(t)$ (or $E(s)$) and does not use the separate information from the two inputs. There may be two principal reasons for an error; one is a change of the reference value (or command signal) $u_c(t)$ and the other a load change or some other disturbance to the system, that will cause a change of the output signal $y(t)$. A change in the reference value is a known disturbance and it is reasonable to think that if the controller can use the advantage of information on the reference change, then the closed loop system would have a better performance. This is what is done in **feedforward control**.

   Let us now consider a controller that has a more elaborate feature than Equation 9.4 and contains two parts. The feedback part $G_{FB}(s)$ is the previous simple controller that reacts on error e. The so-called feedforward part $G_{FF}(s)$ measures the reference value and adds to the control signal a correction term that will more readily correct the total system behaviour according to the reference signal change (Figure 9.2). It is obvious that the control signal $U(s)$ to the process is a sum of two signals,

$$U(s) = G_{FF}(s)\ U_c(s) + G_{FB}(s)\ [U_c(s) - Y(s)] \tag{9.6}$$

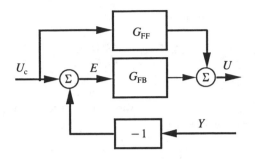

Figure 9.2 A regulator made up by a feedforward loop from the command signal and a feedback loop from the process output.

The controller can be rewritten in the form:

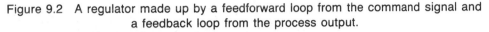

$$U(s) = [G_{FF}(s) + G_{FB}(s)] \, U_c(s) - G_{FB}(s) \, Y(s)$$
$$= G_{FI}(s) \, U_c(s) - G_R(s) \, Y(s) = U_{FFI}(s) - U_{FB}(s) \qquad (9.7)$$

where U_{FFI} is the feedforward part of the control signal and U_{FB} the feedback part. The controller has two inputs $U_c(s)$ and $Y(s)$ (Figure 9.3) and can be represented by two transfer functions $G_{FI}(s)$ and $G_R(s)$. Since the controller (Equation 9.7) has more coefficients to tune than the simple controller (Equation 9.4) it is reasonable to think that the closed loop system can be made to perform better. In particular, the system can react quickly to reference value changes if $G_{FI}(s)$ is properly chosen.

9.2.3 A General Form of the Reference Value Feedforward Controller

By leading the reference value through the feedforward controller it is possible to design a good servo controller, for example, in electric drives, robot systems or machine tools. In these applications it is crucial that the process output has a quick and accurate response to any reference value change.

The controller description can be further generalized. If the transfer functions $G_R(s)$ and $G_{FI}(s)$ are expressed with their numerator and denominator polynomials the

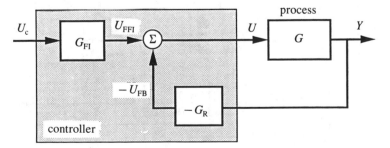

Figure 9.3 Structure of a linear feedforward–feedback controller.

controller can be written in the form:

$$U(s) = G_{F1}(s)\ U_c(s) - G_R(s)\ Y(s) = \frac{T_1(s)}{R_1(s)}\ U_c(s) - \frac{S_1(s)}{R_2(s)}\ Y(s)$$

$$= U_{FF1}(s) - U_{FB}(s) \tag{9.8}$$

where the two parts of the control signal are shown in Figure 9.3. Expressing the transfer functions with a common denominator we get:

$$U(s) = \frac{T(s)}{R(s)}\ U_c(s) - \frac{S(s)}{R(s)}\ Y(s) \tag{9.9}$$

where $R(s) = R_1 R_2$, $T(s) = T_1 R_2$ and $S(s) = S_1 R_1$. This is interpreted as:

$$U(s) = \frac{t_0 s^n + t_1 s^{n-1} + \ \ldots\ + t_n}{s^n + r_1 s^{n-1} + \ \ldots\ + r_n}\ U_c(s) + \frac{s_0 s^n + s_1 s^{n-1} + \ \ldots\ + s_n}{s^n + r_1 s^{n-1} + \ \ldots\ + r_n}\ Y(s) \tag{9.10}$$

where r_i, s_i and t_i are the transfer function parameters, and s the Laplace variable.

The controller can be rewritten in the form:

$$R(s)\ U(s) = T(s)\ U_c(s) - S(s)\ Y(s) \tag{9.11}$$

The process transfer function is also expressed explicitly by its numerator and denominator, i.e.:

$$G(s) = \frac{b_0 s^n + b_1 s^{n-1} + \ \ldots\ + b_n}{s^n + a_1 s^{n-1} + \ \ldots\ + a_n} = \frac{B(s)}{A(s)} \tag{9.12}$$

The closed loop system is illustrated by Figure 9.4 and corresponds directly to Figure 9.3. The closed loop transfer function is:

$$G_c(s) = \frac{Y(s)}{U_c(s)} = \frac{T(s)\ B(s)}{A(s)\ R(s) + B(s)\ S(s)} \tag{9.13}$$

The closed loop transfer function has many degrees of freedom. Note that the A and B coefficients are fixed by the process design. Again, we emphasize that it is not a trivial task to obtain an accurate model of the system, i.e. A and B. However, all the parameters

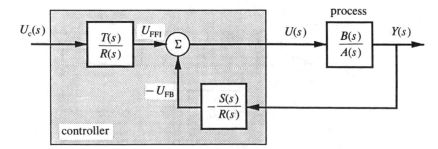

Figure 9.4   The feedforward–feedback controller corresponding to Figure 9.3.

in R, S and T can be tuned. The T and R coefficients belong to the feedforward part of the controller, so by tuning these parameters we can influence how the closed loop system will react on reference value (setpoint) changes. Similarly the S and R coefficients are related to the feedback. By tuning S and R we can affect how the system will recover after a load change or some other disturbance that has influenced the measurement signal $y(t)$.

The transfer function (Equation 9.13) is usually compared with some desired transfer function:

$$G_m(s) = \frac{Y(s)}{U_c(s)} = \frac{B_m(s)}{A_m(s)} \qquad (9.14)$$

and R, S and T are chosen so that:

$$B_m(s) = T(s)\,B(s) \qquad (9.15a)$$

$$A_m(s) = A(s)\,R(s) + B(s)\,S(s) \qquad (9.15b)$$

If the orders of R, S and T are sufficiently high, i.e. if there are a sufficient number of knobs to turn, then the closed loop transfer function (Equation 9.13) can be changed within wide limits. The order n of the controller has to be the same as the process. In particular, by changing R and S the denominator of the closed loop system transfer function (Equation 9.13) can be changed arbitrarily. This means that *the poles of the closed system can be moved to any location*. In Section 3.3.4 we saw that the poles determine the type of transient response of the system, so that the dynamics can be chosen arbitrarily. We are also reminded that the zeros determine the relative size of the different terms in the transient. *There is no possibility of changing the values of the zeros*. One can *insert* new zeros or one can *remove* a zero by cancellation, i.e. place a pole in the same location.

The zeros of the numerator TB are the same as the zeros of T and B. New zeros can be added by the T polynomial. The B zeros, however, are *fixed* and can not be moved. Only if there is a pole in the same location will the zero be cancelled. Such a cancellation, however, has to be made with great caution. For example, if a B zero is located in the right half-plane (which is called a **non-minimum phase system**) then a cancelling pole is also in the right half-plane. This indicates an unstable system where the zero is exactly chosen to cancel the unstable mode. If the cancellation is not exact (which it never is!) the closed loop system will be truly unstable. Consequently a system zero in the right half plane is a plant property that can not be removed by a controller. It can only be minimized by smarter control structures.

The polynomials $R(s)$, $S(s)$ and $T(s)$ can not be chosen arbitrarily. Each of the controller transfer functions (see Figure 9.4) have to be physically realizable. This means that the order of the denominator must be larger than that of the numerator, i.e. the order of $R(s)$ has to be larger than that of both $S(s)$ and $T(s)$, otherwise the controller can not be physically built. The process itself has to be controllable (Section 3.5.1). This is the same as saying that $A(s)$ and $B(s)$ have no common factors.

There are also other limitations as to how the knobs of the controller can be changed. If the knobs are turned too much (e.g. by speeding up the process response time by a factor of ten) then the control signals would probably saturate and the system would no longer be linear. In other words, since the signals have limited amplitudes the closed loop system response cannot be changed arbitrarily.

9.2.4 Feedforward from Load Changes and Process Disturbances

It is intuitively clear that if we knew the disturbances and could measure them it would be possible to correct for them before they actually influence the system output. Such a feedforward can provide dramatic improvements for regulatory control. A couple of examples will illustrate the idea.

In building temperature control systems there is often a sensor to measure the outdoor temperature. When the outdoor temperature changes, a feedforward correction to the hot water valve controller can be made before the outside temperature has influenced the indoor room temperature. The room temperature is continuously measured and fed back to the controller in order to obtain a final adjustment of the temperature.

In chemical process control a feed flow concentration may be measured. This makes it possible to perform corrective actions in the plant before any change has taken place in the output.

The use of feedforward relies on the possibility of measuring the load change or the disturbance. This is not possible or feasible in many applications. If the disturbance can not be measured directly it is often possible to approximate it by some indirect measurement or some estimation. Phosphorus removal by chemical precipitation in wastewater treatment may serve as an example. In order to obtain the right chemical dosage one ought to know the phosphorus content of the influent flow. In practice, it is difficult and expensive to measure phosphorus concentration on-line. Instead, the feed concentration is based on the flow rate and historic records of normal daily or hourly variations of the phosphorus concentration. This type of feedforward still gives an improvement in control system behaviour.

The typical structure of a feedforward link from a disturbance is illustrated by Figure 9.5. In principle the feedforward controller has to produce a signal that will exactly cancel the disturbance signal to the process. Feedforward is normally used in combination with feedback control. The disturbance $W(s)$ influences the process via the transfer function $G_w(s)$, i.e. there is a dynamic relationship between the disturbance and the output $Y(s)$

$$Y(s) = G_w(s) \, W(s) \tag{9.16}$$

Note, that $G_w(s)$ has to be known and is fixed as it is part of the process properties. The idea of feedforward is to make a correction to the control signal via sensor $G_t(s)$ and the feedforward controller $G_{F2}(s)$. Thus the correction via the feedforward loop to the control signal cancels the change in Y caused by W. This means that:

$$G_t(s) \, G_{F2}(s) \, G_v(s) \, G_p(s) \, W(s) + G_w(s) \, W(s) = 0 \tag{9.17}$$

Solving for G_{F2} this gives the ideal feedforward controller:

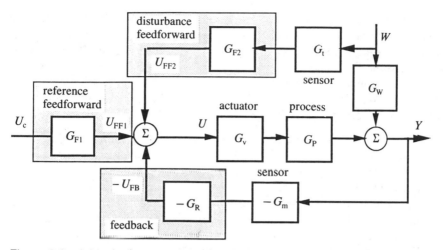

Figure 9.5   A block diagram of the feedforward structure from a disturbance.

$$G_{F2}(s) = -\frac{G_w}{G_t\,G_v\,G_P} \tag{9.18}$$

Note that all the transfer functions in the right hand side are fixed by the process design, so there is no tuning parameter available. In other words, the feedforward signal is completely determined by the system models. If the model is inaccurate, then the feedforward signal would also be inaccurate. In practice, however, the feedforward correction signal may do a good job, even if there is no complete cancellation of the feedforward controller signal and the disturbance signal.

The transfer function of a real physical system is such that the degree of the numerator is larger than the degree of the denominator. For $G_{F2}(s)$ in Equation 9.18, however, the numerator usually has a larger order than the denominator. This means that the disturbance signal has to be differentiated one or more times. This can not be done in practice, so the feedforward control has to be approximated. In computer implementation, a derivative can be approximated by a finite difference so the feedforward signal becomes a function of both the present and the previous values of the disturbance signal.

The feedforward part of the control signal can be written in the form:

$$U_{FF2}(s) = G_{F2}(s)\,G_t(s)\,W(s) = \frac{V_1(s)}{R_3(s)}\,W(s) \tag{9.19}$$

where $V_1(s)$ and $R_3(s)$ are the numerator and denominator polynomials of the transfer function of the feedforward control.

9.2.5 Combination of Feedforward and Feedback

We have noted how heavily the quality of the feedforward control depends on the accuracy of both the disturbance measurements and of the process model. This means that any realistic implementation has to combine feedforward with feedback control. The feedforward action is meant to perform fast corrections due to changes in the reference value or in the disturbance. Feedback gives corrective action on a slower timescale. The real advantage of feedback is that it compensates for inaccuracies in the process model, measurement errors and unmeasured disturbances.

The properties of feedback and feedforward control can be summarized as follows. Feedforward can compensate for some of the limitations of feedback, such as:

- A correction is not made until a deviation occurs in the measured variable. Therefore a perfect control, where the controlled variable does not deviate from the setpoint during load or setpoint changes, is theoretically impossible.
- Feedback can not in a predictive way compensate for known disturbances.
- In systems with long time constants or long time delays, feedback may not be satisfactory. If large and frequent disturbances occur, the process may operate continually in a transient state and never attain the desired steady state.
- If the proper variable can not be measured, feedback is not possible.

The advantage of feedforward is:

- A fast predictive correction can be made, if the disturbance can be measured.

Difficulties with feedforward appear because:

- The load disturbance must be measured on-line. In many applications this is not feasible.
- A model of the process is needed. The quality of the feedforward control depends on the accuracy of the process model.
- The feedforward controller often contains pure derivatives that can not be realized in practice. Fortunately, practical approximations of these ideal controllers often provide effective control.

Feedback is required to complement any feedforward scheme, since:

- Corrective action occurs as soon as the controlled variable deviates from the setpoint, regardless of the source and type of disturbance.
- Feedback requires minimal knowledge of the dynamics of the controlled process, i.e. the process model does not need to be perfectly known.

The controller can now be structured to take care of both feedback signals and feedforward information from the reference value and from process disturbances. Since the systems are linear, all these signals are additive (Section 3.3.3). Referring to Figures 9.3 and 9.4 and Equations 9.8 and 9.19 the control signal into the plant is composed of three terms, the feedforward U_{FF1} from the reference value, the feedback U_{FB} from the output and the feedforward U_{FF2} from the measured disturbance (see Figure 9.5):

$$U(s) = U_{FF1}(s) - U_{FB}(s) + U_{FF2}(s)$$
$$= G_{F1}(s)\, U_c(s) - G_R(s)\, Y(s) + G_{F2}(s)\, G_t(s)\, W(s) \qquad (9.20)$$

$$= \frac{T_1(s)}{R_1(s)} U_c(s) - \frac{S_1(s)}{R_2(s)} Y(s) + \frac{V_1(s)}{R_3(s)} W(s)$$

Expressing the transfer functions with a common denominator, we get:

$$U(s) = \frac{T(s)}{R(s)} U_c(s) - \frac{S(s)}{R(s)} Y(s) + \frac{V(s)}{R(s)} W(s) \qquad (9.21)$$

where $R(s) = R_1 R_2 R_3$, $T(s) = T_1 R_2 R_3$, $S(s) = S_1 R_1 R_3$ and $V(s) = V_1 R_1 R_2$.

9.3 ON/OFF CONTROL

On/off controllers are simple, inexpensive feedback controllers that are commonly used in simple applications such as thermostats in heating systems (cf. Figure 4.18) and domestic refrigerators. They are also used in industrial processes such as simple level control systems or simple dosage controllers for mixers (Section 4.6). For an ideal *on/off* control, the controller output has only two possible values:

$$u = u_{max} \text{ if } e > 0 \qquad (9.22a)$$

$$u = u_{min} \text{ if } e < 0 \qquad (9.22b)$$

where e is the output error (Equation 9.3). *On/off* controllers can be modified to include a deadband for the error signal to reduce the sensitivity to measurement noise. The *on/off* control is also sometimes referred to as **two-position** or **bang-bang control**.

An *on/off* controller causes an oscillation about a constant setpoint, since the control variable jumps between the two possible values. Therefore it produces excessive wear on the final control element. If a valve is used as an actuator this is a significant disadvantage, while it is not a serious drawback if the element is a solenoid switch.

A more advanced type of *on/off* control is used for motor control (Section 4.5.5), where pulse width or other types of modulation are applied to transform the *on/off* control signal to the motor input voltage.

9.4 CONTINUOUS PID CONTROLLERS

9.4.1 The Basic Form of the PID Controller

The **PID controller** is the most common controller structure in process control and in many servo applications. The controller output is the sum of three parts. The first part $u_P(t)$ is proportional to the error of the real system output vs the reference value (setpoint), the second part $u_I(t)$ to the time integral of the error and the third part $u_D(t)$ is proportional to the error derivative. The 'textbook' PID controller looks like:

$$u(t) = u_0 + K \left[e(t) + \frac{1}{T_i} \int_0^t e(\tau)\, d\tau + T_d \frac{de}{dt} \right]$$

$$= u_0 + u_P(t) + u_I(t) + u_D(t) \tag{9.23}$$

Parameter K is the controller gain, T_i the integral time, T_d the derivative time and τ is an integration variable. The value u_0 is a bias value that would give the controller its proper average signal amplitude.

Some controllers, especially older models, have a **proportional band** setting instead of a controller gain. The proportional band PB (in per cent) is defined as $PB = 100/K$. This definition applies only if K is dimensionless. In many systems it is desirable to use engineering units. For example, if the measurement is a flow rate measured in $m^3 s^{-1}$ and the control signal is expressed in volts, then the gain is not dimensionless.

A textbook controller does not show any physical limits on its output. A real controller saturates when its output reaches a physical limit, either u_{max} or u_{min}. In practice, the output of a proportional controller resembles Figure 9.6. If the proportional controller has a very high gain it behaves like an *on/off* controller.

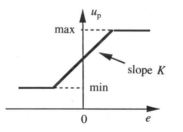

Figure 9.6   Proportional control: actual behaviour.

The integral part of the controller is used to eliminate steady-state errors. Its function can be explained in an intuitive way. Assume that the system is in a steady state so that all the signals are constant, particularly $e(t)$ and $u(t)$. The steady state can only remain if the integral part $u_I(t)$ is constant, otherwise $u(t)$ would change. This can only happen if $e(t)$ is zero.

Note that the integral time coefficient appears in the denominator of Equation 9.23. This makes the dimensions of the controller terms proper and has a practical interpretation. To see this, consider a step change of the error $e(t)$ and its response in a PI (proportional-integral) controller. Immediately after the step the controller output is Ke. After time T_i, the controller output has doubled (Figure 9.7). A *PI* controller is often symbolized by its step response.

The controller can also be described by its Laplace transform. Considering the three terms in Equation 9.23 we obtain:

$$U(s) - U_0(s) = \delta U(s) = U_P(s) + U_I(s) + U_D(s)$$

$$= K \left[1 + \frac{1}{T_i s} + T_d s \right] E(s) = K \frac{1 + T_i s + T_i T_d s^2}{T_i s} E(s) \tag{9.24}$$

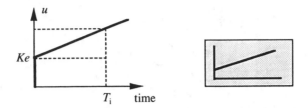

Figure 9.7   Step response of a continuous PI controller and its
process scheme symbol.

where $E(s)$ is given from Equation 9.3, and $U_P(s)$, $U_I(s)$ and $U_D(s)$ are the transforms
of the control signal components $u_P(t)$, $u_I(t)$ and $u_D(t)$, respectively. Note that the degree
of the numerator is less than that of the denominator and the controller gain grows to infinity
for large frequencies. This is a consequence of the derivative term. In practice a derivative
can not be realized exactly, but is approximated by a first-order system with a time constant
T_f:

$$\delta U(s) = U_P(s) + U_I(s) + U_D(s) = K \left[1 + \frac{1}{T_i s} + \frac{T_d s}{1 + T_f s} \right] E(s) \qquad (9.25)$$

Often the filter time constant is normalized to the derivative time:

$$T_f = \frac{T_d}{N} \qquad (9.26)$$

where N is in the order of $5-10$. The gain of the derivative part of the controller (Equation
9.25) then is limited to KN for large frequencies.

The PID controller (Equation 9.25) can be rewritten in the form:

$$T_i s \, (1 + T_f s) \, \delta U(s) = K[T_i s \, (1 + T_f s) + 1 + T_f s + T_i T_d s^2] E(s)$$

This is a special case of the general controller (Equation 9.11). Dividing with $T_i T_f$ it is
obvious that the PID controller can be written in the form (Equation 9.11) with:

$$R(s) = s^2 + \frac{1}{T_f} s \qquad (9.27)$$

$$S(s) = T(s) = K \left(1 + \frac{T_d}{T_f} \right) s^2 + K \left(\frac{1}{T_f} + \frac{1}{T_i} \right) s + \frac{K}{T_i T_f} \qquad (9.28)$$

9.4.2 Derivative of the Measurement Only

In many operating situations the setpoint has abrupt changes from time to time and is
constant in between. A step change in the setpoint value results in a large controller output,
which is sometimes called a 'derivative kick'. This is illustrated in the PID controller step
response, that suggests the symbol that often appears in process control schemes (Figure
9.8).

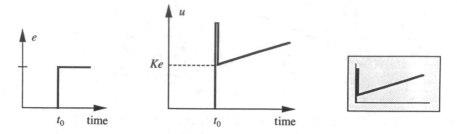

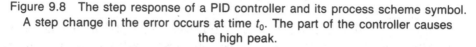

Figure 9.8 The step response of a PID controller and its process scheme symbol. A step change in the error occurs at time t_0. The part of the controller causes the high peak.

In order to avoid the derivative kick in the PID controller, the derivative term is instead based only on the measurement $y(t)$. Since the error derivative can be written:

$$\frac{de}{dt} = \frac{du_c}{dt} - \frac{dy}{dt}$$ (9.29)

we see that the setpoint changes are not calculated by the controller. (It is better to take care of them with a feedforward controller, Section 9.2.2.) The ideal controller becomes:

$$\delta u(t) = K\left[e(t) + \frac{1}{T_i} \int_0^t e(\tau)\, d\tau - T_d \frac{dy}{dt} \right]$$ (9.30)

The derivative is again approximated by a first-order system with time constant T_f:

$$\delta U(s) = K\left(1 + \frac{1}{T_i s}\right) E(s) - K \frac{T_d s}{1 + T_f s} Y(s)$$ (9.31)

This method of eliminating the derivative kick has become a standard feature in most commercial controllers.

9.4.3 Series Representation

Since the PID transfer functions (Equations 9.25 or 9.31) consist of the sum of three terms these can be considered a parallel connection of proportional, integral and derivative actions. The controller can also be written in a *series* form, which can be interpreted as a series connection of a PI controller with a PD (proportional-derivative) controller of the form:

$$G_{PID}(s) = \frac{\delta U(s)}{E(s)} = K'\left(1 + \frac{1}{T_i' s}\right)\left(\frac{1 + T_d' s}{1 + T_f s}\right)$$ (9.32)

The transformation from the parallel to the series form is possible if:

$$T_f \ll T_d \ll T_i$$ (9.33)

The controller gain as a function of frequency is shown in Figure 9.9 and approaches

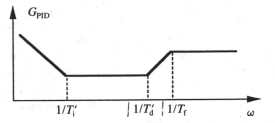

Figure 9.9 The Bode plot of the gain of a series form of a PID controller.

$K' T_d'/T_f$ for high frequencies. From the graph of Figure 9.9 it appears that the PID controller is a combination of a low pass filter in series with a high pass filter (see Section 5.4). This configuration is also called a **lead-lag filter**.

9.4.4 PIPI Controllers

A **PIPI** controller consists of two PI controllers in series, or a PI controller in series with a low pass filter. PIPI controllers are sometimes used in electrical drive systems. The purpose of the extra low pass filter is to limit high frequency signals. In mechanical drive systems there may be resonance oscillations that are suitably dampened by such a filter. The controller transfer function is:

$$G_F(s) = K\left(\frac{1 + T_i s}{T_i s}\right) \left(\frac{1 + T_3 s}{1 + T_2 s}\right) \tag{9.34}$$

where T_i is the integral time and $T_i > T_2 > T_3$. T_3 is usually chosen to coincide with the resonance frequency. The gain as a function of frequency is shown in Figure 9.10.

9.4.5 Other Parametrizations of the PID Controller

In some references a PID controller is parameterized by:

$$\delta u(t) = K_P\, e(t) + K_I \int_0^t e(\tau)\mathrm{d}\tau + K_D \frac{\mathrm{d}e}{\mathrm{d}t} \tag{9.35}$$

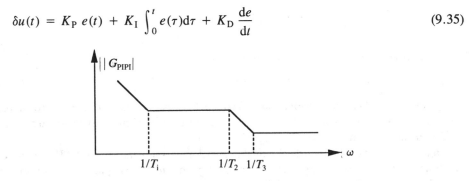

Figure 9.10 The Bode plot gain of the PIPI controller.

This parametrization is equivalent to the form of Equation 9.23. There is an important practical reason why the form of Equation 9.35 is not common practice. In the 'classic' PID controller, Equation 9.23, it is very useful to change the gain of the whole controller by just one parameter K, particularly during start-up or tuning procedures. Moreover, consider the Bode plot (Figure 9.9). In a textbook controller, the shape of the Bode diagram is preserved when K is changed so the gain is altered equally for all frequencies. In the parametrization (Equation 9.35) both the gain and the break points in the Bode diagram will change for any parameter modification.

In the ideal controller the three parameters could be tuned independently, but in practice there are interactions due to the electronic circuitry between the control modes on analog standard PID controllers. Therefore the effective values may differ from the nominal values by as much as 30 per cent. In contrast, in digital control systems the controller settings can be specified as accurately as desired, with no interaction between the terms.

9.5 DISCRETIZATION OF THE PID CONTROLLER

The controller in a digital system must be discretized at some stage in order to be implemented on a computer. In Section 5.5 we discussed the discretization of analog low pass and high pass filters to digital filters. Here the continuous controller will be discretized into a digital form that suits the computer implementation.

In continuous controller design, the controller itself is discretized. Given a sufficiently short sampling interval, the time derivatives can be approximated by a finite difference and the integral by a summation. This approach will be used here (compare with Section 3.4). We will consider one term at a time.

The error is calculated at each sampling interval

$$e(kh) = u_c(kh) - y(kh) \tag{9.36}$$

The sampling period h is assumed to be constant and the signal variations during the sampling interval are neglected (cf. Section 5.2.3).

9.5.1 The Position Form

The time-discrete form of the PID controller is:

$$u(kh) = u_0(kh) + u_P(kh) + u_I(kh) + u_D(kh) \tag{9.37}$$

This is also called the **position form** since $u(kh)$ is the absolute value of the control signal. The control signal offset u_0 has to be added to the control signal. Even for zero error there is usually a non-zero control signal.

In this section we present only the final result of the discretization of the PID controller, while the details of the derivation are left in the appendix to this chapter. The proportional part of the controller (Equation 9.25) is:

$$u_P(kh) = Ke(kh) \tag{9.38}$$

The integral is approximated by finite differences and is given by:

$$u_I(kh) = u_I(kh-h) + K\frac{h}{T_i} e(kh) = u_I(kh-h) + K\alpha e(kh) \tag{9.39}$$

where

$$\alpha = \frac{h}{T_i} \tag{9.40}$$

The integral part forms a recursive expression, i.e. it is updated at every sampling interval. Note that the last term may be small if h is small and T_i is large. Therefore the word length has to be sufficiently large, so that the term $K\alpha$ can be represented with sufficient precision.

The derivative part is also approximated by finite differences,

$$u_D(kh) = \beta u_D(kh-h) - K\frac{T_d}{h}(1-\beta)[y(kh)-y(kh-h)] \tag{9.41}$$

where

$$\beta = \left(1 + \frac{hN}{T_d}\right)^{-1} = \frac{T_d}{T_d + hN} \tag{9.42}$$

Notice that $0 \leq \beta < 1$. If the filter time constant $T_f \rightarrow 0$ (i.e. $N \rightarrow \infty$) then $\beta \rightarrow 0$ and the derivative action becomes a simple difference approximation of the time derivative of the output signal. For $T_d = 0$ we also get $\beta = 0$ which results in $u_D(kh) = 0$, i.e. no derivative action.

9.5.2 The Incremental Form

An alternative approach is to use an **incremental (velocity) form** of the algorithms in which the change in the control output is calculated. If the control output $u(kh-h)$ is subtracted from $u(kh)$, the controller can be written in the form:

$$\Delta u(kh) = u(kh) - u(kh-h) = \Delta u_P(kh) + \Delta u_I(kh) + \Delta u_D(kh) \tag{9.43}$$

The incremental form of the proportional part is easily calculated from Equation 9.38 and the integral part from Equation 9.39:

$$\Delta u_P(kh) = u_P(kh) - u_P(kh-h) = K[e(kh) - e(kh-h)] = K\Delta e(kh) \tag{9.44}$$

$$\Delta u_I(kh) = u_I(kh) - u_I(kh-h) = K\alpha e(kh) \tag{9.45}$$

The incremental form of the filtered derivative part is found from Equation 9.41

$$\Delta u_D(kh) = \beta \Delta u_D(kh-h) - K\frac{T_d}{h}(1-\beta)[\Delta y(kh) - \Delta y(kh-h)] \tag{9.46}$$

where

$$\Delta y(kh) = y(kh) - y(kh-h) \qquad (9.47)$$

The incremental form of the PID controller is useful when the actuator is some kind of adder, such as a stepping motor. From a computing point of view, the calculations are quite simple and simple floating point precision can normally be used. The incremental form of the controller does not exhibit *windup* problems (see Section 9.6). In switching from manual to automatic mode the incremental controller does not require any initialization of the control signal (u_0 in the position form). Presumably the final control element has been placed in the appropriate position during the start-up procedure.

A small disadvantage of the incremental form is that the integral term must be included. Note that the setpoint cancels out in both the proportional and derivative terms, except in one sampling interval following a setpoint change. Therefore, if the incremental form is applied without the integral term it is likely that the controlled process will drift away from the setpoint.

9.6 THE PRACTICAL IMPLEMENTATION OF THE CONTROLLER

9.6.1 Sampling Rate in Control Systems

The choice of sampling rates of continuous signals was discussed in Section 5.2.3. It is not trivial to choose a suitable sampling rate for control; in fact, finding the right sampling frequency still remains more of an art than a science. Too long a sampling period can reduce the effectiveness of feedback control, especially its ability to cope with disturbances. In an extreme case, if the sampling period is longer than the process response time, then a disturbance can affect the process and will disappear before the controller can take corrective action. Thus, it is important to consider both the process dynamics and the disturbance characteristics in selecting the sampling period.

There is an economic penalty associated with sampling too frequently, since the load to the computer will increase. Consequently in selecting the sampling period one has to consider both the process dynamics and the computer capacity at disposal (see Sections 5.2.3 and 10.1). Commercial digital controllers which handle a small number of control loops (e.g. 8–16) typically employ a fixed sampling period of a fraction of a second. Thus the performance of these controllers closely approximates continuous (analog) controllers.

The signal-to-noise ratio also influences the selection of the sampling period. For low signal-to-noise ratios, fast sampling should be avoided because changes in the measured variable from one sampling interval to the next will be due mainly to high frequency noise rather than to slow process changes.

In pure signal processing the purpose is to sample a signal with a computer and to recover it from the time-discrete form. The sampling theorem does not take the computational time into consideration, so that reconstruction from the sampled signal may take a long time. Note that the signal is assumed to be **periodic** (Section 5.2.3). In control applications the signals are usually *not* periodic and the computational time for reconstruction is limited. Therefore the sampling time has more constraints. Many additional rules for the choice of the sampling rate can be found in the control literature.

It is reasonable to assume that the sampling rate is related to the closed loop system bandwidth or the rise time of the closed loop system. Some rules of thumb claim that the sampling rate should be some $6-10$ times the bandwidth or $2-4$ samples per rise time.

In the previous discussion the controller design has been based on continuous system descriptions. One way to calculate a suitable sampling time is to consider the closed loop system as a continuous system with a zero-order sample-and-hold circuit. Such a circuit can be approximated by a time delay of half a sampling interval (cf. Section 5.2.1), which corresponds to a phase lag of $0.5h\omega_c$ radians, where ω_c is the bandwidth (crossover frequency) of the system. If an additional phase shift of $5°-15°$ ($0.09-0.26$ radians) is accepted as a result of the hold circuit we get the rule:

$$h\omega_c \approx 0.15-0.5 \tag{9.48}$$

Usually this rule gives quite high sampling rates causing the Nyqvist frequency to be significantly larger than the closed loop system crossover frequency. This is true in many commercial one-loop or multi-loop PID controllers.

9.6.2 Control Signal Limitations

The controller output value has to be limited, for at least two reasons. The desired amplitude shall not exceed the DAC range, and can not become larger than the actuator range. A valve cannot be more than 100 per cent fully open, or a motor current has to be limited. Thus, the control algorithm needs to include some limit function.

In several control loops it is important to introduce some dead band. If an incremental controller is used, each increment may be so small, that it is not significantly larger than other disturbances. It is of interest not to wear out the actuators. Consequently the control variable increments are accumulated until the control signal reaches a certain value. Naturally the deadband has to be larger than the resolution of the DAC.

9.6.3 Integral Windup

Windup occurs when a PI or a PID controller encounters a sustained error, for example, a large load disturbance that is beyond the range of the control variable. A physical limitation of the controller output makes it more difficult for the controller to reduce the error to zero.

If the control error has the same sign for a long time, the integral part of the PID controller will be large. This may happen if the control signal is limited. Since the integral part can become zero only some time after the error has changed sign, integral windup may cause large overshoots. Note that windup is a result of a non-linear element (the limiter) in the control circuit and can never occur in a truly linear system.

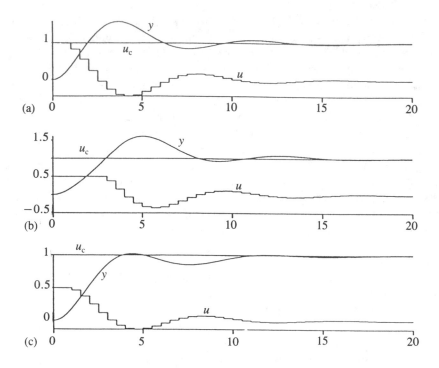

Figure 9.11   Illustration of windup problems of a position servo with PI control.
(a) A step response where there is no actuator limitation so no windup occurs;
(b) the control signal is limited, but no anti-windup is provided; (c) shows the step
response after an anti-windup term is added to the controller. Note the control system
amplitudes in all three cases.

An example will illustrate the problem. Consider a position control system, where a d.c. motor is controlled by a PI controller. The position reference is changed so much that the control signal (the voltage input to the motor) is saturated and limited (i.e. the acceleration of the motor is limited). The step response of the angular position is shown in Figure 9.11.

The integral part of the PI controller is proportional to the area between the step response y and the reference value u_c. The areas provide either positive or negative contributions to the integral term depending on whether the measurement is below or above the setpoint. As long as the error $u_c(t) - y(t)$ is positive the integral term increases. Similarly, as long as there is no control signal limitation there is no windup. When the control signal is limited (Figure 9.11(b)), then the response is slower and the integrator will increase until the error changes sign at $t = t_1$. Even if the control error changes sign, the control signal is still large and positive which leads to the large overshoot of $y(t)$.

One way to limit the integral action is by **conditional integration**. The basic rule is that the integral part is not needed when the error is sufficiently large. Then the proportional part is adequate, while the integral part is not needed until the error is small.

In that instance it is used to remove the steady-state errors. In conditional integration the integral part is only executed if the error is smaller than a prescribed value. For large errors the PI controller acts like a proportional controller. It is not trivial, however, to find in advance the proper error.

In analog controllers conditional integration can be achieved by a zener diode that is coupled parallel to the capacitor in the feedback loop of the operational amplifier in the integration part of the controller. This will limit the contribution from the integral signal.

In digital PID controllers there is a better way to avoid windup. The integral part is adjusted at each sampling interval so that the controller output will not exceed its limits. A PI controller with an anti-windup feature can be described by the following Pascal code:

```
(*initial calculation*)
cl := K*h/Ti;
(*controller*)
 . . .
e := uc − y;
Ipart :− Ipart + cl*e;
v := K*e + Ipart;            (*calculation of the desired control signal*)
u := ulim (v, umin, umax);   (*the function ulim limits the control signal v
                                between the maximum and minimum values*)
Ipart := u − K*e;            (*anti-windup correction of the integral part*)
 . . .
```

If the control signal v stays within the limits, then the integral part will not be changed by the last statement.

To obtain anti-windup for a PID controller the method has to be modified slightly. The integral part is updated with the signal $e_s = u - v$ that describes the difference between the real actuator output u and the desired controller output v. The actuator output can either be measured or calculated from a model. Note that e_s is zero if the actuator produces the desired control signal so that no saturation takes place. The signal e_s is multiplied with a gain $1/T_t$, where T_t is a time constant (called a tracking time constant) for the integral part reset. In the PI controller algorithm shown above, this time constant is equal to h, i.e. an immediate update in the next sampling interval. When a derivative part is used, however, it is advisable to update the integral more slowly. A practical value may be T_t equal to the integral time T_i. The desired control output is then:

$$v(t) = u_P + u_I = K\left[e(t) + \frac{1}{T_i} \int_0^t e(\tau)\, d\tau\right] + \frac{1}{T_t} \int_0^t [u(\tau) - v(\tau)]d\tau \qquad (9.49)$$

where: $v = u_{min}$ if $u \leq u_{min}$
         $v = u$ if $u_{min} \leq u \leq u_{max}$
         $v = u_{max}$ if $u \geq u_{max}$

If the control signal is saturated then the difference $u - v$ will cause the integral part to change until there is no more saturation. Consequently windup is avoided. By taking the derivative of the integral part we obtain:

$$\frac{du_I}{dt} = \frac{K}{T_i} e + \frac{1}{T_t} (u - v) \tag{9.50}$$

which is discretized as:

$$u_I(kh+h) = u_I(kh) + h \frac{K}{T_i} e(kh) + \frac{h}{T_t} [u(kh) - v(kh)] \tag{9.51}$$

and the PI controller is then:

$$v(kh) = Ke(kh) + u_I(kh) \tag{9.52}$$

where $u_I(kh)$ is given by Equation 9.51. Note that the integration has been approximated by forward differences instead of backward differences. This is necessary since $v(kh)$ has to be known before the integral part can be calculated.

9.6.4 Bumpless Transfer

When a controller is shifted from manual to automatic mode the controller output may jump to another value even if the control error is zero. The reason is that the integral part of the controller is not zero. The controller is a dynamic system and the integral part represents one state that has to be known at all regulator mode changes. The sudden jump of the controller output can be avoided and the transfer is then called **bumpless**. We consider two situations:

* Shifting between manual and automatic mode
* Changing regulator parameters

To achieve bumpless transfer for an analog controller going from manual into automatic mode, the control error can be made zero by manually controlling the system until the measurement value is brought to the setpoint. Then the integral part is reset (brought to zero), and since the error is zero a bumpless transfer is obtained. The same procedure can be used for digital controllers.

Another method is to slowly bring the setpoint value up to its target value. Initially it is set equal to the actual measurement value and is gradually adjusted. If sufficient time is allowed then the integral part will be so small that the transfer is bumpless. However, this method may be too slow for many applications.

The PID controller in velocity mode (Equation 9.43) does not need to be initialized at a mode transfer. The operator sets the actuator to a desired position *before* switching from manual to automatic. Then the controller does not change the actuator until an error occurs. It should be emphasized that it is often important to store the actual control signal even when the incremental form of the controller is used. The limits may have to be checked, or the signal value will be used for other reasons.

In digital PID controllers there is yet another way to obtain bumpless transfer. The control algorithm is executed even in manual mode. The measurement y is read into the computer and the error is calculated, which keeps the integral part updated. If the controller is put into automatic mode (and the setpoint is right) the transfer will be bumpless.

The main feature in all bumpless transfer schemes is to update the integral part of the controller to such a value that the control signal is the same immediately *before* and *after* the switching.

Consider a parameter change of a PID controller. Immediately before the change the regulator output can be written as (Equations 9.37–9.42),

$$u(t-) = u_P(t-) + u_I(t-) + u_D(t-) \tag{9.53}$$

and immediately after the parameter change it is:

$$u(t+) = u_P(t+) + u_I(t+) + u_D(t+) \tag{9.54}$$

The parameter changes will influence any of the regulator terms. We want to make $u(t-)$ = $u(t+)$ to obtain bumpless transfer. To do this we can update either the integral part state (Equation 9.39) or the derivative state (Equation 9.41). We choose to update the integral part which gives:

$$u_I(t+) = u_P(t-) + u_I(t-) + u_D(t-) - u_P(t+) - u_D(t+) \tag{9.55}$$

making a bumpless transfer with the difference $u(t+)-u(t-)$ equal to zero.

9.6.5 Rate-of-change Limiting Circuit

In many systems it is necessary to include circuits to limit the amplitude or the rate of change of the command signal. A manual change $u_c(t)$ of the setpoint can be supplied with a limiting protection, so that the process will see the command signal $u_L(t)$ instead.

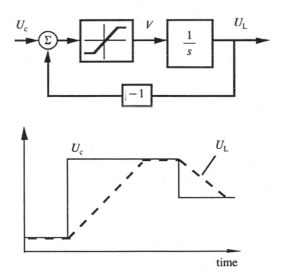

Figure 9.12   Rate-of-change limiter and its effect on a stepwise change of a command signal. A typical response to abrupt set-point changes is shown.

This is common practice, for example, in electric drive systems. A rate limiting function is obtained by a simple feedback system (Figure 9.12). A typical transient is also shown in the figure.

The manual command signal $u_c(t)$ is compared with the permitted command signal $u_L(t)$. The difference is first limited between the values u_{min} and u_{max}. Then the value is integrated. The integral is approximated as a finite sum. The algorithm of the rate-of-change limiter can be written as:

> (*initial calculations*)
> e(kh) := uc(kh) − uL(kh);
> v(kh) := ulim [e, umin, umax];   (*the function ulim limits the error signal e
>   between umin and umax*)
>
> uL(kh) := uL(kh − h) + h*v(kh);
> . . .

9.6.6 Computational Aspects

When discretizing the PID controller, the sequential nature of the computation causes a delay that is not present in the analog version. Also, the anti-windup and bumpless transfer computer schemes depend on the fact that the actuator and controller outputs are executed simultaneously. Therefore it is important to make computational delays as small as possible. Some terms in the controller can be computed before the actual sampling time. Consider the integral part with anti-windup features added (Equation 9.51). The integral part can be written as:

$$u_1(kh+h) = u_1(kh) + c_1 e(kh) + c_2[u(kh) - v(kh)] \tag{9.56}$$

where

$$c_1 = \frac{Kh}{T_i} \qquad c_2 = \frac{h}{T_t} \tag{9.57}$$

Note, that the integral action can be precomputed when *forward* differences are used. The derivative action (Equation 9.41) can be written as

$$u_D(kh) = \beta u_D(kh-h) - K \frac{T_d}{h}(1-\beta)[y(kh) - y(kh-h)]$$

$$= -K \frac{T_d}{h}(1-\beta) y(kh) + \beta u_D(kh-h) + K \frac{T_d}{h}(1-\beta) y(kh-h) \tag{9.58}$$

This can be written in the form:

$$u_D(kh) = -c_3 y(kh) + x(kh-h) \tag{9.59}$$

where

$$c_3 = K \frac{T_d}{h}(1-\beta) \tag{9.60}$$

$$x(kh-h) = \beta u_D(kh-h)+K\frac{T_d}{h}(1-\beta)\,y(kh-h)$$

$$= \beta u_D(kh-h) + c_3 y(kh-h)$$

The state x can be updated immediately after time kh

$$x(kh) = \beta u_D(kh) + c_3 y(kh)$$
$$= \beta[-c_3 y(kh) + x(kh-h)] + c_3 y(kh)$$
$$= \beta x(kh-h) + c_3(1-\beta)\,y(kh) \tag{9.61}$$

Thus $u_D(kh+h)$ can be calculated from Equation 9.59 once the measurement $y(kh+h)$ has been obtained.

The temporary parameters $c_1 - c_3$ have no obvious interpretation. Therefore the basic PID parameters K, T_i, T_d and T_f have to be shown to the operator.

The precision of the calculations should be noted. In the incremental PID algorithm most of the computations are done in incremental terms only, so that a short word length is adequate. In the final calculation, however, the precision has to be higher. The round-off problems in the integral part have been commented on in Section 9.5.

9.6.7 Final Algorithm

A software code for the PID controller is now given. The computation of the controller coefficients c_1-c_3 is done only when some of the controller parameters K, T_i, T_d and T_f are changed. The control algorithm is executed at each sampling interval. An anti-windup feature is added to the integral term. The core of the PID algorithm is as follows.

```
(*calculation of parameters*)
c1 := K*h/Ti;                 (*Equation 9.57*)
c2 :- h/Tt;                   (*Equation 9.57*)
beta := Td/(Td+h*N);          (*Equation 9.42*)
c3 := K*Td*(1-beta)/h;        (*Equation 9.60*)
c4 := c3*(1-beta);

(*control algorithm*)
uc := Ad_input(ch1);          (*read setpoint, analog input*)
y := Ad_input(ch2);           (*read measurement, analog input*)
e := uc-y;                    (*calculate control error*)
p := K*e;                     (*calculate proportional part*)
dpart := x-c3*y;              (*calculate derivative part of Eqn 9.59*)
v := p+ipart+dpart;           (*compute output before limitation*)
u := ulim(v, umin, umax);     (*the function ulim limits the controller
                                 output between umin and umax*)

DA_out(ch1);                  (*analog output, Channel 1*)
```

ipart := ipart+c1*e+c2*(u−v);   (*integral part with anti-windup Eqn 9.56*)
x := beta*x+c4*y;   (*update state Equation 9.61*)
. . .

A commercial digital PID controller looks like the one in Figure 9.13. The figure shows the panel. On the front there are displays of the reference and measurement values. The buttons allow easy switching between manual and automatic mode, and the buttons for manual increase and decrease of the setpoint value are obvious.

9.6.8 A Block Language Implementation

The control algorithms can of course be described in any sequential language. However, in practical implementations more process-oriented high level languages are common. As for sequential control (Sections 8.3 and 8.4), it is common practice to represent controller functions in blocks, where only the input and output signals are marked but the algorithm itself is hidden. The parameters are available for tuning. Figure 9.14 shows a typical structure for PID controllers. The programmer will see the symbols on the screen and

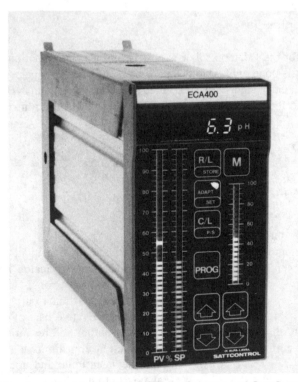

Figure 9.13   An industrial PID controller. (Courtesy of SattControl, Sweden).

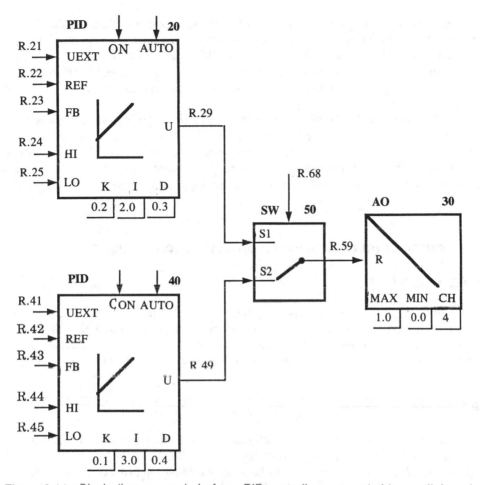

Figure 9.14   Block diagram symbol of two PID controllers connected to a switch and
an analog output unit.

has to label the inputs and outputs with proper variable names to connect the controllers
to other elements.

The diagram shows two PID controllers connected to a switch. One of the two regulator
outputs will be chosen by a binary signal to the switch and sent to the analog output unit.
The *auto* input is a binary variable for manual/automatic mode. The reference value is
sent into REF while the measurement is connected to FB (feedback). The control signal
limits are marked by the two parameters *HI* and *LO*. The tuning parameters *K*, *I* and *D*
correspond to the gain, integral time and derivative time, and their values are displayed
on the bottom line of the symbol. The analog output unit is defined by its range and channel
number.

We note that the control system code is written in a format similar to logical gates.
In most industrial process computer software the controller blocks are incorporated in

the logical programming for easy addition of different logical conditions for the signals. In a modern package there are blocks not only for logical sequences and controllers. The user may define his own blocks from algebraic or difference equations. Thus it is possible to incorporate any of the more general controller structures described in the following sections. The logical programming is further structured by using function charts for sequencing.

The modules may be of different types. A special type of library module can be defined and then copied from the library to several instances. This is reminiscent of reentrant coding (Section 7.6.2), but on a more advanced level, where the user only needs to concentrate on the definition of the module. A system may have a standard library that can be extended with user-defined libraries. Other simpler modules can only appear as single copies.

9.7 CONTROL STRUCTURES BASED ON SIMPLE CONTROLLERS

9.7.1 External setpoints

In the previous discussion, the setpoint u_c has been given explicitly. The setpoint can be read from any register in the computer. The operator can enter the setpoint either via an analog input or a command from the keyboard. Another possibility is to obtain the setpoint from the output of another controller. This is the case in **cascade control**.

9.7.2 Cascade Control

A disadvantage of conventional feedback control is that the correction for disturbances does not begin until after the process output deviates from the setpoint. As discussed in Section 9.2, feedforward control offers large improvements in processes with large time constants or time delays. However, feedforward control requires that the disturbances be measured explicitly and a model must be available for calculating the controller output.

By using a secondary measurement and secondary feedback controller, the dynamic response to load changes may be greatly improved. The secondary measurement point is located so that it recognizes the upset condition sooner than the process output, but the disturbances are not necessarily measured. This is the essence of cascade control. It is particularly useful when the disturbances are associated with the manipulated variable or when the actuator exhibits non-linear behaviour (such as a non-linear valve or the electrodynamics of a motor).

Example 9.1 Control of electric drive systems

A control system for position and velocity control of an electrical drive system was presented in Section 4.5.2. The cascade structure (see Figure 4.19) is the common standard for the

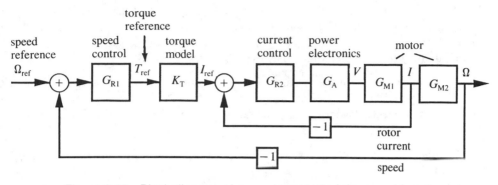

Figure 9.15 Block diagram of cascade control of the speed in an electrical drive system.

control of electric drive systems. In principle the velocity could be controlled by a simple controller, measuring the velocity error and then giving the motor such a voltage that the velocity is corrected. Such a controller, however, would be impractical and extremely complex, because it has to take a large number of features into consideration. The cascade control properties are further illustrated from the block diagram in Figure 9.15.

The velocity controller G_{R1} computes an output signal that is the torque needed to accelerate the motor to the desired speed. The desired current, I_{ref}, that the motor needs to produce the torque is calculated from a mathematical model of the motor. We denote this model simply by a gain K_T, which is adequate for d.c. motors.

The inner loop controls the current that is needed to produce the torque. The output of the controller G_{R2} is a command to the power electronics unit that produces the necessary voltage input to the motor.

Let us calculate the transfer function from the rotor current setpoint I_{ref} to the rotor current I. The power electronics and the electrical part of the motor are represented by the transfer functions G_A and G_{M1}, respectively (the real system is not linear but we can still illustrate the qualitative point). The transfer function G_I of the inner loop is:

$$G_I(s) = \frac{I(s)}{I_{ref}(s)} = \frac{G_{R2} G_A G_{M1}}{1 + G_{R2} G_A G_{M1}} \tag{9.62}$$

If the gain of G_{R2} is large, then the transfer function G_I will approach one and will also be quite insensitive to variations in the amplifier or motor transfer functions. Non-linear behaviour of the motor or amplifier can often be modelled by transfer functions with variable coefficients.

From the output of the velocity controller there are now three quite simple systems in series, the gain K_T, the current control loop G_I, where G_I is close to unity, and the mechanical part G_{M2} of the motor. Thus we can see that the cascade structure eliminates many of the inherent complexities in the power electronics and motor dynamics.

The rotor current feedback serves yet another purpose. Since the rotor current has to be limited, the inner loop functions as a current limiter.

The cascade structure is suitable also for the **commissioning** (first start-up) of the control system. Initially the inner loop is tuned. This tuning does not need to be changed when the outer loop is tuned. Since the inner loop has simplified the dynamics, tuning of the outer loop is much easier. For position control another loop is added outside the velocity loop (see Figure 4.19) and the tuning can proceed in the same manner.

In summary the cascade control has two distinctive features:

- The output signal of the master (primary) controller serves as a setpoint for the slave (secondary) controller.
- The two feedback loops are nested, with the secondary loop located inside the primary control loop.

Note that the dynamics of the secondary loop has to be significantly faster than that of the primary loop.

Windup in cascade control systems needs special attention. The anti-windup for the secondary controller can be solved as shown in Section 9.6.3. To avoid windup in the primary controller, however, one has to know when the secondary controller saturates. In some systems the primary controller is set to manual mode when the secondary controller saturates. For computer control one has to consider the execution order of the two regulators. Due to the different speeds of the control loops they may have different sampling rates. The sampling time for the primary control loop may be significantly longer. If the primary regulator is executed first, then it can present an updated setpoint for the secondary controller. Otherwise, the secondary controller may be fed with an unnecessary old setpoint value. Note that the outer loop controller may receive its setpoint from the operator or some predefined value. The output from the primary controller is not sent to the computer output but to the data area where the secondary controller can pick it up as its setpoint value.

9.7.3 Selective Control

In many process control problems there are more measurements (controlled variables) than manipulated variables. Thus it is impossible to eliminate errors in all the controlled variables for arbitrary setpoint changes of disturbances by using only simple (single-input/single-output) controllers. Then selectors are used to share the manipulated variables among the controlled variables. The **selector** is used to choose the appropriate measurement variable from among a number of available measurements. They are used both to improve the system operation and to protect the system from unsafe operation.

One type of selector device has an output that is the highest (or lowest) of two or more input signals. On instrumentation diagrams the symbol *HS* denotes high selector and *LS* a low selector. For example, a high selector can be used to detect a hot spot temperature. If several temperature transmitters are connected to the selector it chooses the maximum temperature and feeds it to the temperature controller. A median selector calculates an average temperature from several sensors and provides increased reliability due to redundant measurements.

The use of high or low limits for process variables are another type of selective control, called an **override**. At these limits the normal controller operation is overridden by alarm procedures. Anti-windup features in controllers is another type of override.

In summary, selectors are non-linear elements that can be added to the control loops and are easily implemented in digital systems.

9.8 PERFORMANCE LIMITS FOR PID CONTROLLERS

The PID controller is applied successfully to most control problems in process control, electrical drive systems and servo mechanisms. The reason for its success is that most of these processes have a dynamic behaviour that can be adequately approximated by a second-order process. The PID controller is insufficient to control processes with additional complexities such as

- time delays
- significant oscillatory behaviour (complex poles with small damping)
- parameter variations
- multiple-input multiple-output systems

9.8.1 Time Delays

Time delays often occur in process industries because of the presence of distance lags, recycle loops or the dead time associated with composition analysis. The time delay makes information from the true process variable change arrive later than desired to the controller. Generally speaking, all information that is too old causes problems ('the right data too late is wrong data'). They limit the performance of the control system and may lead to system instability.

Any system with time delay that is controlled by a PID controller usually behaves quite sluggishly. The reason is that the gain has to be quite small so that instability is not risked. This can be illustrated by a simple example.

Example 9.2 Control of a system with time delay

Consider a system of two identical chemical mixing tanks in series. The tank is described in Section 3.2.3 (Example 3.7, Figure 3.12). The concentration c of the effluent chemical is measured (y) but the measurement procedure takes a time T. The influence of the delay time is illustrated by PI control of the concentration (Figure 9.16). The PI controller has been tuned to behave well, as if there were no time delay. We note that the closed loop system has the transfer function:

$$\frac{Y(s)}{U_c(s)} = \frac{G_R G_P e^{-sT}}{1 + G_R G_P e^{-sT}} \tag{9.63}$$

where G_R is the regulator transfer function, G_P the process transfer function and e^{-sT} the transfer function of the measurement time delay.

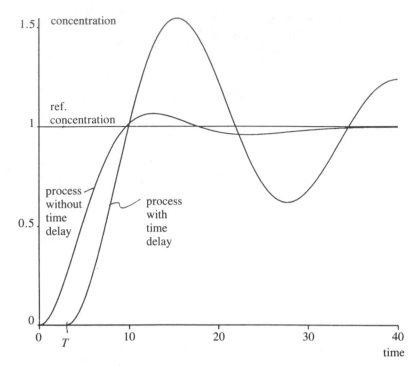

Figure 9.16   A PI control of the chemical dosage with a time delay of 3 time units.

It is intuitively clear that it is difficult to control the system with a simple controller. In the example, the concentration is found to be too low at time 0. The controller then increases the dosage to correct the concentration. Any change due to the control action at time 0 will not be seen until time 3. Since the controller has not recorded any correction at time 1 it increases the dosage further, and continues to do so at time 2. The result is first observed at time 3. If the gain of the controller is large, the concentration change may have been too large. Consequently the controller will decrease the dosage, but will not see the result of this change until time 6, so it may further deteriorate the control at times 4 and 5.

The difficulty with system delays is that necessary information comes too late and creates stability problems. The problem of controlling systems with time delays was solved as early as 1957 by Otto Smith. He suggested a controller where a model of the system is included (Figure 9.17) and the controller consequently is called a **Smith predictor**.

The controller contains a model of both the process and the time delay. Note that the transfer function G_{Pm} in the controller is a model of the true system and is not necessarily the same as G_P. Now assume that G_{Pm} is a perfect model of the process so that G_{Pm} is identical with G_P. Straightforward calculations show that the transfer function of the closed loop system becomes:

$$\frac{Y(s)}{U_c(s)} = \frac{G_R G_P e^{-sT}}{1 + G_R G_P} \tag{9.64}$$

where G_R is assumed to be a regular PID controller, G_P the process model and e^{-sT} the time delay. With the Smith predictor, the denominator of the closed loop system is the same as if the time delay did not exist. In other words, with the predictor the closed loop system transient response looks exactly the same as without the time delay but is delayed by time T.

Let us write the controller equation explicitly. Without the predictor, the control signal is:

$$U(s) = G_R E \tag{9.65}$$

With the predictor we obtain (Figure 9.17):

$$U = G_R(E + G_P \, \mathrm{e}^{-sT} \, U - G_P U) \tag{9.66}$$

The first term is the standard controller output based on the error. The second term is a correction based on a previous control signal $u(t-T)$ that is multiplied with a model G_P of the process. The last term is based on the actual control signal. The important point is that old control values have to be stored. The implementation of the predictor was difficult at the time when Smith suggested the idea, since only analog systems were available. In a digital computer, however, storing the old values is trivial.

It is quite clear from an intuitive point of view that old control signals should be remembered by the controller. Consider again Figure 9.16. If the controller remembers the control signal at Time 0 and knows that the result cannot be expected until Time 3, it is obvious that $u(3)$ should be a function of $u(0)$. Applying the Smith controller on the same process as in Figure 9.16 and with the same controller tuning, the result can be considerably improved. The shape of the transient is the same as if the time delay did not exist. It is only delayed by time T (Figure 9.18). The Smith predictor can also be included in a more general discrete regulator (Section 9.9).

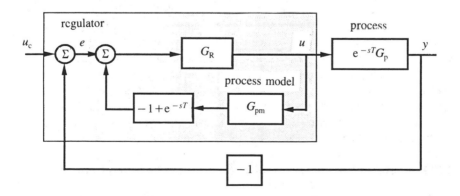

Figure 9.17   Block diagram of the Smith controller.

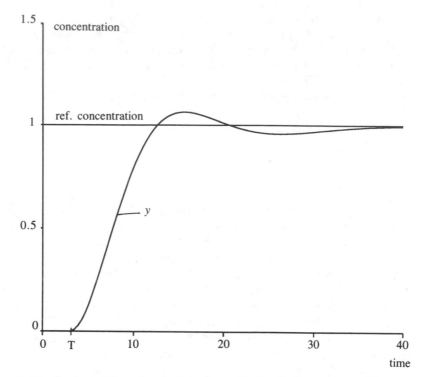

Figure 9.18   Control of the chemical dosage with the Smith predictor. The controller parameters are the same as in Figure 9.16.

9.8.2 Complex Dynamics

Because of the limited number of parameters, the PID controller can not arbitrarily influence a process with high order dynamics. In systems with significant oscillations a higher order regulator is needed. One such example was shown with the PIPI controller in electrical drive systems (Section 9.4.4), where the order has been extended by a low pass filter. The general controller (Equation 9.11) gives the necessary freedom to adjust for complex dynamics, and its time-discrete version will be discussed in Section 9.9.

Some systems have many inputs and outputs, where there are significant couplings between the systems (see Figure 2.13). The control task can not be solved by a simple controller based on one input and one output. Instead one control signal has to be based on several measurement signals. One way to synthesize this type of controller is by state variable feedback (Section 9.10).

9.8.3 Predictable Parameter Variations — Gain-scheduling Control

There are many processes where the process parameters change with operating conditions. A wastewater treatment system serves as an example.

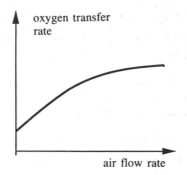

Figure 9.19 Typical behaviour of the oxygen transfer rate as a function of air flow rate.

Example 9.3 Dissolved oxygen control

The dissolved oxygen dynamics of an aerator are non-linear (Section 3.2.3). The transfer rate from gaseous oxygen to dissolved oxygen was modelled in Example 3.8 as $k_La = \alpha u$ with α a constant parameter and u the air flow rate. The k_La term, however, is a non-linear function of the air flow rate (Figure 9.19) and can be considered linear only for small air flow variations.

At a high load, the sensitivity of k_La to air flow changes is smaller than at a low load. Consequently the controller gain needs to be larger at high loads. Moreover, oxygen saturation adds another non-linearity. Since both the air flow rate and the dissolved oxygen concentration can be measured the process gain for different operating conditions can be modelled and the controller gain can be picked from a table.

If the variation of the process gain is known at different operating points it can be stored in a table. The controller can be tuned according to such a table and its tuning calculated in advance. This **gain-scheduling** is a common procedure in many applications, for example, in steam boiler control (different control setting at different power levels) or in aircraft control (different altitude controller settings for different altitudes).

9.8.4 Unknown Parameter Variations — Self-tuning Control

In many systems process dynamics is unknown with constant parameters and in other systems the parameters change slowly with time. There may be many reasons for the gradual changes. The piping in a process may be gradually clogged by material and this may change flow rates or heat transfer properties. In systems such as the air– fuel ratio control in a combustion engine, the sensor changes its gain and bias in an unknown way over time. In a biological fermentor or wastewater treatment plant new organisms may appear and change the pattern of the oxygen uptake rate.

Systems with a low order dynamics are simple to control once the parameters are known and constant. A PID controller is adequate in most cases. However, if the parameters

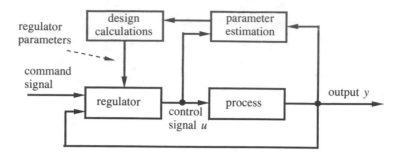

Figure 9.20   The principal parts of an adaptive controller.

are slowly changing the tuning will be quite poor most of the time. One solution to this problem is automatic tuning of the PID controller, a so-called **auto-tuner**. The tuning is initiated by the operator. The auto-tuner then excites some small disturbances to the process to find out its dynamics and computes the PID controller parameters from the process response. The parameters are kept constant until the operator initiates a new tuning.

If the parameters are continuously updated we talk about **adaptive controllers**. There are several commercial adaptive controllers available on the market today. It is outside the scope of this book to describe their features. All of them have two distinctive parts, one **estimation** part and one **controller** part (Figure 9.20).

The estimation part measures the process output and input signals, and continuously updates the process parameters. The controller design algorithm then continuously updates the controller parameters. The system can be said to consist of two loops, one fast control loop that resembles any other control loop and another slow loop that contains the parameter updating procedures.

There are several variants of this general scheme. Instead of updating the process parameters, the controller parameters can be updated directly. It must be emphasized, that even if the essential algorithms may be quite simple from a programming point of view, the adaptive control system needs a large 'safety network' of rules so that misuse is avoided. It is not true — although commonly believed — that the adaptive controller automatically solves all difficult control problems. If it is used with caution and knowledge, however, it gives new possibilities for solving complex control tasks. The controller part of the adaptive controller can be a general discrete controller, as described in the next section.

9.9 A GENERAL DISCRETE CONTROLLER

A general continuous controller was presented in Section 9.2. It was shown to give a lot of freedom permitting the closed loop system to behave in a desired manner. The continuous PID controller was shown to be a special case of the more general controller (Equation

9.11). It is reasonable to assume that a similar controller exists in the time-discrete case. As an introduction to this controller we will consider the time-discrete PID controller.

9.9.1 The Time-discrete PID Controller

By using the shift operator q (see Section 3.4.2) the notation of the PID controller (Equations 9.37–9.42) can be made more compact. With this notation the integral part (Equation 9.39) can be written as:

$$u_I(kh) = q^{-1}u_I(kh) + K\alpha e(kh) \tag{9.67}$$

where α is defined from Equation 9.40. Solving for $u_I(kh)$ we get:

$$u_I(kh) = \frac{K\alpha q}{q-1} e(kh) \tag{9.68}$$

Similarly, the derivative action (Equation 9.41) can be written as:

$$u_D(kh) = \beta q^{-1}u_D(kh) - K\frac{T_d}{h}(1-\beta)(1-q^{-1})y(kh) \tag{9.69}$$

Solving for $u_D(kh)$ we get:

$$u_D(kh) = K\frac{T_d}{h}\frac{(1-\beta)(q-1)}{q-\beta} y(kh) \tag{9.70}$$

Since $0 \le \beta < 1$ the system is always stable. Thus the complete PID controller can be formed from the proportional part (Equation 9.38), and the integral and derivative parts just calculated:

$$u(kh) = K\left(1+\alpha\frac{q}{q-1}\right)e(kh) - K\frac{T_d}{h}\frac{(1-\beta)(q-1)}{q-\beta} y(kh) \tag{9.71}$$

By eliminating the denominator, the PID controller can be written in the form:

$$(q-1)(q-\beta)u(kh) = K(q-\beta)(q-1+\alpha q)e(kh) - K\frac{T_d}{h}(1-\beta)(q-1)^2 y(kh) \tag{9.72}$$

With the definition of q this has the interpretation,

$$u(kh+2h) - (\beta+1)u(kh+h) + \beta u(kh) = K(1+\alpha)e(kh+2h) - K(\alpha\beta-1+\beta)e(kh+h)$$

$$+ K\beta e(kh) - K\frac{T_d}{h}(1-\beta)[y(kh+2h) - 2y(kh+h) + y(kh)] \tag{9.73}$$

By translating the time two sampling intervals backwards we can rewrite the expression in the form:

$$u(kh) = (\beta+1)u(kh-h) - \beta u(kh-2h) + Ke(kh) + K(\alpha-1-\beta)e(kh-h)$$

$$+ K\beta(\alpha-1)e(kh-2h) - K\frac{T_d}{h}(1-\beta)[y(kh) - 2\,y(kh-h) + y(kh-2h)] \qquad (9.74)$$

9.9.2 A General Time-discrete Controller with Reference Feedforward

The discrete PID controller can be considered as a special case of the more general digital controller, the continuous case. The general time-discrete controller can be written in the form:

$$u(kh) = \frac{T(q)}{R(q)} u_c(kh) - \frac{S(q)}{R(q)} y(kh) = u_{F1}(kh) - u_{FB}(kh) \qquad (9.75)$$

which is illustrated in a block diagram in Figure 9.21. In analogy with Equation 9.11, the control signal is composed of the two terms $u_{F1}(kh)$, the feedforward part, and $u_{FB}(kh)$, the feedback part. The controller can be written in the form:

$$Ru(kh) = Tu_c(kh) - Sy(kh) \qquad (9.76)$$

where R, S and T are polynomials in q:

$$R(q) = 1 + r_1 q + \ldots + r_n q^n$$
$$S(q) = s_0 + s_1 q + \ldots + s_n q^n \qquad (9.77)$$
$$T(q) = t_0 + t_1 q + \ldots + t_n q^n$$

The controller form (Equation 9.76) is interpreted as:

$$u(kh) = -r_1 u(kh-h) - \ldots - r_n u(kh-nh)$$
$$+ t_0 u_c(kh) + t_1 u_c(kh-h) + \ldots + t_n u_c(kh-nh)$$
$$- s_0 y(kh) - s_1 y(kh-h) - \ldots - s_n y(kh-nh) \qquad (9.78)$$

The structure is similar to the continuous case and is a combination of feedforward from the command signal and feedback from the measurement value.

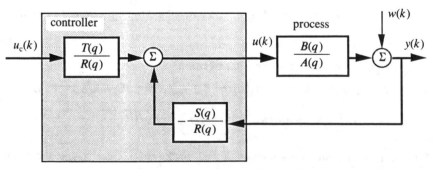

Figure 9.21   A general time discrete controller with feedforward from the set point and feedback from the measurement.

As described in Section 3.4.2 the time-discrete process transfer operator $H(q)$ is (Equations 3.68 and 3.71):

$$\frac{y(kh)}{u(kh)} = H(q) = \frac{B(q)}{A(q)} \tag{9.79}$$

where the polynomials A and B are defined as:

$$A(q) = q^n A(q^{-1}) = q^n + a_1 q^{n-1} + \ldots + a_n \tag{9.80}$$
$$B(q) = q^n B(q^{-1}) = b_0 q^n + b_1 q^{n-1} + \ldots + b_n$$

With this general controller, the input/output relationship for the total closed loop system in Figure 9.21 can be expressed by:

$$y(kh) = \frac{TB}{AR+BS} u_c(kh) + \frac{AR}{AR+BS} w(kh) \tag{9.81}$$

where the first term denotes the transfer operator from the setpoint and the second term that from the disturbance $w(kh)$ to the output y. This corresponds directly to the continuous case (Equation 9.13).

The A and B parameters are fixed by the process design, while the R, S and T parameters can be tuned as in the continuous case. By changing the R and S parameters the poles of the closed loop system can be changed arbitrarily provided the system is controllable. Consequently the dynamic response to either reference value or disturbance changes can be arbitrarily chosen. The response behaviour in terms of amplitude can be changed by adding closed loop zeros via the T polynomial. As for the continuous case, there are algebraic conditions that have to be satisfied. Also, since the signals are limited the poles can not be moved without limits.

With regard to computer implementation, we can see that the controller has to store old control signals as well as old setpoint and measurement values. Storing old control signals makes it possible to compensate for time delays, as with the Smith predictor (Section 9.8).

By comparing the PID controller structure (Equation 9.72) with the general controller it is clear that:

$$R_{\text{PID}}(q) = (q-1)(q-\beta) = q^2 + (1+\beta)q + \beta \tag{9.82a}$$

$$\begin{aligned} T_{\text{PID}}(q) &= K(q-\beta)(q-1+\alpha q) \\ &= K(1+\alpha)q^2 - K(1+\beta+\alpha\beta)q + K\beta \end{aligned} \tag{9.82b}$$

$$\begin{aligned} S_{\text{PID}}(q) &= -K(q-\beta)(q-1+\alpha q) + K\frac{T_d}{h}(1-\beta)(q-1)^2 \\ &= K(-1-\alpha+\gamma)q^2 + K(1+\beta+\alpha\beta+2\gamma)q + K(-\beta+\gamma) \end{aligned} \tag{9.82c}$$

where

$$\gamma = \frac{T_d}{h}(1-\beta)$$

If the derivative action is based on the control error, then the polynomial R is still the same, while the T polynomial becomes identical to the S (compare with the continuous case in Section 9.2). This means that the controller contains another zero in the transfer function. The different realizations of the PID controller consequently will include more or less zeros in the feedforward loop. This will influence the closed loop behaviour.

Note that there is quite a complex relationship between the R, S and T parameters and the initial PID controller parameters. The polynomial parameters do not have an apparent physical meaning and should consequently be hidden for the operator. The operator will tune the initial PID parameters and the computer will transform them to the R, S and T parameters, which are considered internal variables.

9.9.3 Feedforward from Process Disturbances

The discrete controller can be extended with another term from any measured disturbance. In Section 9.2, the continuous controller could be made to include both reference and disturbance feedforward terms. Naturally it can be discretized and the general discrete controller will correspond to the controller (Equation 9.21). The continuous feedforward controller from disturbances (Equation 9.18) has a discrete version, that can be written in the form:

$$H_{F2}(q) = \frac{H_w(q)}{H_t(q)H_v(q)H_p(q)} \tag{9.83}$$

where the transfer operators $H(q)$ are the discrete versions or the transfer functions $G(s)$ in Equation 9.18. The feedforward part of the control signal can be expressed in the form (we refer also to Figure 9.5):

$$u_{FF2}(kh) = H_{F2}(q)\, H_t(q)\, w(kh) \tag{9.84}$$

The controller is expressed in its numerator and denominator, i.e.:

$$u_{FF2}(kh) = \frac{V_1(q)}{R_3(q)}\, w(kh) \tag{9.85}$$

9.9.4 Feedforward from Both Reference and Disturbances

In analogy with the continuous case (Equation 9.21) the general discrete controller can be expressed by three terms, the feedforward from the reference value, the feedback from the output and the feedforward from the measured disturbance:

$$u(kh) = u_{FF1}(kh) - u_{FB}(kh) + u_{FF2}(kh)$$

$$= \frac{T_1(q)}{R_1(q)}\, u_c(kh) - \frac{S_1(q)}{R_2(q)}\, y(kh) + \frac{V_1(q)}{R_3(q)}\, w(kh) \tag{9.86}$$

Expressing the transfer functions with a common denominator we get:

$$u(kh) = \frac{T(q)}{R(q)} u_c(kh) - \frac{S(q)}{R(q)} y(kh) + \frac{V(q)}{R(q)} w(kh) \qquad (9.87)$$

where $R(q) = R_1 R_2 R_3$, $T(q) = T_1 R_2 R_3$, $S(q) = S_1 R_1 R_3$ and $V(q) = V_1 R_1 R_2$.

It is interpreted by analogy with Equation 9.78 as:

$$\begin{aligned}
u(kh) = &-r_1 u(kh-h) - \ldots -r_n u(kh-nh) \\
&+t_0 u_c(kh) + t_1 u_c(kh-h) + \ldots + t_n u_c(kh-nh) \\
&-s_0 y(kh) - s_1 y(kh-h) - \ldots - s_n y(kh-nh) \\
&+v_0 w(kh) + v_1 w(kh-h) + \ldots + v_n w(kh-nh)
\end{aligned} \qquad (9.88)$$

9.9.5 Different Criteria for the Discrete Controller

The general form of the discrete controller can satisfy different kinds of criteria. By stating the closed loop performance as a desired model we get a very general criterion on the system behaviour. Yet this model or criterion does not explicitly model the disturbances to the process. The classical process control criterion is to keep the measurement 'as close as possible' to the reference value. This criterion may be formulated as:

$$J_{mv} = \frac{1}{N} \sum_{k=1}^{N} y^2(k) \qquad (9.89)$$

and letting $N \to \infty$. The variable y is the deviation from the desired reference value. The criterion is known as a **minimum variance criterion**. The parameters in the regulator (Equation 9.76) can be tuned so that J_{mv} is minimized.

In the minimum variance criterion there is no limitation on the control signal u. In many practical situations u has to be limited because large control actions may cost as well as wear out the actuators. The corresponding criterion is then:

$$J_{lq} = \frac{1}{N} \sum_{k=1}^{N} [y^2(k) + \rho \, u^2(k)] \qquad (9.90)$$

There is a penalty cost for large control signals. The control law that minimizes J_{lq} is called a **linear quadratic control law** and can be expressed in terms of the general controller (Equation 9.76).

In principle, all the regulators mentioned, including the adaptive controllers, can be formulated in the general framework (Equation 9.76). From a software structure point of view, the system can be prepared for any controller complexity. The parameters are then chosen according to a relevant criterion.

9.10 STATE FEEDBACK

So far in this chapter dynamic systems have been described in continuous or discrete time by their transfer functions or transfer operators. These give only the relations between input and output. Similarly the controllers are formulated in input/output forms. The internal couplings in the process have been hidden.

There are many occasions where it is advantageous to describe the process with its state/space model. The internal model description leads to other structures of the controllers known as **state feedback**.

A linear time-discrete state/space structure was shown in Section 3.4.1 (Equation 3.61), and can represent a system with several inputs and outputs. The system parameters usually have a physical interpretation since the equations derived from force, momentum, mass or energy balances.

The state model can be the basis for a multi-input/multi-output state feedback design of the form:

$$\mathbf{u}(t) = \mathbf{M}\mathbf{u}_c(t) - \mathbf{L}\mathbf{x}(t) \tag{9.91}$$

where $\mathbf{M}$ and $\mathbf{L}$ are matrices and $\mathbf{u}_c$ the reference signal (Figure 9.22). In the case of a single control variable, the control law resembles:

$$u(t) = mu_c(t) - l_1 x_1 - l_2 x_2 - \ldots - l_n x_n \tag{9.92}$$

where m, l_1, $\ldots$, l_n are constants. In principle, state feedback consists of a sum of proportional controllers, one from each state variable. If a state variable is not known or measured directly, then it may be estimated (Section 3.5.2). The controller still has the same form but the state variable is replaced by its estimated value.

Assuming the states are known, the closed loop system with state feedback is described by:

$$\begin{aligned}
\mathbf{x}(kh+h) &= \Phi\mathbf{x}(kh) + \Gamma[M\mathbf{u}_c(kh) - L\mathbf{x}(kh)] \\
&= (\Phi - \Gamma L)\,\mathbf{x}(kh) - \Gamma M\mathbf{u}_c(kh)
\end{aligned} \tag{9.93}$$

The matrices Φ and Γ are given by the plant design, while M and L are available for tuning.

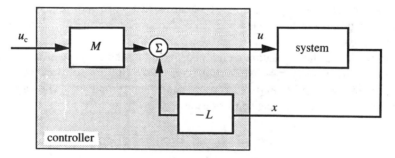

Figure 9.22 Structure of the state feedback controller.

The controller output is constant between the sampling instances and is implemented by matrix and vector operations on the available measurements at time kh.

The closed loop system (Equation 9.93) has a dynamics that is described by the system matrix $\Phi - \Gamma L$. The eigenvalues of this matrix determine the dynamic behaviour. The controller parameters L can change the closed loop system eigenvalues in an arbitrary manner as long as the system is controllable. As for the input/output controller, the system dynamics (poles and eigenvalues are identical concepts) can be changed arbitrarily only if the control signal can be unlimited. Therefore there are practical limits as to how much the dynamics can be changed.

We have seen that the eigenvalues (poles) of the system can be changed with state feedback, just as an input/output controller could do. The difference by using state feedback is that the internal description sometimes gives a favourable insight into the system and controller structure.

9.11 SUMMARY

Feedback is of fundamental importance in any process control. All the principal issues are the same for continuous and time-discrete systems. The structures of the continuous and time-discrete linear controllers are similar and differ only in the parameter values. From a computer implementation point of view many different control types can be included in a single general controller.

A computer implemented controller can be derived in two different ways:

- Design a continuous controller and discretize it afterwards.
- Make a time-discrete model of the process and design a discrete controller based on this model.

The first method has been discussed here. It has the drawback that the sampling interval generally tends to be smaller than that for the discrete design.

Feedforward is an important concept, and can be used to increase the controller capability. In high performance servos it is important to feed forward the reference value. In process control it is crucial to compensate as early as possible for disturbances and measurable load changes or disturbances can be forwarded directly to the controller. In principle the transfer function from the setpoint to the output should have a high gain for all relevant frequencies (in a servo) while the transfer function from a disturbance to the plant output should have as low a gain as possible.

The PID controller is a dominating controller structure in industrial control. The reason why this type of controller is so successful is that a majority of processes can be approximated by low order dynamics for which the PID controller represents a practical and inexpensive solution. A time-discrete version of the PID controller has been derived and can be more versatile than the continuous version. For example, in the discrete controller it is easier to include features for anti-windup and bumpless transfer, and to obtain adequate filtering for the derivative action. The control signal and its rate of change can also be easily limited. PID controllers can be used in cascade, for example, in electric drive system applications.

In systems with more complex dynamic features, PID controllers are no longer adequate. The most apparent problems appear in systems with dead times, with highly oscillatory behaviour or with parameter variations. The most general discrete-type controller can handle these problems and satisfy more elaborate criteria. The general discrete controller can be programmed in quite a straightforward manner, once the parameters are given, and can include both feedback from the process output and feedforward from the reference value and measurable process disturbances.

For known parameter variations it is possible to use gain-scheduling techniques while adaptive controllers can be applied in processes with unknown parameter variations. If the system dynamics is of low order, so-called auto-tuning PID controllers can be employed successfully.

FURTHER READING

A complete introduction to control theory with attention to the state/space description and special aspects of digital sampling is given in Kuo (1991). Chemical process control and applications of control structures is discussed in detail in Seborg, Edgar and Mellichamp (1989) and Stephanopoulos (1984). Shinskey (1967) describes process control problems from a more practical point of view.

Time-discrete control is extensively described in Åström and Wittenmark (1990) and in Franklin and Powell (1980). Also in Seborg, Edgar and Mellichamp (1989) there are a number of guidelines for the selection of sampling rates for PID controllers. The dead time compensator was first published by Smith (1957).

Adaptive control has been the focus of extensive research during recent decades. Åström and Wittenmark (1989) present a comprehensive description of the theory and applications of adaptive controllers. Åström and Hägglund (1988) give a practical and very readable description of autotuners.

APPENDIX: DISCRETIZATION OF THE PID CONTROLLER

The PID controller is represented in Laplace transform as the sum of three terms (Equation 9.25):

$$\delta U(s) = U_P(s) + U_I(s) + U_D(s)$$

$$= KE(s) + K \frac{1}{T_i s} E(s) - K \frac{T_d s}{1 + T_f s} Y(s) \tag{9.A1}$$

The error is calculated at each sampling interval:

$$e(kh) = u_c(kh) - y(kh) \tag{9.A2}$$

where the signals are considered constant during the sampling interval h, which is assumed fixed. The proportional, integral and derivative actions will be considered separately.

9.A1 Proportional Action

In the proportional part of the controller:

$$U_P(s) = KE(s) \qquad\qquad (9.A3)$$

the continuous signals are simply replaced by their sampled values at the kth sampling instant:

$$u_P(kh) = Ke(kh) \qquad\qquad (9.A4)$$

9.A2 Integral Action

The integral term is:

$$U_I(s) = K \frac{1}{T_i s} E(s) \qquad\qquad (9.A5)$$

which corresponds to the integral expression:

$$u_I(t) = K \frac{1}{T_i} \int_0^t e(\tau)d\tau \qquad\qquad (9.A6)$$

Approximating the integral by a summation, the integral part at time kh (corresponding to t) can be expressed as:

$$u_I(kh) = K \frac{h}{T_i} \sum_{n=0}^{k} e(nh) \qquad\qquad (9.A7)$$

If the last term of the summation is extracted we obtain:

$$u_I(kh) = u_I(kh-h) + K \frac{h}{T_i} e(kh) = u_I(kh-h) + K\alpha e(kh) \qquad\qquad (9.A8)$$

where

$$\alpha = \frac{h}{T_i} \qquad\qquad (9.A9)$$

Another way to obtain the discrete form of the integral action is to take the derivative of Equation 9.A6:

$$\frac{du_I}{dt} = \frac{K}{T_i} e(t) \qquad\qquad (9.A10)$$

The time derivative is approximated with a backward difference approximation that immediately leads to Equation 9.A8.

Notice that if a forward difference approximation is used, then:

$$u_I(kh) = u_I(kh-h) + K\alpha e(kh-h) \tag{9.A11}$$

Now the integral part can be precomputed before the measurement signal is obtained at time kh.

9.A3 Derivative Action

The derivative part can be discretized directly from the ideal derivative in Equation 9.30 using backward differences. The derivative action $u_D(kh)$ at the kth sampling interval is

$$u_D(kh) = -K\frac{T_d}{h}[y(kh)-y(kh-h)] \tag{9.A12}$$

A more elaborate derivation of the derivative part is made by considering its filtered transfer function (Equation 9.A1).

$$U_D(s) = -K\frac{T_d s}{1+T_f s} Y(s) = NK\left[1 - \left(\frac{N}{T_d}\right)\Big/\left(s+\frac{N}{T_d}\right)\right]Y(s) \tag{9.A13}$$

where T_f has been replaced by T_d/N. The quantity $U_D(s)$ can be expressed as:

$$U_D(s) = NK[-Y(s) + Z(s)] \tag{9.A14}$$

where $Z(s)$ is introduced as a state variable and defined as:

$$Z(s) = \left[\left(\frac{N}{T_d}\right)\Big/\left(s+\frac{N}{T_d}\right)\right]Y(s) \tag{9.A15}$$

The corresponding differential equation is:

$$u_D(t) = NK[-y(t) + \zeta(t)] \tag{9.A16}$$

$$\frac{d\zeta}{dt} = \frac{N}{T_d}[-\zeta(t) + y(t)]$$

where $\zeta(t)$ is the state variable of the derivative part and $u_D(t)$ the derivative part of the controller output. With the derivative approximated by a backward difference the sampled form is:

$$u_D(kh) = NK[-y(kh) + \zeta(kh)] \tag{9.A17}$$

$$\zeta(kh) = \zeta(kh-h) + h\frac{N}{T_d}[-\zeta(kh) + y(kh)]$$

The last equation can be rewritten in the form:

$$\zeta(kh) = \beta\zeta(kh-h) + (1-\beta)\,y(kh) \tag{9.A18}$$

where

$$\beta = \left(1+\frac{hN}{T_d}\right)^{-1} = \frac{T_d}{T_d+hN} \tag{9.A19}$$

If Equations 9.A17 and 9.A18 are combined and ζ eliminated, the derivative action can be written in the recursive form:

$$u_D(kh) = \beta u_D(kh-h) - K\frac{T_d}{h}(1-\beta)\,[y(kh)-y(kh-h)] \tag{9.A20}$$

Notice that $0 \le \beta < 1$. If the filter time constant $T_f \to 0$ (i.e. $N \to \infty$) then $\beta = 0$ and the derivative action is a simple difference approximation of the time derivative of the output signal (Equation 9.A12). For $T_d = 0$ we also get $\beta = 0$ which results in $u_D(kh) = 0$, i.e. no derivative action.

A third way to approximate the derivative action (Equation 9.A13) can be shown by a more straightforward approximation of the derivatives. The transfer function (Equation 9.A13) corresponds to the differential equation (note that the representation (Equations 9.A17 and 9.A18) does not contain any derivative of y):

$$\frac{T_d}{N}\frac{du_D}{dt} = -u_D-KT_d\frac{dy}{dt} \tag{9.A21}$$

Approximating the derivatives with backward differences we get:

$$\frac{T_d}{N}\frac{[u_D(kh)-u_D(kh-h)]}{h} = -u_D(kh)-KT_d\frac{[y(kh)-y(kh-h)]}{h} \tag{9.A22}$$

which can be rearranged to:

$$u_D(kh) = \frac{T_d}{T_d+hN}u_D(kh-h) - \frac{KT_dN}{T_d+hN}[y(kh)-y(kh-h)] \tag{9.A23}$$

With β (Equation 9.A19) inserted, this is identical to Equation 9.A20. Note that the backward difference approximation is numerically stable for all T_d.

9.A4 The Position Form of the PID Controller

The final form of the PID controller is:

$$\delta u(kh) = u_P(kh) + u_I(kh) + u_D(kh) \tag{9.A24}$$

One realization is based on the approximations of Equations 9.A4, 9.A7 and 9.A12. The integral term is expressed by its recursive form (Equation 9.A8), i.e.

$$\delta u(kh) = K\left\{e(kh)-\frac{T_d}{h}[y(kh)-y(kh-h)]\right\} +u_I(kh) \tag{9.A25}$$

Taking the derivative filtering action into consideration, then the PID controller terms in Equation 9.A24 are expressed by:

$$u_P(kh) = Ke(kh) \tag{9.A4}$$

$$u_I(kh) = u_I(kh-h) + K\frac{h}{T_i}e(kh) = u_I(kh-h) + K\alpha e(kh) \tag{9.A8}$$

$$u_D(kh) = \beta u_D(kh-h) - K\frac{T_d}{h}(1-\beta)[y(kh)-y(kh-h)] \tag{9.A20}$$

while α and β are defined from Equations 9.A9 and 9.A19, respectively.

10

Digital Communication

Aim: To introduce the principal concepts in the communication process and describe practical industrial applications with reference to the most important standards

Back in the introduction we saw that communication is fundamental to the function of any organized system. Thus, this chapter and Chapter 11 are dedicated to important aspects of communication in relation to control systems: the transfer of information within a computer system and to its exterior (a technical problem), and between the computer and its human users (a psychological problem). For both purposes a general model of the communication process is necessary.

This chapter, dedicated to electrical information transfer, does not pretend to be exhaustive. The issues related to communication are so many, the standards so numerous and the equipment so manifold that a somewhat comprehensive approach would require many hundreds of pages. Instead, we will first concentrate on basic concepts and on the most important issues for communication in industrial environments.

Section 10.1 presents a general model for communication following which is a description of the Open Systems Interconnection (OSI) scheme in Section 10.2. The OSI framework is taken as a reference throughout the chapter to examine common procedures and standards. Section 10.3 is dedicated to the physical links, Section 10.4 to communication protocols and Section 10.5 to local area networks (LANs). In these sections the reader will recognize many known standards, e.g. the ubiquitous printer interface EIA-232. The Manufacturing Automation Protocol (MAP), described in Section 10.6, is a new general concept aiming for interconnectivity (openness) in industrial communication. Section 10.7 is dedicated to fieldbuses, local networks for data exchange in factory environments. Public data networks are included in this chapter (Section 10.8) because a basic knowledge of the most important related concepts is also useful in industrial applications.

10.1 INFORMATION AND COMMUNICATION

10.1.1 What is Information?

Information is a concept for which we have a more or less intuitive understanding. It is a fundamental quantity that cannot be expressed by other fundamental quantities in the same way as, for instance, speed can be expressed by the ratio of length and time. Information is an intrinsic property (a structure contains information in itself), and can be transmitted at little energy expense: the drawings of a house contain as much structural information as the house itself, but are easier to carry around. A remarkable and very important property of information is that it can be duplicated at will without degrading.

Mathematically, information is defined as the measure of the state of order of a collection of entities. If a variable quantity can take a certain number of states, learning its current state is equivalent to getting some information about it. The more states a variable can take, the more information we get by learning its current state.

The information **I** associated with the knowledge of one out of **N** possible states that a variable can take, is the logarithm of **N**. For logarithms in base 2, the associated measure is in **bits**:

$$\mathbf{I} = \log_2 \mathbf{N} \text{ bits} \tag{10.1}$$

If a variable can assume only two states, its information content is $\log_2(2) = 1$ bit. The information content of a decimal figure $0-9$ is $\log_2(10) = 3.32$ bits. Other logarithm bases are seldom used in practice. For logarithms in base e the information unit is the **nit** and for logarithms in base 10 the **dit** or **Hartley**. Equation 10.1 holds under the assumption that the probabilities for the different states are equal.

The bit as a unit of information is very practical in digital logic, in which it is easily represented by two different power levels in an electric circuit. To transport or store the information related to a variable, a number of elementary circuits (switches) equal to or greater than the associated information is needed. In the case of a figure $0-9$ with an information content of 3.32 bits, at least four such switches are needed (with four switches it is possible to represent $2^4 = 16$ different states).

A continuous variable, i.e. a variable that can assume infinite values in a given range, has infinite information content. Digital applications can use only quantized values and some loss of information is accepted. For instance, in the digital representation of a voltage level varying between 0 and 10 V, approximation to 0.1 V means that for the description of the 100 resulting states 7 bits are sufficient ($2^7 = 128$).

The use of the logarithm as a metric for information is justified by a number of reasons:

- It is an increasing function of the number of possible states.
- When only one state is possible, the value is zero (i.e. the information content of a constant is zero).
- It makes information an additive measure. In comparison, the combination of the states of independent variables is the product of the states that each variable can assume.

Otherwise stated, the number of possible states for each symbol is **v** and therefore the number of combinations for **n** symbols is v^n. If the information content of every symbol is **I, n** symbols must have a total information content **n*I**. The function log(**n**) satisfies this property. An important consequence of the logarithmic measure of information is that information is always positive.

10.1.2 Communication

Communication is information moving in space. Communication plays a fundamental role in all organized systems. As a basic natural process, communication takes place within living entities via chemical and electrical signals; external communication through speech and vision helps sustain life in the surrounding environment. In developed societies, communication has a formidable technological infrastructure with the telephone network, the press and television.

Communication theory is an important tool in many fields. The marketing expert who tries to convince a large number of people to buy a certain product uses concepts similar to those of the communication engineer who designs radio links. In control systems, communication is a central issue: information about the state of a system continuously moves to a central processor and control signals from the processor out to the physical system.

Communication theory was first born as an engineering topic to solve transmission problems, but moved later to other non-technical fields such as psychology and linguistics. Here new results were found, to be later fed back to engineering. The basic issues and results are quite similar and lead to generalized models. For a long time, engineers were more interested in 'low level' information transfer while linguists and psychologists paid more attention to the purpose of the communication process, i.e. how to *reach a goal*. Now, the different aspects are gradually converging so that in engineering also the **semantic** (meaning-related) and **pragmatic** (purpose-related) aspects of communication can no longer be overlooked. Data transmission is then not a goal in itself but a means of achieving specific purposes.

10.1.3 A Model for the Communication Process

A general model for the communication process is shown in Figure 10.1. This model has general validity, i.e. it is not restricted to engineering applications but can be related to other types of communication, provided that its elements are appropriately identified.

All communication processes involve a **sender** and a **receiver**. The sender transmits a **message** (sequence of symbols) to the receiver along a **channel** or **medium**, which is common to both sender and receiver. The message is about some external **object** and the total amount of information carried is the sum of the information content of each one of the symbols composing the message. Because information is dimensionless, the sender has to change some parameters of the channel according to a **code** in order to convey

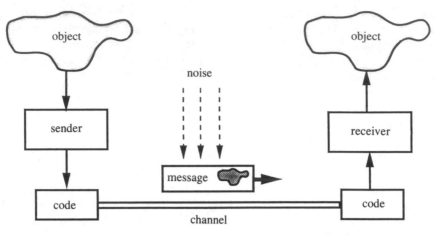

Figure 10.1   The general model for communication.

the message and the receiver uses the same code to extract the information from the message. The channel is subjected to **noise** which distorts the message and makes it difficult for the receiver to understand the message correctly.

In most cases the message deals with something external to the communication process, but sometimes it can be related to some of the communication entities. A familiar example is the 'hello' of phone conversations, which is in fact a message about the channel itself. Its purpose is to test whether the channel is open and operating correctly.

The basic issue in data communication is to move in a given time a message from location A to location B and minimise the influence of noise, or alternatively recover and reconstruct a message corrupted by noise. The question may also be put another way: how to let the receiver at B fulfill the purpose for which A is sending the message (this is obviously very important in advertising). In automation systems either unit A wants to pass information (its current state) to another unit B or A wants B to do something.

Example 10.1 Printed information

The writer (sender) conveys information to a reader (receiver) on paper (medium, channel). The message is the content, the code is the language, the symbols are the letters and the words. The reader can decode (understand) the message only by using (knowing) the same language as the sender.

Example 10.2 Data transmission

Here the medium is an electric cable and the information to transmit is digital data. The coding is straightforward: 0 and 1 correspond to different voltage levels in the cable.

Example 10.3 Electrical text transmission

A text is coded (written) in some language, the symbols are the words and letters. The text may be stored on an electronic medium, where each symbol has to be coded again, for instance, with different magnetization patterns. Frequently used is the **ASCII** (American Standard Code for Information Interchange) code, where each character (letter, figure, other symbol) is coded with a unique 7-bit (or in some applications, 8-bit) pattern.

In electrical text transmission, each bit leads to changes in some physical parameter of the channel and the text appears as a sequence of 0s and 1s with no apparent meaning. At a high abstraction level, the meaning is understandable if the same language is used, e.g. English or French. From this standpoint it does not matter if the lower level, internal machine coding is done in ASCII or in another code like **EBCDIC** (Extended Binary-Coded Decimal Interchange Code), with bit patterns different from ASCII.

The model for text transmission is quite subtle because several levels are involved. Coding is done in different ways depending on the level. And in real situations, much more complicated interactions are commonplace. In order to bring some clarity, at least where digital communication is concerned, the **Open Systems Interconnection (OSI)** model has been introduced. OSI is treated in detail in Section 10.2.

Independent of the reason and purpose for communication, in dimensioning a communication process it is possible to play with several factors, from the physical properties of the channel to the way the messages are coded. Communication design requires that different aspects are considered at several levels: the basic hardware, the processing software and the structure of the messages. The same functional result can often be achieved in different ways: what matters is the common operation of the selected methods.

An important factor that negatively influences communication is noise. Noise can be just a nuisance, but if its power level is too high compared with the power level of the original message, the message might be distorted to the point of becoming unintelligible. The receiver might decode the message wrongly and take an action other than the one envisioned by the transmitter. Noise protection and the recovery of corrupted messages are therefore very important issues in communication. Noise is a hard reality of the real world. In dimensioning communication channels, noise is the issue on which most compromises have to be made. In principle, noise can be eliminated but at great expense; in practice there are ways to cope with it and make it (almost) harmless.

10.1.4 Basic Quantitative Aspects of Communication

The basic parameter to describe a communication channel is its **capacity**, the amount of information it can transport per time unit measured in bit s^{-1}. When a communication channel is correctly dimensioned, the capacity is related to the amount of information that has to be transported. Basically one wants to carry as much information as possible at as low a cost as possible, but channel capacity in general costs money, the higher the capacity the more expensive the channel.

Example 10.4 Transmission of printed text

A sheet of paper (the channel) may contain 2000 letters; if each letter is coded in 7-bit ASCII, the total amount of information carried by the sheet is 14 000 bits. If the paper is sent through the mail and reaches its destination in one day, the equivalent channel capacity is 14 000/(24*3600), about 0.16 bit s^{-1}. A 200-page book counting approximately 300 000 letters and travelling at the same speed, has a capacity 150 times greater, i.e. 24.3 bit s^{-1}.

Note that in Example 10.4 the distance between source and destination is not relevant, assuming that the mail service requires the same time to deliver the correspondence independent of the destination. In fact, in communication distance plays a role only in those cases where there is a sensible delay in propagation, for example, in satellite links in which the radio signals need a total of about 270 ms to reach a geosynchronous satellite and then bounce back to earth. The real issue of communication is not the length but the **width** (i.e. the capacity) of a channel.

In industrial applications where different devices are connected together with electrical cables, there are propagation delays. These delays are not actually due to transmission along the transport medium but to the reaction speed of electronic equipment along the path. Communication delays might play a role and have to be taken into consideration in those applications where fast reaction times are required.

Another basic concept in communication is the **bandwidth** of an electrical channel. Bandwidth is the range of frequencies, i.e. the difference between the maximum and the minimum frequency that the channel can carry with an attenuation of less than 3 dB (50 per cent in power level). This concept is similar — although not identical — to the instrument bandwidth we have seen in Section 4.1.3. The bandwidth is measured in hertz or some of its multiples. For example, a normal telephone link carries signals in the frequency range 300 to 3400 Hz and the channel bandwidth is 3 kHz. A television channel has a typical bandwidth of 5.5 MHz.

The relation between the frequency bandwidth **W** (in hertz) of an electrical channel and the maximum data rate $\mathbf{R_{max}}$ (bit s^{-1}) that the channel can carry was found in 1924 by the American Harry Nyquist. In Nyquist's relation the way the signal is coded, and thus its ability to carry information, plays an important role. With **V** signal levels it is possible to carry $\log_2(\mathbf{V})$ bits and the total capacity of a channel becomes:

$$\mathbf{R_{max}} = 2*\mathbf{W}*\log_2(\mathbf{V}) \tag{10.2}$$

This relation is close to the problem of signal sampling and reconstruction described in Section 5.2. We can get enough information about a signal with bandwidth **W** by sampling it at a frequency **2*W**; sampling it at higher frequencies would not add anything to the information needed to reconstruct it later. The maximum data rate is obtained when each sampled signal level is different from the previous one.

In the technical literature, the concepts of channel capacity and bandwidth are often mixed. The confusion is due to the use of the same name for two different things. Channel capacity is a general concept valid for every kind of channel and every type of communication, and is not restricted to channels carrying electric signals. The bandwidth of a channel, on the other hand, has only to do with the frequency range for carrying electrical signals with no appreciable attenuation. Considered alone, bandwidth does not imply anything about channel capacity.

In binary data transmission $V = 2$ and the logarithm term of Equation 10.1 is 1. This has led to the common misconception that the capacity of a channel in bits per second is equal to twice its bandwidth in hertz. According to Equation 10.2, there is no limit to the amount of information that the channel can carry provided that a sufficiently large number of symbols is used.

In practice, problems are encountered rapidly because of the influence of noise. If the voltage levels for the different symbols are many and close to each other, they become difficult to separate and identify univocally. The smallest voltage spike due to noise on the line would erroneously be interpreted as a different level and thus as a different symbol.

The problem of communication in the presence of noise was studied by the American mathematician Claude Shannon. In 1948, Shannon proposed the following relation to describe a channel with bandwidth W (Hz) subjected to noise. According to Shannon, the maximum capacity R_{max} (bit s^{-1}) of such a channel is:

$$R_{max} = W * \log_2 (1 + S/N) \qquad (10.3)$$

Shannon's relation shows the maximum data rate which can be transmitted without errors in presence of noise. S/N is the ratio of average signal power to the average noise power. Signal power varies, in fact, with time depending on the symbols and the coding. Noise follows a random pattern. The limit indicated by Shannon's relation is considered a fundamental physical limit that is impossible to reach in practice. The concept is equivalent to the thermodynamic limit on the conversion of heat into work. But as in thermodynamics, Shannon's relation is a good reference to define how well a real communication process operates.

In fact it is very difficult to even approach Shannon's limit; a data rate one third of the maximum is generally considered to be more than satisfactory. To go above that limit, special multilevel signal coding is needed, which requires time for processing. In the end, what is gained in transmission rate is lost again in signal coding and decoding in the transmitter and the receiver. If data is put on a channel at rates above Shannon's limit, the errors due to noise will distort the signal in such a way that it is impossible for the receiver to decode it correctly.

Pay attention to the fact that Equation 10.2 is *not* a special case of Equation 10.3 in the absence of noise (i.e. for $S/N \to \infty$). Nyquist's relation (Equation 10.2) is a function of the number of coding symbols and leads to a theoretically infinite capacity for all channels. Shannon's relation is a function of the S/N ratio. According to Equation 10.3, in the absence of noise and with any bandwidth $W > 0$ it is possible to carry as much information as desired provided the coding is chosen appropriately. In the case of a telephone link with bandwidth 3000 Hz and a typical S/N ratio of 30 dB (corresponding to a power ratio of 1000), the Shannon limit is about 30 kbit s^{-1}. Nyquist's relation indicates that a coding scheme with $V = 32$ different levels must be used to carry this amount of information.

10.2 THE OPEN SYSTEMS INTERCONNECTION (OSI) SCHEME

10.2.1 The Changing Needs for Data Communication

When the first attempts to transfer digital data along existing telephone lines were made, attention was concentrated on the lowest communication level, the physical link. At that time programming was done in assembler and programmers worked at bit level anyway, so that more abstract data representations were not needed. Today's technology offers the means for transmitting large amounts of data at low cost, and attention tends to focus more and more on different applications: databases, process control, computer-aided production systems, etc. Connectivity among systems must be ensured at several levels, from the digital bits to the data and functions they represent. Generally accepted standards are the key to interconnectivity.

To overcome the difficulties of having to deal with a large number of incompatible standards, the International Organization for Standardization (ISO) has defined the **Open Systems Interconnection (OSI)** scheme. OSI itself is not a standard, but offers a framework to identify and separate the different conceptual parts of the communication process. In practice, OSI does not indicate what voltage levels, which transfer speeds or which protocols need to be used to achieve compatibility between systems. It says that there *has* to be compatibility for voltage levels, speed and protocols as well as for a large number of other factors. The practical goal of OSI is optimal network interconnection, in which data can be transferred between different locations without having to waste resources for conversion purposes with the related delays and errors.

The conceptual simplicity of OSI does not imply that its description is also simple: the related documents are several thousands (!) of pages long. OSI was initially published by the International Organization for Standardization (ISO) in 1984, in a document set called ISO 7498. The other major international standards organization, the CCITT (Comité Consultif International de Télégraphie et Téléphonie) published in the same year a comparable recommendation called X.200. The recommendation was redefined, with only minor changes, in 1988 and is open for revisions every 4 years.

10.2.2 Open Systems Interconnection Basics

OSI introduces a conceptual model for communication similar to the different levels of operating systems, where the operations have different abstractions, ranging from machine code and assembler programming to high level languages and applications. The multilayered structure of the operating systems is fundamental in the realization of functional interfaces by providing services while hiding their implementation details from the higher layers. The case of the OSI protocol and service structure is not much different.

Seven functional layers are defined in OSI (Figure 10.2). Each layer communicates directly only with the layers above and below it, requesting services from the layer below and providing services to the layer above. OSI service calls are similar to operating system calls: the requesting layer passes data and parameters to the layer below it and waits for

Layer

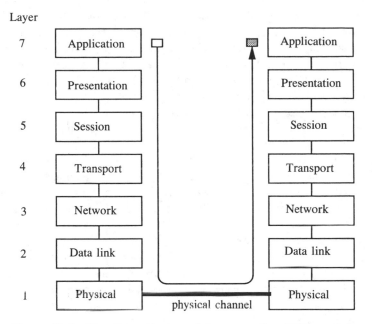

Figure 10.2   The Open Systems Interconnection (OSI) model.

an answer, but ignores the details of how the request is carried out. Modules located at the same layer and at different points of the network (i.e. running on different machines) are called **peers**; they communicate via **protocols** that define message formats and the rules for data exchange.

OSI defines the services each layer must put at the disposal of a higher level. The services (what to do) are strictly separated from the protocols (the actual implementation). Interconnectivity builds on the fact that different systems are structured around similar services and that at each layer the protocols are the same. The basic philosophy of OSI can be summarized with 'do not mix up unrelated entities' and 'let related entities communicate with each other'.

The layers defined in OSI are the following.

1. **Physical link layer** This consists of electrical, mechanical and optical interfaces with related software device drivers for the communication ports. At this level all details about transmission medium, signal levels, frequencies and the like are handled. The physical layer is the only real connection between two communicating locations (**nodes**).
2. **Data link layer** This level provides for the verification that bit sequences are passed correctly between two nodes. If errors occur due, for example, to line noise, this level may request the retransmission of a corrupted bit sequence. The data link layer presents to the higher layers an error-free data link between the nodes.
3. **Network layer** This sets up a complete path and oversees that messages go all

the way from source to destination node, even when the path is composed of different legs traversing several nodes.

4. **Transport layer** This layer provides end-to-end communication control and is the interface between the application software that requests data communication and the external network. It has the responsibility of verifying that data from one machine to another is transmitted and received correctly.

5. **Session layer** This enhances the transport layer by adding services to support full sessions between different machines. One example is the login via a network in a remote computer.

6. **Presentation layer** Data encoding and conversion, in which the raw binary data is related to its meaning: messages, texts, figures or other, take place at this layer.

7. **Application layer** The highest layer deals with application system management tasks such as file transfer, distributed database operations and remote control.

The physical layer is the only one with a concrete physical appearance. All the other layers are sets of rules and descriptions of functional calls and are therefore implemented in software or firmware. The lower three layers are network or communication layers. The three highest layers are implemented in the application software of a target computer. Layer four, the transport layer, is the link between the application and the communication-oriented layers.

The basic principle of OSI operation is not difficult to understand. Two peer entities are connected via a **virtual (logical) link**. To the peers, the virtual link appears to be real, although only at level 1 are virtual and physical links the same. The peer modules exchange data according to a protocol specified for that level. In practice, the entities request the services of the layer immediately below via procedure calls (Figure 10.3). The protocol internals are not visible to the entity requesting the service and can be changed at any time without the outside entity knowing it. There is no direct link, real or virtual, between modules on the same machine at a distance of more than one level from each other or between modules running on different machines and that are not located at the same level. For instance, a module at Level 4 on a machine can communicate only with Levels 3 and 5 on the same machine and with Level 4 on a remote computer.

A protocol is a set of rules on how to initiate, carry on and lay down a communication process. In OSI, protocols are used for communication among peers. In some cases the peers must exchange protocol-related information, which is appended to the original message. Each layer has its own protocol and therefore adds its protocol data to the original message. The result is somewhat like Russian 'matrjoshka' dolls or Chinese boxes, each is placed inside a larger one (Figure 10.4).

Other protocol data may be exchanged between peer modules at different levels for operational **signalling**, e.g. to establish or bring down a link. Signalling information is transmitted separately from application messages.

Not all layers require direct exchange of data. Figure 10.4 gives an indication of the basic concept, but reality need not be this complicated. Protocols that require active data exchange are needed only for some of the peer layers and interconnectivity can function even if some of the layers are bypassed — with the agreement of the communicating

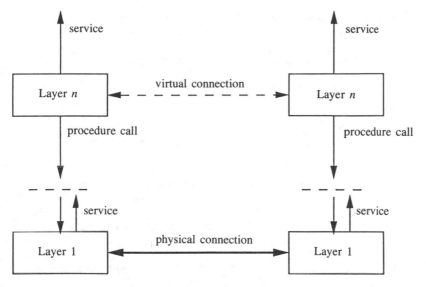

Figure 10.3 Principle for virtual peer-to-peer communication in OSI.

partners. Procedure calls are described in OSI and in the related documents with a generic syntax for each function and its parameter list.

For each of the OSI layers there are one or more sets of standards issued by the major standardization organizations (see Section 12.1.4). At the physical and data link levels some earlier standards were incorporated in OSI. For the other levels new protocols had to be defined following the indications of the OSI model.

OSI's full compatibility between the different layers suggests that, in principle, one could put together a working application by mixing products from different sources. As usual, reality looks rather different. OSI intermediate levels are not sold as separate software products and manufacturers and software developers offer instead packages for the levels 3–4 to 6–7. The internal interfaces do not need to follow OSI requirements, and in practice that is not very important. Instead of supporting all layered OSI protocols, software is designed to be as efficient as possible.

The OSI model is not exempt from criticism. It has been observed that it lacks some important functions such as data encryption (which should be part of layer 6) and that the structuring and division of layers 4–7 has a somewhat academic and pedantic flavour. As the layered software runs on the same machine anyway, its internal structure could be left to the program developers as well. Protocol definition at the higher levels is also inefficient. For these reasons, OSI is usually not implemented in its entirety, but computer manufacturers sell products making use only of the necessary layers, avoiding all features that are not needed for a specific application.

It might seem that the OSI model is of more interest in long-distance data communication than in automation applications. Actually the opposite is the case. Effective automation requires that different computers running different applications be interconnected

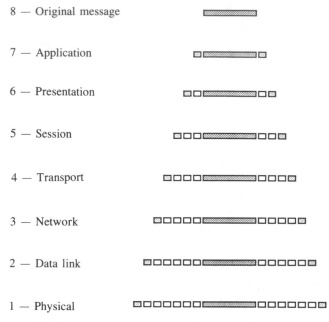

Figure 10.4   Aspect of data in layered communication protocols.

with no major effort and OSI provides the framework for such interconnection. The newest communication standards in industrial and office environments, the **Manufacturing Automation Protocol** or **MAP** (Section 10.6), and the **Technical and Office Protocol (TOP)** are based on the OSI layered model.

A future perspective is the integration of local control systems in major geographically-distributed constellations. OSI would guarantee that applications such as warehousing, production and statistics, in plants and offices would run together exchanging data in a general approach to factory automation, logistics and planning, regardless of where the plants and offices are located. OSI provides the framework to set up a rational structure for data communication.

In the following description, we will examine the two lowest layers and layer 7 (application) in more detail, paying attention to the current implementations and trends for communication in factory environments and process control applications. A more detailed description of the other OSI levels is outside the scope of this book; the interested reader will find stimulating material in the works of Black (1989) and Tanenbaum (1989).

10.2.3 Virtual Devices

Virtual device is a concept frequently referred to in OSI (and not only there). A virtual device is the set of all functions and parameters, described with a specific syntax, which are relevant for a particular, real, device. The set of procedures composing the virtual

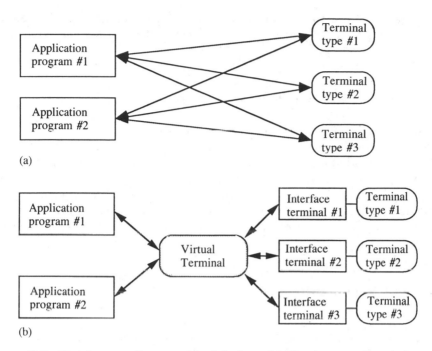

Figure 10.5 Direct connections vs virtual devices. (a) Direct connections between application programs and physical devices; (b) interfacing physical devices via a virtual terminal.

device takes care of all the low level details and accepts - or delivers — only 'clean' data. For instance, a virtual terminal is a collection of commands for placing a cursor on the screen, writing a string of characters with particular highlights such as boldface or underline, accepting characters from the keyboard and so on. Without virtual devices, a programmer writing a text editor would have to care about the different control sequences for performing the same function on different terminals. With a virtual terminal such a nightmare is avoided; it is sufficient to write commands to control the virtual terminal. Other interface routines convert the commands for the virtual device in control sequences for the real terminals (Figure 10.5).

A major advantage of the virtual device concept is that new devices may be added without having to change existing application programs. It is sufficient to provide the new physical device with an interface to the virtual one. The application program does not even *need* to know that a different physical device is being used.

An important virtual device supported under OSI is the virtual filestore, with protocols called **File Transfer Access and Management (FTAM)**. FTAM applications are important not only in process control but also in distributed databases, as found in financial transaction systems, airline seat reservation systems and similar applications. FTAM describes with an abstract notation file properties such as creation date, access control and many others, and supports operations in a multitasking environment with functions such as open/close, read/write and set locks. The mapping software between FTAM and the specific file system hardware is part of the operating systems provided by the computer manufacturers.

10.3 PHYSICAL LINKS (OSI PHYSICAL LAYER)

The most common way to move digital information is to send it on electrical cables. Simplicity, low cost and an established technological base make cables the most appropriate medium to carry information in limited areas, while optical cables and radio links are more cost-effective for moving large amounts of information between distant geographical locations. In the following, we will only deal with the issues related to the transport of digital information on electrical cables that are of specific interest in industrial automation. For a more detailed discussion of the other media, refer to Black (1989) or other relevant publications.

10.3.1 Electrical Conductors

The most common types of conductors used for communication are the **twisted pair** and the **coax cable** (see Section 4.7.2). Both types of cable are insensitive to disturbances, although the best screening is, of course, provided by the shield of the coax cable. The bandwidth of the twisted pair is limited to a few megahertz, which means that it cannot

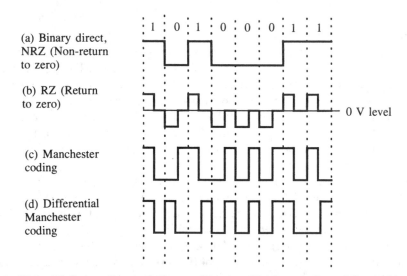

Figure 10.6   Digital coding techniques. Observe that knowledge of the 0 V level is relevant only in RZ coding; in the other schemes the absolute signal levels are not relevant to identify the related data. (a) Binary direct, NRZ (non-return to zero); (b) RZ (return to zero); (c) Manchester coding; (d) differential Manchester coding.

support data rates higher than a few Mbit s^{-1} over distances of more than a couple of kilometres. However, because of its simplicity and its low cost, the twisted pair is a common choice for data communication when the performance does not have to be too high.

Coax cables allow the transmission of a bandwidth up to about 500 MHz and are commonly used to carry high frequency radio and TV signals. Thanks to their broad bandwidth, in data communication coax cables can sustain much higher data rates than the twisted pair.

A functional distinction is made between baseband and broadband coax cable. The difference does not have that much to do with the type of cable (which actually are quite similar), but with the type of signals carried. **Baseband** coax is used for digital communication with only one carrier frequency, usually 5, 10 or 20 MHz. This is the standard solution for most industrial applications. **Broadband** is used to support several kinds of signals at different frequencies, and is slowly gaining in importance, especially on campuses, but so far few industrial users seem to be interested in mixing telephone, office data, TV signals and factory automation controls on the same physical conductor. The installation and the maintenance of twisted pair as well as coax cable is generally unproblematic. The industrial area networks that will be described later in this chapter use twisted pair or coax cables as the physical medium.

10.3.2 Bit Coding

There are basically two possibilities for transmitting a bit sequence along a physical channel.

- Putting the bits on the line as they are or with some kind of coding, but maintaining the aspect of digital data.
- Modulating a carrier in amplitude/frequency/phase and transmitting the modulated signal.

The most direct way to move digital data is to put it directly on the electrical line. With **direct coding**, a voltage level of 0 V would represent a logical '0' and +10 V a logical '1' (Figure 10.6(a)). In widely used terminology binary '0' indicates **space** and binary '1' is **mark**. Quite often, the bits are **reverse-coded**: '0' (or space) would keep the line at high voltage level and '1' (or mark) at a low level. Also common is **polar coding**, where digital '0' and '1' have opposite levels in respect to a common reference. Direct, reverse and polar coding are called **non-return to zero (NRZ)** codings because there is no forced transition back to zero level. A sequence of 1s would keep the line potential level constantly high or low according to the selected encoding scheme.

NRZ coding is straightforward but quite sensitive to noise and distortion. To compensate for line attenuation and distortion, a threshold may be defined at the receiver site so that anything less than 2 V is interpreted as '0' and levels higher than 5 V as a logical '1'. A Schmitt-type trigger set on those levels is then used to reconstruct the digital signal (Figure 10.7).

There is, however, another big problem with straight NRZ coding. Just putting bits on the line does not enable the receiver to detect where each bit starts and ends and, at

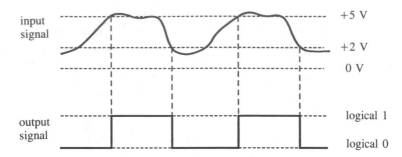

Figure 10.7   Threshold for a Schmitt trigger.

the beginning of a transmission, the actual data transfer speed if different speeds are possible. In other words, with pure NRZ coding there is no way to distinguish 'no message' from a sequence of information-carrying 0s. Should an incoming pulse be interpreted as one long or two short 1s? A possible solution would be to require each message to have a preamble, e.g. a sequence of alternating 0s and 1s, to give the correct timing. But there still would be a risk of the receiver losing synchronization with the transmitter during data transfer, leading to a false interpretation of the incoming data. Finally, the capacitances distributed along a communication line might lead to d.c. potential build-up if the data pulses always have the same polarity.

All these problems may be solved with **return to zero (RZ)** coding (Figure 10.6(b)), which combines the original data with a synchronization signal. In RZ coding two potential levels are defined of opposite polarities with respect to zero. Each bit starts at high or low potential level and in the middle of each pulse there is a transition to the zero level; the transition edge is used to synchronize the receiver. RZ coding requires twice as much bandwidth as NRZ and has more complicated interface electronics, but its advantages definitely offset the disadvantages.

A different and widely used solution is **Manchester coding**, also known as **biphase level** (or **Bi$^\phi$-L**). In Manchester coding each bit is coded with two voltage levels and a transition in the middle of each pulse. In straight Manchester coding, bit '0' is represented by a transition from a low to a high voltage level and bit '1' from a high to a low level (Figure 10.6(c)). A similar scheme is **differential Manchester coding** in which bit '0' is represented with a level transition at the beginning of each new period and bit '1' with the absence of a transition (Figure 10.6(d)). In the differential Manchester scheme the coding for a bit depends, therefore, on the level of the second half of the preceding bit.

Manchester and differential Manchester coding contain the synchronization reference and offer better noise immunity than RZ coding. As for RZ coding, the Manchester schemes require twice as much bandwidth in respect to NRZ coding. An advantage of Manchester coding is that it works with two voltage levels instead of three and the processing hardware is simpler than for RZ coding. Manchester coding is widely used in local area network applications.

Several other coding schemes have been proposed, but they are more important in long distance and satellite communication than in industrial control and will not be dealt with here.

10.3.3 Carrier Modulation

Carrier modulation is used to match the characteristics of the signal for transmission with those of the line carrying it. **Modulation** is the change of some of the parameters (amplitude, frequency, phase) of a high frequency sine wave **carrier** as function of the input signal (Figure 10.8). The receiver extracts the original signal from the modulated carrier.

Amplitude modulation is relatively little used in digital transmission while **frequency** and **phase modulation** are more common. A normal application of carrier modulation is found in modems for data transmission along analog telephone lines (Section 10.8.1). For instance, medium-speed modems operate with carrier frequencies ranging between 800 and 2500 Hz. Carrier frequencies for transmission on broadband cables may reach up to 500 MHz.

Carrier modulation need not strictly follow the envelope of a digital input signal. A change in some carrier parameter may be associated with a sequence of several bits and not just one bit, so that it is possible to carry more information at the same carrier frequency. For instance, in phase modulation the direct relation bit 0 implies a phase shift of $0°$ and bit 1 implies a phase shift of $180°$ could be employed. Alternatively, the bit sequences 00, 01, 10, 11 could be associated with phase shifts of $0°$, $90°$, $180°$ and $270°$, respectively, so that the same signal carries twice the amount of data. An even more sophisticated technique is **quadrature amplitude modulation (QAM)**, which combines amplitude and phase modulation to carry several bits for each change in the envelope of the carrier signal.

The added data transport capacity with multibit encoding and QAM technique is not free, however. Referring to Equation 10.2, the number of symbol **V** was increased from 2 to 4 when four phase shifts were employed. To process the related signals, more complicated circuitry is needed and the transmitted signals become more sensitive to noise. For a given line of a known bandwidth and a constant noise factor and for a fast carrier frequency, there is an optimum data rate. Exceeding that rate does not improve the throughput of the channel because data correction information also has to be carried together with the original data, and additional processing is needed at both transmitter and receiver. But as technology becomes cheaper and allows the construction of more complex processing circuitry, increased use is being made of combined modulation techniques in order to make the best use of the available physical channels. The theoretical maximum transmission speed for a channel, however, does not depend on technology and is only a function of the channel bandwidth and of the signal to noise (S/N) ratio (Equation 10.3).

The number of times per second at which the carrier changes in some of its parameters is the **signalling** or **baud rate**. There is often confusion between bit and baud rate; they are the same only when a change in a modulation parameter corresponds to one bit carried, i.e. when the carrier may take two states only. In QAM modulation, if one combined change in amplitude and phase corresponds to four bits, the bit rate is four times the baud rate.

10.3.4 Time Synchronization

In order for a communication process to operate correctly, it is necessary that transmitter and receiver agree on a time reference, i.e. that they are **synchronized**. The synchronization

carrier $u_c(t) = u_0 \sin(\omega_c t + \phi)$

signal $u(t)$

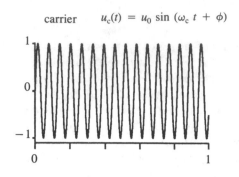 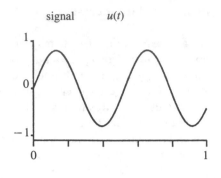

AM

$u_{AM}(t) = (u_0 + ku(t)) \sin(\omega_c t + \phi)$

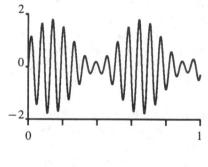

FM

$u_{FM}(t) = u_0 \sin[(\omega_c + ku(t))t + \phi]$

PM

$u_{PM}(t) = u_0 \sin[\omega_c t + ku(t) + \phi]$

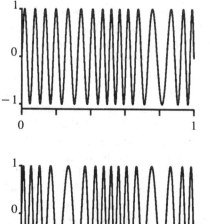

Figure 10.8 Amplitude modulation (AM), frequency modulation (FM) and phase modulation (PM) of a carrier.

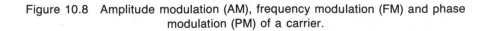

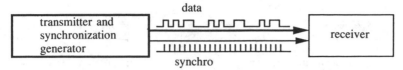

(a) the transmitter generates the synchronization reference

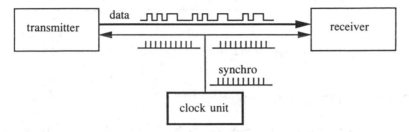

(b) an external unit generates the synchronization signal

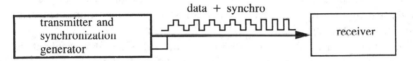

(c) the transmitter generates the synchronization signal
and superimposes it on the data

Figure 10.9 Three methods for the distribution of a common synchronization reference. (a) The transmitter generates the synchronization reference; (b) an external unit generates the synchronization signal; (c) the transmitter generates the synchronization signal and superimposes it on the data.

reference is a pulsed signal with defined frequency, generated by one of the communicating units or some other external unit (Figure 10.9). The synchronization signal is either distributed with a dedicated conductor or transmitted with the digital information, as, for example, in RZ and Manchester-type coding (Section 10.3.2). In the first case, additional cabling is required, in the other additional bandwidth.

Data communication may or may not be related to a time reference. If the bits for transmission are generated together with the synchronization pulses (the fixed time reference), transmission is **synchronous**; if there is no relation to a time reference the transmission is **asynchronous** (Figure 10.10). There is a similarity in synchronous data communication and synchronous data transfer in bus systems (see Sections 6.2.6 and 6.2.7), as both require a common clock reference.

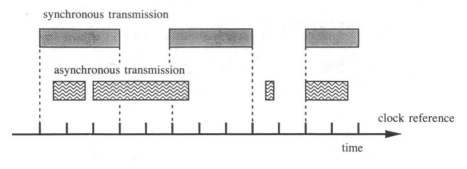

Figure10.10   Synchronous and asynchronous transmission.

10.3.5 The EIA-232-D Interface Standard

The interface standard EIA-232-D is probably the best known and most used of all interfaces for serial data communication. The operation of the EIA-232-D and of the other data communication interfaces is not complicated. The real challenge is to keep track of all the reference documents issued by the standards organizations, what they do (and do not) specify and how they refer to each other in a fashion closely reminiscent of the double-linked list data structure.

The standard EIA-232-D was previously called RS-232. It was introduced in 1969 by the Electrical Industries Association (EIA) to define the interface between computers or terminals and modems. The latest revision, when the name was changed from RS to EIA, dates from December 1987. The original RS-232 specifications were used, with minor changes, by the international standards organizations CCITT and ISO to issue their own sets of specifications. The loop is now closed with EIA-232-D, which refers back to the CCITT and ISO standards.

EIA-232-D was initially defined as an interface between data processing and data communication equipment connected to an external network, but today the standard is commonplace for many other applications, e.g. the connection of computers with terminals, printers and other external equipment. The original definition has led to the terminology **data terminating equipment (DTE)** for the generic processor (computer, terminal) and **data communication equipment (DCE)** to indicate the communication interface.

The physical connector has 25 pins and is normed as ISO-2110. The EIA-232-D connector with the most important pin definitions is shown in Figure 10.11. EIA-232-D defines circuits for communication on two channels and DCE testing, but in practice only the circuits for the first communication channel are used. We will limit our discussion to these.

The electrical interface of EIA-232-D follows the CCITT V.28 recommendations. The signal levels are between $+3$ and $+15$ V for a logical '0' and -3 and -15 V for a logical '1' (reverse coding). The input electronics should withstand surges up to ± 25 V. Maximum data rate is 19 200 bit s^{-1} for a cable length of up to 15 m; for lower data rates, the cable length may be increased.

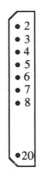

Figure 10.11   The EIA-232-D connector with related signals. The physical connector is normed as ISO 2110. The EIA-232-D circuits for the first communication channel are the following:

Protective ground (GND), Pin 1 — connection to the equipment frame or to an external ground.

Signal ground (SG), Pin 7 — 0 V reference for all other signals; it is common practice to connect together SG and GND at the modem site.

Transmitted data (TD), Pin 2 — data transmitted from the terminal device or computer (DTE) to the modem. When no data is transmitted, the line is kept at logical 1 (mark) state.

Received data (RD), Pin 3 — data received from the modem (DCE) to the terminal device or the computer. When no data is received, the DCE keeps the line at mark condition.

Request to send (RTS), Pin 4 — signal from DTE to the DCE to request clearance for transmission.

Clear to send (CTS), Pin 5 — reply from DCE to DTE to indicate readiness for transmission after an RTS request.

Data set ready (DSR), Pin 6 — indication that the DCE is connected to the telephone or communication line and is able to start data transmission. This signal indicates that state of the DCE and is independent of any requests by the DTE (DSR has been renamed DCE Ready in EIA-232-D).

Data terminal ready (DTR), Pin 20 — Indication that the DCE equipment is ready for operation (DTR has been renamed DTE Ready in EIA-232-D).

Carrier detect (CD), Pin 8 — indication that the DCE recognizes a carrier signal from the remotely connected DCE.

The definition of the pin signals and the procedural operations (protocol) of EIA-232-D follow the CCITT V.24 recommendations. The EIA-232-D circuits for the first communication channel are shown in Figure 10.11.

The signals in EIA-232-D have been defined to interface with a modem connected to a communication line (Figure 10.12(a)); for this reason, most of the signals are not relevant in other applications. To connect other equipment directly and without going over an external communication line, so called 'null modem' cables are used. There are different

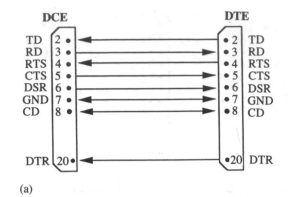

(a)

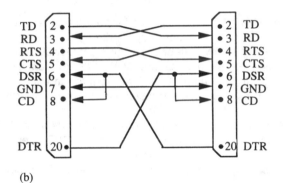

(b)

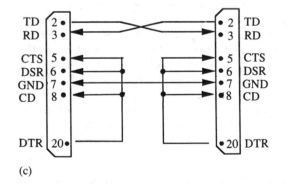

(c)

Figure 10.12 Direct connection DCE–DTE. Examples of null-modem cables with and without handshaking. (a) Direct connection between DCE and DTE; (b) null-modem cable with handshaking (both units operate as DTEs); (c) null-modem cable with no handshaking (both units operate as DTEs). (TD — transmit data; RD — receive data; RTS — request to send; CTS — clear to send; DSR — data set ready: GND — ground; CD — carrier detect; DTR — data terminal ready.)

types of null modem cables. In one solution (Figure 10.12(b)) the pins are connected so that the control signals from the communicating devices can handshake. A different configuration is to connect the handshake pins to a constant voltage, so that they are permanently asserted (Figure 10.12(c)). The equipment at one end of the line assumes that the one at the other end is always ready to exchange data.

The standard EIA-232-D does not imply anything about the type of transmission, which can be asynchronous as well as synchronous (the circuits of pins 15 and 17 carry the timing signals for transmission and reception synchronization). The digital data may use any kind of coding.

The major drawback of EIA-232-D is the limitation on the maximum operating speed to about 20 kbit s^{-1}. To overcome this limitation, a new standard, RS-449, has been defined which extends the functions of EIA-232. RS-449 defines more comprehensive signalling than EIA-232-D in order to exploit new capabilities in the public data network services. The new standard has two connectors, one 37-pin connector for the basic circuits and an optional 9-pin connector for the secondary channel.

RS-449 does not directly specify the electrical signal levels but refers to two other documents: RS-422 for balanced and RS-423 for unbalanced transmission (the standards RS-422 and RS-423 deal purely with electrical specifications and not with the rest of the functional requirements for a complete communication interface). The main connector of RS-449 provides two additional pins for the return circuits of balanced transmission. According to RS-422, a data rate of up to 2 Mbit s^{-1} can be reached, while using unbalanced circuits and a common return line the maximum defined is 20 kbit s^{-1}. Provisions are made in RS-449 for interoperability with EIA-232-D using relatively uncomplicated equipment. So far, few devices are available that make use of the RS-449 interface standard.

10.3.6 The Multipoint Electrical Interface RS-485

The standards we have seen so far allow data exchange only between two communicating devices, but many applications require that several units are connected together for common data transmission and reception. The standard RS-485 defines the electrical interface when several units are connected together. This standard is only concerned with the electrical interface parameters and does not specify anything about signal quality, timings, protocols, pin assignments, etc. It supports binary data exchange at signalling rates up to about 10 Mbit s^{-1}.

In the RS-485 model several units are connected via a twisted pair, balanced electric cable. The units can be transmitter (generator), receiver or transmitter/receiver combined. The cable has two terminating resistors of at least 60 Ω each at both ends (Figure 10.13).

The interface operation is similar to bus tristate logic (see Section 6.2.3): the generators can be in the active or passive state. In the passive state they present a high impedance to the network; in the active state they drive the network with a differential voltage (measured between the two output pins) of between 1.5 V and 5.0 V. Binary '0' is related to one polarity, one output pin is positive with respect to the other, and binary '1' has

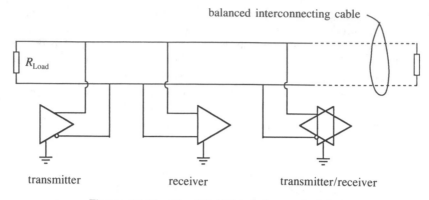

Figure 10.13   The RS-485 interface principle.

reversed polarity with respect to level '0'. The differential threshold for the receivers is set at 0.2 V, with an allowed range of input voltages varying from -7 to $+12$ V with respect to receiver ground. In this configuration no conductor is at ground potential. Reversing the connections from a generator or to a receiver is equivalent to inverting the bit values.

The input impedance of a receiver and the output impedance of a generator in passive (idle) state is measured in unit loads, exactly defined in the RS-485 document. A generator must be able to drive up to 32 unit loads and the two terminating resistors at a total equivalent load of 54 Ω on the line. The generators must also be able to withstand the power dissipated when two or more generators are active, some sourcing and some sinking current at the same time.

10.4 COMMUNICATION PROTOCOLS

10.4.1 The OSI Data Link Layer

The establishment of a workable physical link is only the first step for building reliable communication. At Layer 2 in the OSI scheme we find data link control, that is, the set of procedures and protocols necessary to guarantee that messages arrive intact at the receiver node.

At the physical layer little can be done to ensure that data is not distorted by noise and to recover data which has been corrupted. Verification of the validity of received data is the responsibility of the layers above the physical layer; of primary importance is the operation of the data link layer. The method used transmits data according to specific protocols and together with additional information allows verification of data integrity.

There is a large number of data link protocols, but their basic operating principles are quite similar. We will examine the HDLC protocol in more detail because it is standardized and has a structure that is more or less the same as all the other protocols.

10.4.2 Virtual Channels and Multiplexing

Because of their limited bandwidth, communication channels must be considered as protected resources that have to be used as efficiently as possible. If at a given moment more information has to be carried than the communication channel can handle, a selection principle for access to the channel becomes necessary. The division of a physical channel among several users is called **multiplexing** (Figure 10.14). Multiplexing is transparent for users, who are not aware of how the channel is managed in detail. Each user 'sees' a virtual channel with a fraction of the capacity of the original physical channel, in analogy to the concepts of virtual CPU and virtual memory described in Section 7.2. The selection principles for channel access are similar to those used for bus systems or CPU allocation.

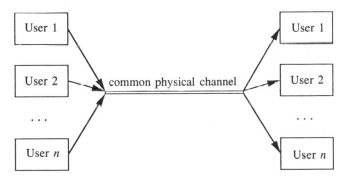

Figure 10.14   The principle of multiplexing.

Multiplexing can be carried out with reference to time or to frequency. In **time division multiplexing (TDM)** the channel is divided in periodic time slots and each user has access only to their assigned slots (Figure 10.15(a)). In **frequency division multiplexing (FDM)** the channel bandwidth is divided into frequency bands, each allocated to one virtual channel (Figure 10.15(b)). FDM requires that the original data modulates a carrier wave at the central frequency of the allocated band. With TDM the entire channel bandwidth is available, but only for a fraction of the time. In FDM a fraction of the original bandwidth is available all the time. In this sense, TDM and FDM are complementary methods.

The two types of multiplexing are also equivalent in a practical sense, because they require that the data is processed at both ends of the channel. TDM has, however, two advantages over FDM: all processing is digital, and there is no need to install and service high-frequency equipment. In addition, with TDM it is possible to control the allocation of the packets to the different users who request the channel. If a user does not send data for a moment, the empty slots may be given to another who needs that capacity.

Under the assumption that a fraction of the users will not need to access the channel at a given time, it is even possible to share a channel among more users than the channel would physically support. This technique is known as **statistical multiplexing** and is used, for example, in data terminal concentrators. The channel is allocated only to those unit(s)

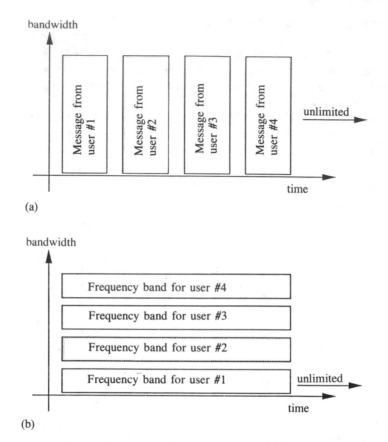

bandwidth

Message from user #1

Message from user #2

Message from user #3

Message from user #4

unlimited

time

(a)

bandwidth

Frequency band for user #4

Frequency band for user #3

Frequency band for user #2

Frequency band for user #1

unlimited

time

(b)

Figure 10.15   Time division multiplexing (TDM) and frequency division multiplexing (FDM). (a) Time division multiplexing; (b) frequency division multiplexing.

that request it at a given moment. On the other hand, if the number of units that need the channel at a certain time is higher than the channel can support, some of these units will have to wait before they get access. The net result is that the channel is used by more entities than if every slot were always assigned to the same user.

If transmitter and receiver are located at the opposite ends of a channel and always send data in the same direction, the transmission mode is called **simplex**. But the units on both sides of a channel may need to both transmit and receive, so that the channel has to be multiplexed for communication in both directions. If only one unit at the time disposes of the channel, the transmission is called **half duplex**. This is equivalent to TDM and requires that the units exchange control signals according to a protocol to tell each other when it is time to take over the transmission. In **full duplex** communication the units communicate all the time in both directions. Full duplex is realized with FDM, where each side has its own dedicated channel. Two different frequency bands are allocated to the messages going in opposite directions.

Multiplexing is not the only way to divide channel capacity among different users. The more general problem is known as **medium access** and is particularly important in local area networks, where a high number of units (which may count in the hundreds) shares a common physical channel and exchanges messages in all directions. In Section 10.5 some of the methods for medium access control are examined in relation to the local area networks that use them.

10.4.3 Error Detection and Correction

A physical link supports data transfer between different locations, but cannot ensure that the data arrives in exactly the same form as at transmission time. Line noise may have corrupted part of the message. To ensure data protection, two error correction strategies may be employed, both require the active participation of the receiver. **Error detection** entails appending some information to the original message in order to enable the receiver to find out whether there have been transmission errors. **Error correction** involves appending sufficient information to the original message to enable the receiver to reconstruct a corrupted message on the basis of the received data.

Contrary to what might seem to be the case, errors tend to come in bursts rather than singularly. In any data message it is more likely that several bits in succession do not take their correct value rather than that just one bit at random is wrong. This is due to the fact that discrete noise sources (different from the continuous white noise always present on a channel) generate pulses a few milliseconds long. At common data transmission rates, some tens of bits that would be affected by the disturbance are transmitted during this time.

In order to determine whether any errors have occurred during data transmission, checking and correction methods have been defined. Error checking is done in practice by computing a **checksum** with the help of some algorithm. The data for transmission is divided into **blocks** of known length (from a few bytes to a couple of thousand bytes). The sender computes the checksum, which has a length from up to some tens of bits and appends it to the data block. The receiver computes again the checksum from the incoming data on the basis of the same algorithm. If both checksums are equal the block is accepted, otherwise the receiver requests the sender to retransmit the whole block.

The checksum is called **cyclic redundancy check (CRC)** and also **frame check sequence (FCS)**. The name derives from the fact that the operations to compute the CRC may be performed by shifting the bits of the incoming data block through a register. The original string of data for transmission is divided by a binary number one or two bytes long, commonly expressed in the polynomial form:

$$x^n + x^{n-1} + \ldots + x^2 + x^1 + 1$$

The polynomial form is not an equation, it is just a notation. The check polynomial is one bit longer than the resulting CRC and begins and ends with 1s. For instance, the standard CRC-CCITT polynomial is expressed as:

$$x^{16} + x^{12} + x^5 + 1$$

which corresponds to the binary sequence 1000100000010001.

The CRC complements the division of the rest of the original data block by the check polynomial. The original data block must be longer than the check polynomial so that the division makes sense. When the CRC is appended to the original data, the resulting bit sequence is a multiple integer of the check polynomial. A new division performed on the whole sequence must yield a result of 0 if the data has not been corrupted. A different result indicates that the original data block was corrupted and that retransmission is needed.

The efficiency of the checksum in error recovery decreases with very long data fields. With checksums, it is possible to detect all error bursts shorter than the CRC and more than 99% of the longer error bursts.

Error correction schemes transmit redundant information together with the original data. The receiver uses all the information received to compute back the original data block. The necessary redundant data requires quite a large share of the message (typically 10–20% of the original block length), so that in general it is more economical to use a simple error detection scheme and request retransmission for the corrupted data blocks. Error correction procedures are important when communication takes place in one direction only or when full duplex handshaking with data block retransmission is not feasible.

10.4.4 Character-oriented Protocols

A common method for asynchronous transmission is to transmit each byte independently (**character-oriented transmission**). Every byte is preceded by a start bit to allow the receiver to be synchronized on the incoming transmission. The byte is terminated with a parity bit and a stop pulse of length 1, 1.5 or 2 bits (Figure 10.16). Reverse coding is common, so that an idle line is kept at a high level. Due to the presence of the start and stop bits, asynchronous transmission is also known as **start-stop transmission**.

The value of the **parity** bit depends on the related communication parameter. For **even parity** the total number of mark '1' bits plus parity bit must be even; for **odd parity** the sum must yield an odd number. For parity = none, the state of this bit is disregarded. If it detects a parity error the receiver can do no more than communicate it to the higher layers in control of communication that have to decide, for instance, whether to send a control character back to the transmitter to request retransmission of a wrong byte. The

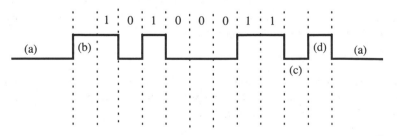

Figure 10.16   Asynchronous byte transmission. (a) Idle line; (b) start bit; (c) parity bit
(even parity, total number of 1s is even); (d) stop bit.

combinations of the number of bits, parity type and length of stock bits are many; the protocol 8N1 (8 bit, No parity, 1 stop bit) is finding wide acceptance.

Character-oriented, asynchronous transmission is uncomplicated but also inefficient because of the additional bits required to delimit each byte. Gaps between characters contribute to time losses. Asynchronous transmission is used, in practice, for up to 20 kbit s^{-1} and only with short connecting cables. In its most common application, the use of modems on telephone lines, asynchronous transmission works at a typical maximum speed of 9600 bit s^{-1}.

10.4.5 Bit-oriented Protocols

There are many types of data-link layer protocols that ensure the integrity of the transmitted data or allow the reconstruction of a message. The protocols can generally be divided in **bit-oriented** (the content of each transmitted message is a string of bits of variable length) and **byte-oriented** (a string of bytes with some delimiting control character is transmitted).

The most famous of the bit-oriented protocols is **synchronous data link control (SDLC)** by IBM. Several other protocols which are similar — but incompatible — to SDLC, have been subsequently defined. Particularly important in industrial automation is **high-level data link control (HDLC)**. This protocol is standardized by ISO and serves as a reference for other protocols, e.g. the Ethernet data frame (Section 10.5.2).

The fields in the HDLC frame are predefined and have fixed length (Figure 10.17); only the data field has variable length. The delimiting **start** and **end flags** have the unique pattern 01111110. Of course it may happen that the data to be transmitted also contains the pattern 01111110 somewhere. A technique called **zero insertion** or **bit stuffing** is used to avoid a potential conflict in such a situation. When there are five consecutive 1s in the original data, the sender inserts a 0 immediately after them. On its side, the receiver deletes a 0 following five 1s, so that the original data pattern is reconstructed. The 0 is obviously not inserted during the transmission of the start and end flags.

The **address** field is relevant only when several units can receive the same message. This is the case in local area networks where all units are connected to the same physical medium and must identify which messages are addressed to them.

The **control** field indicates the type of message and contains related information. In HDLC three types of messages are foreseen: information, supervisory and unnumbered. An **information frame** carries data; its control field indicates the current frame number and whether more frames are going to follow. The **supervisory frame** is used in the receiver to communicate with the sender and request retransmission of lost information frames,

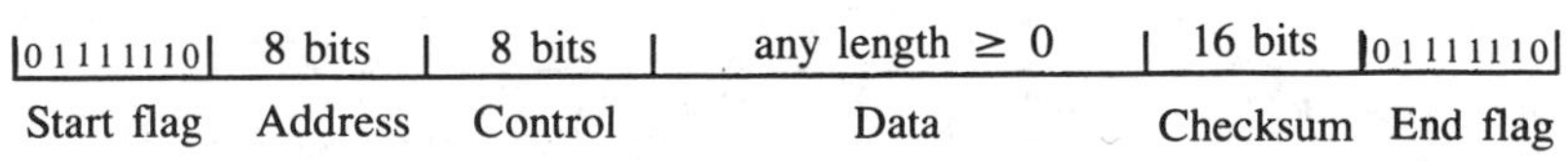

01111110	8 bits	8 bits	any length ≥ 0	16 bits	01111110
Start flag	Address	Control	Data	Checksum	End flag

Figure 10.17 Example of the basic high-level data link control (HDLC) frame structure.

for instance, if there has been a mismatch in the frame numbering sequence. **Unnumbered frames** are used for link initialization and status reporting.

The **data** field may have any length and the **checksum field** is computed with a CRC-CCITT polynomial. The **end flag** contains the pattern 01111110 like the start flag.

The general HDLC format provides the means for full duplex communication with error control while allowing implementation of different communication protocols. For instance, the sender may continue to transmit frames without waiting for their acknowledgment, or the receiver could have to either acknowledge all of them or indicate only which frames need to be retransmitted.

10.4.6 Block-oriented Protocols

In **byte-** or **block-oriented protocols**, the bytes to transmit are grouped in blocks and each block is delimited by special control characters such as SOH (start of header), SYN (synchronization), STX (start of text), ETX (end of text) and EOT (end of transmission). Each transmission starts with the SOH and STX characters and terminates with ETX, EOT. A few SYN characters at the beginning of the data block are used to synchronize the receiver on the incoming transmission (Fig. 10.18).

It is common practice never to keep a line idle. Transmitter and receiver exchange synchronization characters when there is nothing else to communicate. Depending on how the actual communication protocol is implemented, the ACK (acknowledge) and NAK (not acknowledge) characters are used to control the transmission flow. In one solution, the receiver might have to explicitly acknowledge every transmitted block. In a different procedure, the receiver requests only the retransmission of the corrupted blocks.

Data block transmission is used to transmit blocks of several characters, up to some hundreds, in an uninterrupted sequence. A data block starts with a few synchronizing characters to warn the receiver of the impending transmission and ends with a special end-of-block character. The pulses corresponding to the data bits are in phase with the synchronizing reference signal.

Synchronous data communication is, in general, more efficient than asynchronous, as it makes better use of the line and avoids idle times. It is used for data rates higher than about 2 kbit s^{-1}. On the other hand, the equipment for synchronous data transmission is more complicated, and therefore more expensive, than for asynchronous transmission. Synchronous transmission is efficient for long byte sequences transmitted in bursts, such as file and message transfers. For applications that do not require high speed and where the characters are sent one at a time, as the connection of a terminal to a computer or of a computer to a printer, asynchronous transmission is usually adequate and generally used.

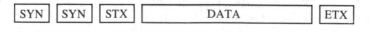

Figure 10.18 Synchronous block transmission. (SYN — synchronization character: STX — start of transmission character; ETX — end of transmission character.)

10.5 LOCAL AREA NETWORKS

10.5.1 Communication Networks

So far, we have seen how to connect together two different points. This is still not sufficient for practical applications, as communication often involves more participants. The installation of a dedicated line between all possible communication partners is not only impractical but also expensive, because of all the necessary cabling.

Local area networks (LANs) are used to connect together several locations (called **nodes**) so that they can all communicate with each other. A LAN is basically a shielded cable, twisted pair or coax, installed in such a way that it comes into the proximity of each unit to be connected. Depending on the operational method, a LAN may have to be built as a ring or be open-ended. LANs have a typical length of some tens of metres to a couple of kilometres and allow data rates of between 1 and 10 Mbit s^{-1}.

When several nodes are connected via a LAN, different topologies (that is, geographical configurations) are possible. Two nodes may be connected together either directly or the communication path may have to pass via another unit, which relays the data to its destination. The topology and the layout of the communication channels depend on several factors, for instance, the type of traffic, the distance between the nodes, the kind of disturbances which may affect the data channel and the desired speed, throughput and accuracy in data transmission. Different network topologies are shown in Figure 10.19.

In the selection of a network topology, it is important to choose a structure which guarantees efficient data transport from the source to the destination and provides redundant paths in case a direct link should be broken. Alternatively, a LAN may consist of a cheap and strong medium, so that its maintenance is facilitated.

LANs are covered by Layers 1 and 2 in the OSI model. Layer 2 is divided in two sublayers: **medium access control (MAC)**, the actual interface to the physical medium, and **logical link control (LLC)** which is responsible for the coordination of network access.

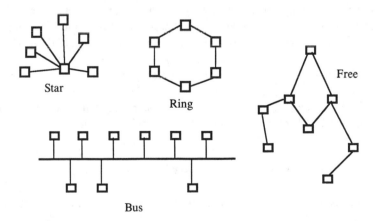

Figure 10.19   Network topologies.

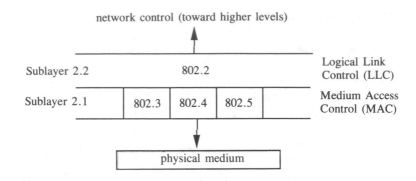

Figure 10.20   IEEE/ISO network standards structuring for OSI layer 2.
Standard IEEE 802.2 (ISO 8802-2): Logical Link Control.
Standard IEEE 802.3 (ISO 8802-3): Carrier sensing multiple access/collision
detection (CSMA/CD).
Standard IEEE 802.4 (ISO 8802-4): token bus.
Standard IEEE 802.5 (ISO 8802-5): token ring.

The concept is illustrated in Figure 10.20. We see here an important consequence of layering: the physical medium may be the same for different LAN types (coax cable or twisted pair). The signals, modulation type and medium access protocols differ for different LANs. Finally, the higher control of the LAN operations is again the same. At LLC and higher layers it is not important to know which type of LAN is used.

The most comprehensive documentation and specification of LANs has been devised by the professional engineering society IEEE (Section 12.1.4). The IEEE organization has published a set of documents known as the 'IEEE 802' standard, which describes both general principles and particular types of LANs. The IEEE 802 specifications have been taken over by ISO, which publishes them as standard ISO 8802. The LAN standards are subject to periodic revisions, but their major features are now stable.

The same thing has happened with LANs as with other products for communication and automation technology. In the beginning it was hoped that only one, general LAN standard for office and industrial applications would be selected and that everybody would follow it. When it was time to take a decision, there were three competing — and mutually incompatible — technologies, each backed by powerful companies. The committee charged with defining the standard could not agree on any one of them and in the end it was decided that three different standards would be better than no standard at all. This has led to today's standards CSMA/CD ('Ethernet'), Token Bus and Token Ring.

10.5.2 The IEEE 802.3 Network (Ethernet)

Ethernet is a widely used LAN for both industrial and office applications. Jointly developed by the companies Xerox, Intel and Digital Equipment, Ethernet was introduced to the market in 1980. Ethernet follows the IEEE 802.3 specifications. Although there are a few

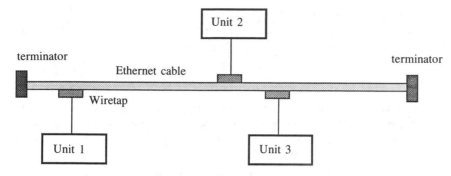

Figure 10.21 Ethernet configuration with wiretaps.

differences between the data packet description of IEEE 802.3 and the packet used in industrial Ethernet, their principles of operation are identical. We will refer in the following to the industrial standard.

Ethernet has a bus topology with branch connections. At the physical level, Ethernet consists of a screened coax cable to which peripherals are connected with 'taps' (see Figure 10.21). Two types of cable are available: 'thick' Ethernet (diameter about 2.5 cm) and 'thin' Ethernet (diameter 0.5 cm); the latter also known as ThinWire Ethernet or Cheapernet (the name is self-explanatory). The connection to thick Ethernet is made with wiretaps, i.e. cable fasteners with a nail that punches through the plastic coat to reach the inside conductor (Figure 10.22). 'Thin' Ethernet uses radio-frequency RG-58 coax cable and

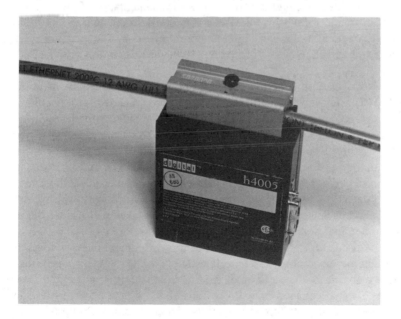

Figure 10.22 Ethernet H4005 wiretap. (Courtesy of Digital Equipment Corporation.)

8 Byte	2/6 Byte	2/6 Byte	2 Byte	46-1500 Byte	4 Byte
Preamble	Destination Address	Source Address	Length of data field	Data	Checksum

Figure 10.23   Ethernet packet.

BNC connectors; in some cases even twisted pair cable can be used. Thick and Thin Ethernet cables may be joined together with special connectors. The maximum length between the most remotely connected points is about 2500 m; the IEEE specifications suggest that it should not exceed 500 m.

Ethernet does not have a network controlling unit and all devices decide independently when to access the medium. Consequently, as the line is entirely passive, there is no single-failure point on the network. Ethernet is a baseband medium; and the digital data is put directly on the line using Manchester coding. Ethernet supports communication at different speeds, as the connected units need not decode messages that are not explicitly directed to them, with a maximum data transfer rate of 10 Mbit s^{-1}.

For the data link and medium access layer in Ethernet a packet format is defined similar to the HDLC packet (see Figure 10.17). The Ethernet packet (Figure 10.23) starts with an 8-byte preamble with a 010101 ... pattern, followed by the destination address and source address, each 6 bytes long. The destination address may be structured to cover only one destination unit, a defined subset, or all the units in a group. A particular code indicates 'broadcasting', when a message is addressed to all connected units. The destination address is followed by a 2-byte type field, whose definition does not follow particular requirements and depends on the actual system implementation. Then the data, defined to be between 46 and 1500 bytes long (messages shorter than 46 bytes have to be padded with empty characters), follows this preliminary information. The total packet length varies thus between 72 and 1526 bytes. The packet is terminated by a 4-byte checksum and the end flag.

The method used in Ethernet for medium access coordination, a protocol called **CSMA/CD (Carrier-sensing Multiple Access/Collision Detection)**, is quite interesting. According to the CSMA/CD protocol, the units decide independently when to access the line and start transmission. They only have to check that there is no other transmission, but if the line is free, a new transmission can begin. The transmitting unit constantly controls the quality of the signal on the line, listening to what it is transmitting itself. If any other unit had also sensed a free channel and had attempted transmission at the same time, both units would detect that their signals are garbled and immediately stop transmitting. To ensure that all units detect that a message collision has taken place, after the interruption both transmitters put a short noisy 'jam' signal on the line.

The time needed for an electric signal to travel the maximum cable length is about 50 μs; in Ethernet operations this time is called 'time slot'. If a transmitting unit does not detect a collision for the duration of the first time slot, it is safe. But if two units have to wait after a signal collision, they do so for a random time period of 0 or 1 time slots (0 or 50 μs) before they attempt a new transmission. The probability of a new collision is now 50%. If a new collision occurs, the range from which the waiting period is selected at random is increased by powers of two to 0-1-2 slots, then 0-1-2-3-4 slots and so on, up to a maximum of 0 to 1023 time slots (50 ms). In the case that a collision still takes place, the units assume that the problem has a different cause and report the situation to the higher layers.

Another situation where error handling is not included in the transmission protocol but left to higher layers occurs when the checksum computed by the receiver does not match the one received with the message (Section 10.4.3). These higher layers can decide whether to issue a request for retransmission, which in practice is done most often. In order for the acknowledgment messages not to compete with normal network traffic, the first time slot after a successful transmission is reserved for immediate acknowledgment from the receiver to the transmitter.

Ethernet's concept is flexible and open. There is little capital bound in the medium, and the medium itself does not have active parts such as servers or network control computers which could break down or act as bottlenecks and tie up communication capacity. Some companies offer complete Ethernet-based communication packages which may also implement higher layer services in the OSI hierarchy.

A disadvantage of a network based on the CSMA/CD principle is that the probability of collision increases with both the number of connected units and the length of the line; as the collisions increase so does the time that is lost in trials. This means in practice that there is no upper bound to the time it may take to access the medium and transfer a message. This is a serious drawback for industrial real-time applications, where it is necessary to know exactly 'worst-case' performances in advance.

10.5.3 The IEEE 802.4 Network (Token Bus)

The LAN-type **Token Bus** is described in the IEEE 802.4 specifications. Together with Ethernet, it is the most widely used type of local area network in industry. In Token Bus all units are connected to the network in a bus fashion (figure 10.24). The physical conductor is, as for Ethernet, either coax cable or twisted pair and supports data rates in the range $1-10$ Mbit s^{-1}. Only one unit at the time can send messages on the LAN. The right to send is given by the **token**, a special bit pattern that is passed from one unit to the other in a round-robin fashion. A unit that receives the token gets the right to transmit for a specified interval, and must then pass the token to the following one. If a unit does not have anything to transmit, it just passes the token to the next. The circular pattern in which the token is passed makes the Token Bus a **logical** ring, although its physical topology is a bus.

The devices connected to the token bus can be active or passive. Active devices circulate the token and may transmit whenever they hold it. The **passive (or slave) stations** may issue messages only when they are addressed by an **active (master) station**; their reply or acknowledgment must be immediate. Active devices are usually process computers, robot controllers and other advanced equipment; passive units are sensors, PLCs and other low-end devices which do not need to deliver information unless requested to do so by the end user of the data.

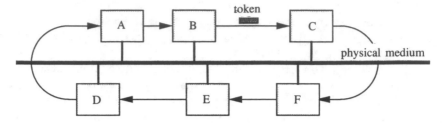

Figure 10.24   Token bus operating principle. The stations A–F circulate the token
and can communicate only when they hold it.

   The Token Bus specification has provisions for network management that require the
active participation of all devices. A new token must be generated at power-on or in the
case that the station which currently holds it should crash and destroy the token. Other
examples of network management functions are configuration changes, addition and removal
of stations, recognition and isolation of faulty stations, etc.
   The Token Bus has a computable worst-case delay for a unit to gain access to the
medium. On the other hand, a unit cannot freely initiate a transmission but must first wait
for the token. Compared to Ethernet, there is a computable worst-case waiting time (the
time for the token to circulate through all other units and for these to keep it as long as
they are allowed to), but this also implies enforced waiting before initiating a transmission.
Transmission time is also limited, longer messages have to be divided and sent in separated
blocks. Ethernet, on the other hand, allows a transmission to be immediately initiated,
but does not promise anything about its success. The advantage of the Token Bus concept
in applications with tight timing requirements is just that luck does not need to be an explicit
communication parameter.

10.5.4 The IEEE 802.5 Network (Token Ring)

The LAN solution **Token Ring** was introduced by IBM. The operating principle of Token
Ring is similar to Token Bus; the difference is that the ring is not only logical but also
physical, with the stations connected in a circular path (Figure 10.25). The token is
continuously circulated on the ring and each station receives it, regenerates it and puts
it on the line again. When a station holds the token, it may transmit a message. Messages
can be sent by passing them onto the ring instead of the token (the sender does not pass
the token further until it has terminated its transmission). Alternatively messages can be
appended to the token or the token itself can be appended to the message. The cable used
is a shielded twisted pair with a capacity of $1-4$ Mbit s^{-1}.
   Each station checks the destination field of incoming messages. If the message is not
intended for that station, it is just sent further. If the message is intended for the station,
it is retained. According to the actual protocol type, the message can also be circulated
until it reaches the station that transmitted it, as an acknowledgment that it actually reached
its destination. When a station is not powered in Token Ring, its input and output bus

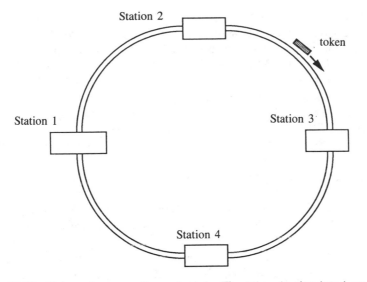

Figure 10.25 Token ring operating principle. The token is circulated among the stations. A station can send messages only when it holds the token. All stations identify and collect the messages directed to them and relay further the others. A message stays on the ring until it is returned to the transmitting station.

connections are shorted together via a relay so that other messages can circulate on the bus without hindrance.

Token Ring is slower than Ethernet and Token Bus, and requires more circuitry, so that it is hardly a good candidate as a general bus standard. In introducing it, IBM probably intended to play the same game as with the IBM PC and set *de facto* standards that everybody else had to follow — whether they wanted to or not. Today's industrial applications are based in practice either on the Ethernet or the Token Bus principle.

10.5.5 The Higher OSI Layers (Layer 3 to Layer 7)

The layers in the OSI model above Layer 2 are of relatively little importance for industrial applications (with the exception of Layer 7), so we will just mention them briefly. The higher layers are used for remote operations, file exchange and database applications over public data networks.

In the OSI scheme, **Layer 3 (Network)** is responsible for establishing and operating a virtual link between any two nodes that wish to communicate. Layer 3 describes how the messages are routed, i.e. passed from one node to another that is directly connected, and from here to another, until it reaches its destination after some hops.

OSI Layers 1 to 3 are called **external, network** or **communication layers. Layer 4 (Transport)** is the actual interface between the machine software and the external network. Although there are several protocols and methods at the lowest layers to guarantee safe data transmission, it is only at Layer 4 that it is possible to verify that the data was actually sent from the source to the destination machine. **Layer 5 (Session)** is responsible for establishing and for bringing down a connection. The session layer also provides extensions to the services of Layer 4 with particular concern for the functions for file transfer from one machine to another.

Layer 6 (Presentation) is concerned with data coding and structuring. Here two standard specifications play an important role. ISO 8824, known as **Abstract Syntax Notation no. 1 (ASN.1)** gives guidelines about how to structure data for transmission. The specification ISO 8825 for **Basic Encoding Rules (BER)** indicates how the data is to be coded in binary, i.e. the relation between the content of a message and the 0s and 1s to be transmitted. It is here that the coding in, for example, ASCII or EBCDIC takes place.

Layer 7 (Application) covers several fields related to the utilization of the transmitted data. Here can be found the **File Transfer Access and Management (FTAM)** protocol, an abstract notation to describe the different features of a file, e.g. its creation date, access control, protection passwords and multiple access information. Another Layer 7 specification is the **Message and Handling System (MHS**, CCITT X.400), a set of procedures for electronic mail transfer in the form of messages. The **Virtual Terminal (VT)** concept with hardware-independent commands to control terminal output (Sec. 10.2.3) also forms part of Layer 7. We will examine these ideas in more depth in the next section.

10.6 THE MANUFACTURING AUTOMATION PROTOCOL (MAP)

10.6.1 A New Concept for Industrial Communication

The need for a practical and general way to connect together different units in production lines and process control is widely recognized. In particular, the American car-manufacturing giant General Motors (GM) realized early on that the incompatibility of computer systems was a major hindrance to the integrated automation of their production plants and began studies on how to interconnect its production computers. At GM, they noticed how significant the rewiring costs were related to the retooling for every new car model. Moreover, the costs pointed no way but up. According to estimates made at the beginning of the 1980s, by 1990 there should be about 100,000 different units to interconnect at GM plants, e,g, robots and PLCs. The costs just for the interconnections would take a major share of the total company investments in automation. GM decided therefore to develop a comprehensive and standard approach to plant-floor communication. The idea quickly won the interest of major computer manufacturers and other companies in the industrial automation business, and has led to what is known today as Manufacturing Automation Protocol.

Manufacturing Automation Protocol (MAP) is not a standard, an interface or a type of electric cable, but a comprehensive concept to realize interconnections between the different equipment at plant-floor level and higher planning and control systems.

However, realization of the conceptually simple principle that has different units communicating together using common protocols has taken about 30 years and is far from being completed. There may have been two major culprits: the lack of a general frame for data communication (which is now provided by OSI with the related standards) and the fact that major corporations did not consider it to be in their interests for their products to be compatible with those of their competitors. Now — at last — compatibility has turned out to be a good sales argument. The principal goals of open communication are **interoperability** (all information should be understood by the addressed units without the need for conversion programs) and **interchangeability** (a device replaced with another of a different model or manufacturer should be able to operate without changes in the rest of the connected system).

MAP follows the OSI layering scheme. For each OSI layer there is a defined standard within the MAP scheme. The standards at Levels 1 to 6 are also used in applications other than MAP; the MAP-specific part is the **Manufacturing Message Specification (MMS)**, described later in this section. The foreseen MAP standards for each OSI layer are as follows:

Layer 7:   ISO 9506 Manufacturing Message Specification (MMS)
Layer 6:   ISO 8824 Abstract Syntax Notation no. 1 (ASN.1) and
           ISO 8825 Basic Encoding Rules
Layer 5:   ISO 8326/8327
Layer 4:   ISO 8072/8073
Layer 3:   ISO 8348/8473 (CLNS) and ISO 9542 (ES/IS)
Layer 2:   ISO 8802.2 Logical Link Control and ISO 8802.4 Token Bus
Layer 1.   Broadband/Carrierband link

Put another way, a MAP application must have a physical connection which follows the LAN Token Bus standard with Logical Link Control according to IEEE 802.2; it must code data following ASN.1 (ISO 8824) and the Basic Encoding Rules of ISO 8825; and it has to exchange MMS messages (ISO 9506). Any other combination, even if it is technically feasible, is not consistent with the MAP scheme. For instance, a solution where Ethernet is used instead of Token Bus for the data link and physical connection is not a MAP application.

At the physical level, MAP can be implemented with different media and signal types. The initial GM requirements to transport data at 10 Mbit s^{-1} require two adjacent channels with 6 MHz bandwidth if broadband AM-PSK is used. For baseband-based MAP two data capacities are specified and FSK modulation is used. At 5 Mbit s^{-1} the FSK frequencies are 5 and 10 MHz; at 10 Mbit s^{-1} the frequencies are 10 and 20 MHz.

A scheme similar to MAP but more suited for integrated process-oriented and administrative data communication is the **Technical and Office Protocol (TOP)**. TOP follows the same structure as MAP and is for the most part equivalent, referring to the same standards as MAP; the only difference between MAP and TOP is at OSI Layers 1 and 2. MAP uses the Token Bus local area network while TOP is based on Ethernet. At Layer 7, TOP is richer in the choice of application interfaces: the Virtual Terminal (VT), the Message Handling Systems (MHS) and the File Transfer Access and Management

(FTAM) protocol. The TOP concept has been supported by Boeing Corporation, which for a long time has had an Ethernet-based solution to connect their plant equipment with their production planning system. The basic issues related to MAP and TOP are quite similar; in what follows we will limit our discussion to MAP because it has been specifically developed for use in factory environments with real-time processing requirements.

The reason behind the selection of particular standards for MAP, and most of all for the choice of broadband network and the Token Bus medium access method, was that these had already been tested in real plants and that Token Bus devices were already available. Token Bus also has a specified and computable worst-case time limit for message transmission which Ethernet lacks (we have already seen how in Ethernet there is no upper bound to the time it may take to access the medium, it all depends on the number of connected devices and on luck). Some real-time applications cannot be designed with this uncertainty factor. Not surprisingly, MAP and TOP were supported by the respective companies GM and Boeing, which have quite different requirements because of the way their production is organized, production lines versus fixed-position mounting. Compatibility at higher levels ensures the interconnectivity of MAP and TOP applications.

In factory automation there are, generally speaking, three operational levels: general management, process control or production line control, and field control. MAP supports the central levels of communication, it coordinates the operations of multiple cells on a production line and of several lines at plant level. MAP is not apt for communication and control down to the sensor level. MAP is a very 'heavy' product because of all the layers involved with their related protocols, and this does not match the need for simple, fast and cheap technology required at the lowest factory automation levels. Here a different technique, the fieldbus, is used (described in the following section). Neither is MAP apt for management support at the highest levels where strategic factory decisions are made. The software tools for use at that level do not have to satisfy special real-time requirements and may be developed with the normal programming techniques for statistical processing and the analysis of large quantities of data. However, MAP remains the key concept for the practical realization of computer-integrated manufacturing (CIM) applications. Factory communication at large will be treated in more detail in Chapter 12.

10.6.2 The Manufacturing Message Specification (MMS)

The **Manufacturing Message Specification (MMS)** is a collection of abstract commands, (in practice, a language) for remote monitoring and control of industrial equipment. MMS defines the content of monitoring and control messages as well as the actions which should follow, the expected reactions, acknowledging procedures, etc. MMS is an ISO standard (ISO/IEC 9506) structured in the following documents:

Part 1:   Service definition
Part 2:   Protocol definition
Part 3:   Robot control messages
Part 4:   Numerical control (NC) messages

Part 5:   Programmable logic controller (PLC) messages
Part 6:   Process control messages

MMS provides a large number of different services and options. Functions that are of general type, e.g. reading and writing the values of variables on remote devices, starting and stopping the execution of programs and transferring files between different units, are described in Part 1 and Part 2.

The MMS companion standards (Parts 3–6) are oriented to specific, real, devices. Although MMS tries to cover all the functions that are needed in factory automation applications, it is still necessary to leave a certain margin for flexibility and future expansions. This is the background to the current document structure. Each device does not need to understand all the MMS commands, but it is sufficient for it to understand the appropriate subset, as described in the relevant part. Each subset of the whole standard is defined and revised independently of the others.

MMS is based on object-oriented programming concepts, where classes of objects are defined together with the operations which can be performed on them. A central concept of MMS is the **virtual manufacturing device (VMD)**. A VMD is a collection of all possible commands for some type of device. A real machine will understand and react to VMD

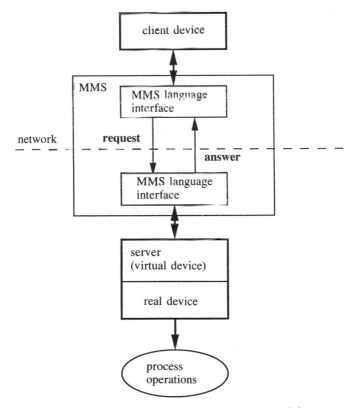

Figure 10.26   The MMS client–server model.

commands by carrying out the specified function in a standard, predefined way. MMS and VMD must be very comprehensive in order to cover different types of similar devices; real devices in general can execute only part of the VMD commands.

The MMS services are based on a 'client–server' model (Figure 10.26). The client requests a service from the server. The server executes the service and gives an answer to the client in order to acknowledge and specify the result of the operation.

A VMD represents the function of a real device, but from the point of view of the client there is only the virtual server. A VMD command sent to a virtual robot to move its arm 30° around the z axis will lead different, real robots to perform the same action. The actual commands generated by the robots to control their actuators can be quite different, depending on the electrical and mechanical design. In the case that a robot is not able to execute the command, either because it has reached the end of the turn radius or because it has no Z axis, it will answer with an operation result code stating the reason.

An important feature of the client–server model is that devices acting as servers do not have a state defined in a local model unknown to the client. All requests from the client lead to self-contained answers and do not imply the use of information that was sent previously. In other words, information from the server devices (VMD) can be considered to be its own database, where updated values can be collected, or stored, at any time. The VMD independence of historical data to be collected by the client and the specification that all requests be acknowledged by the VMD help to avoid errors and inconsistencies due to lost or delayed messages.

10.7 FIELDBUSES

10.7.1 A Solution for Low Level Plant Communication

The emphasis given by OSI and MAP to interconnectivity at several levels with all the necessary protocols to accomplish this goal does not always go hand in hand with the requirement for fast and effective communication that is needed in industrial real-time applications. As already indicated for the OSI model, not every layer is needed for all applications. When all communicating units are located in a close work-cell and are connected to the same physical bus there is no need for multiple end-to-end transfer checks as if the data were routed along international networks. To connect computers in the restricted environment of a factory plant, the data exchange definition of OSI Layers 1 and 2 and an application protocol, such as that provided by MMS, are more than enough. There is little use for other layers, so they can be skipped altogether.

In industrial applications, the largest share of the work (and of the costs) for data collection and processing is due not to central processing, but comes from the field where the devices are installed. In order to gain the advantages of digital technology, there is a growing demand in the industrial world for a new digital standard for low level communication; this standard is known as **Fieldbus**. There is no single Fieldbus as yet, but different solutions have been presented by industry and by research institutions. In the course of time, what has been proposed and what is operating in the field will crystallize around one or maybe a few technologies that will then become part of a more general Fieldbus.

A Fieldbus standard must gain wide acceptance, as has been the case with the 4−20 mA current loop. With an accepted standard, costs are kept down and problems with incompatible components are minimized. An obvious advantage of digital versus analog technique is the saving in the amount of cabling: a single digital loop can replace a large number of 4−20 mA conductors.

The possibilities opened by fieldbuses are notable. A large share of the intelligence required for process control is moved out to the field. (In Chapter 12 we will examine the implications and the advantages with distributed architectures.) The maintenance of sensors becomes much easier because operations such as test and calibration can be remotely controlled and require less direct intervention by maintenance personnel. And, as we have already pointed out, the quality of the collected data influences directly the quality of process control.

The International Electrotechnical Commission (IEC) is working on an international fieldbus standard. Similar to MAP, the fieldbus standard should ensure interconnectivity between different devices connected to the same physical medium. Various national projects have already begun to define how the future standard will look. A final agreement has not yet been reached, but nobody wants to wait until a general standard is introduced. Some companies have already defined their products and are marketing them, and projects have been carried out in France and Germany to define national fieldbus standards. In the end, all experiences and proposals may come together into a single and widely accepted standard, but it might turn out that the different   already existent — proposals will live their own, parallel, lives.

We will briefly examine here the main features of the fieldbuses FIP from France and PROFIBUS from Germany as well as the industrial Bitbus developed by Intel. The interest in these solutions lies not only in their importance for practical applications but also in the different technical operating principles.

10.7.2 FIP

A group of French, German and Italian companies has worked on a Fieldbus design called **FIP**, where the acronym may be considered to be in English **Factory Instrumentation Protocol** or in French **Flux Information Processus**.

FIP uses a twisted pair conductor and the transmission speeds are 31.25 kbit s^{-1}, 1 Mbit s^{-1} and 2.5 Mbit s^{-1}. The cable length for a segment is 500 m for operations at 1 Mbit s^{-1} and varies depending on the actual speed. Several segments can be connected together with repeaters, the maximum number of stations connected to a segment is 32 and for a whole system is 256. All the interface functions are contained in a special integrated circuit called FULLFIP (a trademark of CEGELEC). This integrated circuit is connected with the CPU of the communicating device (PLC, intelligent sensor or other) and on the other side with the FIP bus.

FIP uses a so-called source addressing polling method. Control over access to the bus is responsibility of a device, the 'bus arbitrator', that cyclically polls all the other devices that can be data 'producers' or 'consumers' or both. The arbitrator contains a list of all variables used at all connected stations and each variable is identified by a unique 16-bit address (that means 65 536 different identifiers). When the bus arbitrator places an address on the bus, the device that recognizes it answers by sending back the content of the related variable. At the same time, the consumer units that have also recognized the address store the value they read from the bus (Figure 10.27). Several units can act at the same time as consumers and receive the same message. The arbitrator polls continuously all the variables from the connected devices. The net effect is a distributed database where every variable can either be accessed with its identifier or has its value appearing periodically on the bus.

The data update with the arbitrator has an advantage for data consistency. The operations can be so defined that a fixed (and known) time is used for variable polling and a variable part of the time is left for asynchronous operations. This ensures that the collected data always has the same phase relation (no phase jitter).

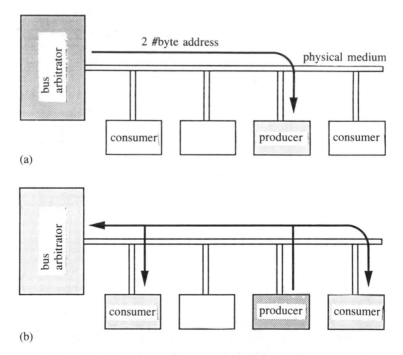

Figure 10.27 The FIP messaging principle. The bus arbitrator transmits an identifier (16 bit logic address) and the addressed unit (producer) replies with the actual data. The consumers receive the data. Information on what data is to be used by which devices is contained in local tables prepared and downloaded by the bus arbitrator. (a) Transmission of an address; (b) reply with current data.

The messages in FIP contain address or data and a code to identify the type. The frame structure is similar to the HDLC format with preamble, checksum and end flag (Section 10.4.5). In the basic operation mode, there is no need to acknowledge messages because the real-time data is periodically updated at very short time intervals anyway. In FIP there are functions for network operations, e.g. the start and stop of remote tasks.

10.7.3 PROFIBUS

PROFIBUS (Process Fieldbus) is supported by a group of German manufacturers; it was initially developed under the coordination of the German Federal Ministry of Research and Technology and is now a German standard (DIN 19245). Similarly to the other fieldbuses, PROFIBUS combines specifications at OSI Layers 1, 2 and 7 (physical medium, data link and application). PROFIBUS follows existing standards at the different layers and attention has been paid to compatibility with the hardware already in use. The most important bus operations do not require specialized processors but can be carried out with programs running on common CPUs.

The PROFIBUS physical medium is a screened twisted pair cable according to RS-485 specifications, with a maximum length of 1200 m (up to 4800 m may be reached with the help of repeaters). Data transmission speeds are 9.6, 19.2, 187 and 500 kbit s^{-1}.

PROFIBUS uses a hybrid version of the IEEE 802.4 Token Bus model with master and slave stations. Up to 127 stations, divided into **active** and **passive**, can be connected to the bus. Active stations circulate the token and may transmit when they hold it. Passive stations do not circulate the token and must be addressed by an active station to be able to reply with a message (Figure 10.28).

In PROFIBUS are foreseen different types of messages, from simple data requests and acknowledgments to longer data packets. The messages have two different levels of priority (high/low) and their maximum length is 256 bytes. PROFIBUS messages are based on a subset of the MMS language which includes the 'virtual device' client–server concept and a set of commands to perform network operations, remotely start and stop tasks, etc.

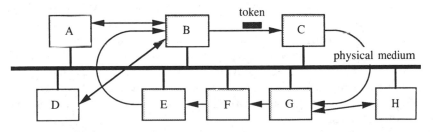

Figure 10.28 The PROFIBUS operating principle. The master stations B, C, E, F and G circulate the token and can communicate when they hold it. The slave stations A, D and H are only allowed answer to direct requests from one of the master stations that holds the token.

An innovative feature of PROFIBUS is the definition of management functions, e.g. to change the bus configuration, add or take away stations as well as detect and isolate faulty units.

PROFIBUS has been designed to support the connection on the same bus of intelligent sensors together with more complex units such as PLCs, regulators and small process computers. The message-based communicaton scheme allows flexibility in the type and amount of data which can be exchanged. However, a price must be paid in terms of speed. PROFIBUS cannot guarantee that all sensor data are collected at a constant rate. Whether this is relevant or not for a process must be defined from case to case.

10.7.4 Bitbus

Bitbus is the name of a fieldbus introduced by Intel in 1984 and is the basis of industrial products of other companies. Bitbus consists of the two first OSI layers: physical (1) and data link (2).

At the physical level, Bitbus uses twisted pair cable according to the RS-485 standard. A maximum of 28 devices can be connected to a single bus and several buses can be joined together with help of repeaters. Transmission speeds are 62.5 kbit s^{-1}, 375 kbit s^{-1} and 2.4 Mbit s^{-1}. At the lowest speed, the distance between repeaters can be up to 1200 m.

Bitbus is structured hierarchically. One of the connected devices acts as master, all other devices are slaves. The master is thus always in control of the communication process. The communication protocol is strictly causal: the master sends a message to a slave device and the slave must reply; a slave cannot initiate communication on its own, but must wait for a request from a master. A slave must reply to a master request within a specified time. If the slave fails to respond, the master can try again several times. If no request is successful, the master writes off the slave station as unreachable. When several Bitbuses are in operation and interconnected via a network, the master units also act as network communication units. The Bitbus protocol does not foresee multiple masters and there is no arbitration method to transfer master right. In general, a 'strong' unit will act as master and the slaves will be simpler devices, with less electronic circuitry.

The strict division in master/slaves has an advantage in real-time operations, as it makes reaction times exactly computable in advance. The Bitbus packets follow the general HDLC structure (see Section 10.4.5); they are handled and sent as intertask messages between the tasks on the master unit and the tasks on the slave units. The application programmer, however, will not work directly at the packet level because the bus functions can be accessed via defined procedures in the Bitbus support software. These procedures include functions to read and write from the master unit in the memory of local units (and thus to/from the process to be controlled), to download tasks and data, to start and stop tasks in the local units, etc.

Bitbus is a mature industrial product supported by several companies. It is apt for use in smaller production cells or processing units with relatively limited data traffic, where there is no need for several masters to control communication. In larger scale applications, the master units are connected together with high speed local area networks and transfer the concentrated data to a central process computer at higher levels.

10.7.5 Towards a Fieldbus Standard?

The solutions that we have seen are by far not the only ones that exist. Every major company in the industrial automation business has some kind of proprietary data bus more or less based on — and compatible with — existing standards. The different fieldbus implementations of today are not necessarily a disadvantage where functionality is concerned. PROFIBUS has the flexibility of the token, any unit can start its own data search when it owns the token. FIP and Bitbus with their central unit concept are more effective as long as they follow predetermined schedules and address simpler devices, but they support asynchronous operations with more independent devices less well. In this sense, FIP and Bitbus are strictly hierarchical with only one master, PROFIBUS supports the concurrent activity of several intelligent units.

It is difficult to foresee whether the idea of an integrated fieldbus standard will see the light of day, or if different standards will have to live side by side. Given that national chauvinism and market forces each play a role at least as important as that of standardization committees, the standards that we have seen are almost certainly here to stay. This does not necessarily have to be a disadvantage. The loss in uniformity at component level will be offset by the freedom to select a communication method that can best match a specific application. Real-time oriented fieldbuses like FIP or Bitbus can be used at the lowest level to collect sensor data in a cell database; PROFIBUS may connect together several independent devices and intelligent sensors in a hybrid configuration.

10.8 COMMUNICATION OVER PUBLIC NETWORKS

Local area networks are useful for connecting together the equipment within a building or an otherwise limited geographical area. On the other hand, LANs are of little use when the distances are larger, in a range of more than about one kilometre, or if the physical link has to cross public property. In many countries, notably in Europe, one can connect all the equipment in a large plant with a high performance LAN, but run into trouble if the connection is to extend to an office building located on the other side of a public road. The issue of communication monopoly is a complicated and still unresolved question, and we will not deal with it here but take it as a fact of life. In practice, it means that there are situations where data has to be carried over networks owned by a public PTT administration (the abbreviation means post, telegraph and telephone, and is used to denote national telecommunication organizations) or some other telecommunications company.

Using a public network to transfer data is not necessarily a disadvantage. The telephone network, that can also be used to carry data, is installed and operates, in principle reaching out to the whole world. On the other hand, its bandwidth is limited and allows only low data transmission rates. The following description is a brief introduction to the main features of public networks from the point of view of the user interface to the networks. It will not describe the internals of how public networks operate.

10.8.1 The Telephone Network and Modems

The telephone network can rightfully be considered one of the wonders of the modern world. It connects together some half a billion telephones in every country on Earth and works reasonably well, albeit in many cases with deficiencies and shortcomings. It is a good example of how a very complex system gradually evolved around a simple design instead of it being planned from the beginning. In fact, it took over 100 years for the telephone system to evolve to its present state. But the world telephone system also reflects the huge differences among world countries. According to UN estimates, two-thirds of the world population still do not have access to telephone services.

From a data transmission point of view, a telephone connection is a poor medium. Its bandwidth is about 3000 Hz, enough to carry a human voice intelligibly, but poor for data communication. Note that the limitation in bandwidth is not due to the circuitry in the telephone set or to the twisted pair cable connection, which could support data rates of some hundreds of kbit s^{-1}, but to filters installed at telephone exchanges and along the line.

The signals are limited in bandwidth to allow their multiplexing for long distance transmission. Analog and digital technology are combined in the telephone system. The voice signals from the subscriber are carried to the local toll station via the twisted pair connection cable (analog communication). At the toll exchange they are digitized and transported via high speed multiplexed lines (cables, microwave links, satellites, fibre optics, etc.) to another exchange. Here they are demultiplexed and carried again to their destination via a twisted pair.

Digital data cannot be transmitted directly along a telephone line because the limited bandwidth would cut out the d.c. components and distort rapidly changing signals. To overcome this problem it is necessary to use **modems** to match the digital data to the characteristics of the telephone link. Modem is the abbreviation of **mo**dulator/**dem**odulator. Modems are also denoted as **data communication equipment** or **DCE** (see Section 10.3.5). A modem generates a carrier wave at a frequency of between about 1000 and 2000 Hz and modulates it according to the data to be transmitted. The frequencies and the modulation type used (amplitude, frequency or phase) vary according to different standards.

Modem operation can be full or half duplex. In half duplex, only one device at a time transmits and the other receives. A protocol indicates when it is time to switch over transmission. In full duplex mode different frequencies are used, so that the modems can send data simultaneously. Some of the most modern modems are able to recognize and adapt to the communication speed and other parameters. A common interface between the computer or terminal and the modem is the EIA-232-D (see Section 10.3.5).

Modems operating with sophisticated coding schemes today reach about half of the maximum theoretical speed as indicated by Shannon's equation (Equation 10.3). In the presence of noise, their performance drops quickly and the need for error correction schemes makes their throughput data rate actually less than for slower modems. Typical current modems for dial-up lines operate at 1200 or 2400 bit s^{-1}, some modems and the modems installed in fax machines reach 9600 bit s^{-1} and can automatically switch back to a slower speed if poor line quality requires it.

10.8.2 Digital Networks

The most important digital services provided by the PTTs for direct data transmission are:

- circuit switching
- leased lines
- message switching
- packet switching

Circuit switching is a different denomination for the use of modems on the telephone network. A telephone link is dialled up and remains connected for the whole duration of the communication.

A leased line is a permanently open connection between two fixed points. Lines leased for data transmission need to be **conditioned**, i.e. specially calibrated and shielded in order to offer a greater bandwidth and a better S/N factor, which gives higher data transmission capacity. The rental costs for leased lines is fixed on a per month or year basis and does not depend on the amount of voice or data traffic. Data transmission on leased lines uses modems with typical speeds varying from 19.2 to 56 kbit s^{-1}. Higher speeds are possible with special types of lines.

Another method of data communication is **message switching**. The whole digital message is put on the network and transferred from node to node until its final destination in what is known as a store-and-forward fashion. Message switching is used in telex and for electronic mail transfer, but is, in general, of no interest in industrial automation.

The state of the art in digital communication is **packet switching**, which offers a **virtual** network connection between the communicating nodes. The data from the sender is divided into packets of limited length (up to some hundreds of bytes each). Each packet contains protocol information such as destination address and sequence numbering. The packets in the network are routed toward their destinations on virtual channels. Each packet is routed independently of the others, so that packets addressed to the same destination actually take different physical routes. Under particular network load conditions, some packets could even reach their destination before packets that had been sent earlier. Thanks to sequential numbering, the last network node before the destination can keep track of the right order and reorganize the packets before forwarding them to their destination.

Packets from different users are multiplexed on long-distance channels, so that the network capacity is used more effectively than with direct connections and circuit switching. The network can also be more evenly loaded because digital users do not require full transmission capacity all the time, but transmit high-capacity data bursts for short time periods. In this way, users tend to compensate for each other loading the channel. Another big advantage of packet switching is that, in case of a node crash, packets can be routed on a different path without the end users noticing it and having to intervene. Thanks to these features, packet-switched networks normally show high availability factors.

Packet networks are generally accessed via an X.25 interface; the corresponding CCITT recommendations cover the first three OSI layers. The physical level interface between DTE and DCE is X.21; this interface defines the electrical connection and a handshaking protocol oriented to the network operations ('dial number', 'line busy', 'communication established', etc.). The X.21 interface makes use of a 15-pin connector of which only 8 contacts are used. However, the X.21 standard is seldom used and therefore a provision, called X.21 bis, is made to allow the DTE—DCE connection via EIA-232-D.

At the data link layer of X.25 a specific protocol is defined, called **Link Access Procedure-Balanced (LAPB)**, with a frame format similar to HDLC (see Section 10.4.5). At the network layer, X.25 provides for commands for establishing, managing and ending the virtual connections. If communication always takes place to the same destination, the end address may be stored in the network computers so that the virtual path can be activated immediately on request. This feature is known as fast number selection.

10.8.3 The Integrated Systems Digital Network (ISDN)

Analog transmission is satisfactory for voice communication but is a poor medium for data transmission which has to be converted to analog form first. For data transfer, it would obviously be better to have a digital link all the way. This is precisely the purpose of the **Integrated Systems Digital Network (ISDN)**, which has been defined in accordance with the OSI model. The basic idea of ISDN is that all communication is either digital (data, fax, remote monitoring, etc.) or can be transformed to digital form (voice sampling), so that the best way to transport it is with digital links all the way, from one end point (the subscriber's telephone set or other device) to another end point. ISDN is the main concept for the telephone system development at present and will be until well into the twenty-first century.

In the basic service for normal subscribers, ISDN provides the end user with two 64 kbit s^{-1} channels, called, respectively, B and D channels. The two B channels are used for two parallel communication services, voice and telefax or voice and E-mail, at the same time. A maximum of eight communication devices may be connected together under the same user number, although only two will be able to operate at any one time. The 16 kbit s^{-1} D channel is used for operational signalling to carry additional information, for instance, shorter messages to be displayed on a telephone set window while its user is calling. In ISDN, the voice is sampled at 8 kHz and reconstructed at the opposite end; all other kinds of transmission, which are digital in nature, are put directly on the line. The three ISDN channels are multiplexed on the already installed subscriber twisted pair line, with a bandwidth capacity that is sufficient to carry the data.

Digital customers with high data traffic may use primary access services. In North America and in Japan, primary access consists of 23xB + 1xD channels, for a total of about 1.5 Mbit s^{-1}. In Europe the service provides 31xB + 1xD channels, totalling about 2 Mbit s^{-1}. ISDN functions are requested with packets called **Link Access Procedure, D-channel (LAPD)**, with a structure similar to LAPB and HDLC. Another service of interest is Broadband ISDN with a capacity of 150 Mbit s^{-1}. The innovation behind Broadband ISDN is that the service is of switched type, i.e. can be routed like conventional telephone calls and is not limited to fast point-to-point connections. Optic fibres are the physical medium used for Broadband ISDN.

The two channels of the basic service have been introduced for marketing reasons.

With ISDN, a normal customer interested only in voice services would notice a better signal quality and a faster connection time compared with the conventional, analog voice network, but this alone is hardly a sales argument. The advantage of providing two services at the same time, as would be the case with voice and fax, is for many users the deciding factor. Another argument in favour of ISDN is that it does not need to rewire the end user connections; connection to the new digital toll exchanges can be implemented with the older cabling. In many countries ISDN is already available as an optional communication service, although it will not be fully operative on a large scale until well into the twenty-first century.

ISDN supporters envision a world where all communiction is digital. With ISDN no special data networks would be necessary, as ISDN is a digital network on its own. Thanks to the fact that ISDN was defined in detail before its implementation, all telecommunication administrations from different countries can build ISDN following the same guidelines, thereby keeping compatibility problems for equipment and network to a minimum. The industry could produce ISDN hardware for a world market, so that it would not be necessary to support several different versions of the same product for use in different countries, as is the case today.

Some critics point out that with ISDN it is much easier for the PTT administrations to wiretap and record telephone calls, and they warn that, under undemocratic governments, the technology might be used to such a purpose. Undoubtedly the legal and personal integrity aspects of ISDN still have to be resolved. ISDN is still evolving, but at the moment while there is a search for unifying concepts in the chaotic realm of data communication, it remains one of the key reference concepts.

10.9 SUMMARY

Communication plays a central role in the interconnection of different devices. The key parameter to describe communication capability is the capacity (in bit s^{-1} or some multiple) of the communication channel. The capacity is a function of the channel bandwidth, of the noise affecting the channel and of the coding method used.

The Open Systems Interconnection (OSI) scheme has been introduced to bring order to what is meant by 'communication' and 'compatibility'. In OSI seven layers are defined for the different aspects of communication, from the physical layer (cabling, plugs and electric signal levels) to the application layer, where programs can exchange information on predefined objects (files, production equipment) using standard languages.

To achieve as much compatibility as possible, it is necessary to follow a few, established, standards. Today there are probably more 'standards' than would be desirable; in fact, work is done to introduce progressively wider concepts that cover several of the communication layers and integrate different standards.

Two comprehensive concepts for information exchange in industrial processes are MAP (Manufacturing Automation Protocol) and TOP (Technical and Office Protocol). They are largely compatible with each other and are oriented to different aspects of industrial processes (production versus administration). Both MAP and TOP are resource-intensive products and support the interconnection of a large number of devices in medium-sized to large plants.

In low level (or 'factory-floor') industrial automation, the important OSI layers are 1, 2 and 7. Fieldbuses cover those levels with products especially designed to operate in industrial environments. It is unclear whether a general standard for fieldbuses will be introduced. By analogy with what happened with local area networks, two fieldbus principles will probably coexist: one hierarchical-deterministic (e.g. FIP and Bitbus) and the other more dependent on the amount of traffic (PROFIBUS).

FURTHER READING

The classic reference for the foundations of communication theory is Shannon and Weaver (1963). This is a reprint of their famous articles from 1948 about the statistical analysis of the communication process.

Tanenbaum (1989) tells you almost everything that is to be told about computer communication at a very high level and yet is not boring. As has already been pointed out, Tanenbaum is one of the few authors who can break established writing patterns without leaving safe ground. Black (1989) is a modern and comprehensive guide to several types of communication, not only those related to data. Both books treat everything from the physical line to network communication, and follow the OSI model quite strictly. A general introduction to OSI is given in Voelcker (1986).

DEC (1982) describes data networks in general and provides a specific introduction to Ethernet.

Kaminski (1986) describes the major ideas behind the MAP concept. In a parallel article in the same journal issue, Farowich (1986) introduces the TOP protocol.

An overview of the proposals and expectations for a general Fieldbus standard is given in Wood (1988). The issues related to open communication for fieldbus devices are treated in Warrior and Cobb (1988).

FIP documentation can be obtained from the French standardization organization CEGELEC or through other national standards organizations (see Standards for the addresses to contact). A general introduction to PROFIBUS is given by Göddertz (1988, 1990). The complete standard is available as DIN (German standardization organization) document (DIN 19245, Part 1 and 2).

Held (1989) describes data communication from the point of view of the devices: modems, multiplexers, interfaces. It is quite advanced reading, and illustrates in detail communication via public data networks.

11

Man—Machine Communication

Aim: To understand the basic principles relating to man—machine communication and to apply them to the design of user interfaces in order to make the transfer of information and control of a process easier and more effective

The design of the **man—machine interface** (also called a **user interface**) is an essential aspect of computer systems. As the name shows, the purpose of the user interface is to facilitate the exchange of information between the user and the machine (computer or system) to be controlled. A well designed interface not only makes work conditions more pleasant, but also helps considerably to reduce errors and thus limit the extent of possible damage.

The chapter begins with a general presentation of the topic (Section 11.1) followed in Section 11.2, by an introduction to the psychological background for human cognition that is the basis of information exchange. Section 11.3 is dedicated to the hardware for the user interface and Section 11.4 to general interface design principles, with the emphasis on the structuring of terminal screen layouts, command syntax and menu organization. Practical suggestions and hints on how to define user-friendly computer communication in the form of layouts and man—machine commands are also given.

11.1 WHAT IS MAN—MACHINE COMMUNICATION?

The study of how human capabilities are best utilized in the work environment and how this environment can be configured to adapt best to human workers is a comprehensive discipline known as **ergonomics**. It integrates fields as different as engineering, physics, physiology and psychology.

Some aspects of ergonomics are of direct importance in control system applications. The control system engineer faces the problem of the man—machine interface as user or as designer. In both cases some knowledge of the related ergonomic principles will be profitable. The **user** should know how to approach a system, what to look for, what to expect and how to quickly recognize the general command principles for a new machine. In fact, if a system is built following consistent and logical rules, the user will be able to master it in a short time. The **designer** of a control system has to define how data is presented on terminals and control panels, and the aspect of commands given by the user. Here it must be ensured that output data is easy to understand, even in complex situations, so that the user can make the correct operational decisions.

If you discover that in a system basic principles of ergonomic design are violated, you may assume that the system as a whole is messy. A fancy interface could be a cover-up for bad functionality. On the other hand, if the user interface is well structured and easy to use and understand, then the underlying system is probably also well structured. Some knowledge of ergonomic principles helps in analysis when you 'feel' something is wrong, but you are not really sure of what it is.

There are no predefined, handbook-written rules for the design of a man—machine interface, but some principles can be followed in order to avoid major mistakes. If a solution is not necessarily the best possible, at least it does not have to be bad. Experience has already shown principles and trends, but ergonomics and computer user psychology remain open research fields.

In the definition of a new interface, analysis leads to a specification document and programming is based on this. In some cases implementation may be done directly by the process engineer with help of screen editors and command dictionaries. In such a case it is easy to improvise, but nevertheless the aspect of the interface should be correctly defined beforehand. It should be noted that the software making up the man—machine interface may well represent as much as 50—75% of the total control system software. Nevertheless, it is an investment which can pay off many times over.

11.2 PSYCHOLOGICAL CONSIDERATIONS

It is estimated that the total amount of information entering the body is one billion bits per second, of which only about 100 bit s^{-1} are processed consciously. The brain tends to further reduce the amount of information to process. If too much information is presented at the same time, acting capacity is lost and the attention tends to concentrate only on part of the input data. As a common example, we know that the brain is highly skilled at 'filtering out' an image or a sound from a noisy background.

To define effective ways for the design of man—machine communication, a model of human cognition and information processing is needed. Psychologists have dealt for a long time with this problem and have proposed several models. In the cognitive process, modern psychology differentiates between functional blocks: sensory storage, short term (or work) memory and long term memory (Figure 11.1). The stages in information processing conducted by the brain are perception, storing in short and long term memory, planning and conversion in control action. A human being can concentrate only on one thing at a time, although attention may rapidly switch from one thing to another.

Perception is either visual or acoustic; the other senses are less important and are not relevant in the man—machine information exchange. The information collected by the

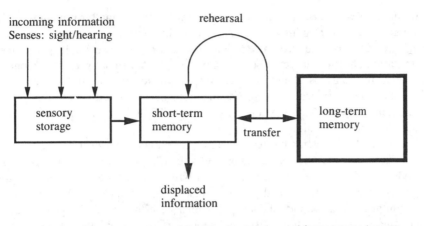

Figure 11.1 The memory model with short-term and long-term storage.

sense organs (**sensory storage**) is transferred to **short term memory** where we can consciously pay attention to it. From the short term memory, and in most cases only with a voluntary effort, information can be transferred to the **long term memory**. Short term memory is fast to recall (and forget), but we can 'see' all the information contained in it at the same time and react quickly based on this. Short term memory is our consciousness, it holds whatever we think about at a given instant and provides us with a base for action. Long term memory has an almost infinite storage potential, but memorizing and recalling takes longer. Information in long term memory comprises a person's entire knowledge and includes everything from the use of language to childhood's memories, from multiplication tables to the name of the King of Ruritania. Information in short term memory has a retention time of seconds, in long term memory it can last a lifetime.

Different studies suggest that in short term memory there is place for 7 ± 2 information items, also called '**chunks**', which can contain quite a lot of information. New information coming in will erase or displace the existing chunks. Items not thought about decay quickly and are lost from consciousness. A relevant fact is that the items in short term memory are at about the same abstraction level, or show at least some homogeneity. If we are thinking about football scores, it is not difficult to compare other football scores at the same time, but we can not, say, dedicate our attention to football scores and plan a journey at the same time (at least, not most people).

Organization and the relationship with previous knowledge can help us to handle new information more easily. Take, for example, the number sequence 14154569220. It just looks like an arbitrary sequence of eleven figures. Most people would not be able to recall such a sequence without some effort and would probably quickly forget it. But regrouping the sequence as 1-415-456-9220 makes it more managable, it is even more so if it is identified as a San Francisco[1] telephone number. Regrouping has reduced the number of chunks from 11 to 4, an amount that most people can handle without major effort.

---

1 Bay Area residents are obviously facilitated in this task. The SF area code is stored in their long term memory.

In most cases, information is stored and recalled from long term memory only with a voluntary effort. Storing information is easier if the new data can be put into an existing 'frame', i.e. if the data can be related to information already present in long term memory. Memorization of different facts also works better if these are not presented alone but are put in causal relationships. Similarly, recalling is facilitated by 'cues' hinting at some aspect of the data to be retrieved. Human memory does not work with direct cell addressing as computers do (what state is your neuron #2023965 in ?); rather it works on the base of analogies and associations.

In user-centred systems the relationship to previous experience is established using symbols from everyday life. On a computer screen, a pen will symbolize something to write with and an eraser something to cancel with. The symbols are not real, they are metaphors for real life and relate the operation of the known objects with similar operations for the computer. This important concept is called **visibility**: symbols and tools give clues to their function and operation. Visibility is most effective when it is related to everyday experiences and objects.

With any item or machine the user tends to build a conceptual model of how it works to find out later the appropriate controls for leading the machine to take the desired action. The mental model for driving a car is that turning the steering wheel clockwise will lead the car to the right. We can also think of a car where turning the steering wheel to the right leads the car to the left. It would take considerable effort to learn how to drive such a car because the clues to its operation would run contrary to our normal expectations.

Visibility has direct practical applications in man—machine communication. Simple models, possibly related to common, everyday experience should be selected. For example, a value of 66 alone does not mean anything. If we call it a temperature and compare it with a maximum allowed value of 70°C, we have already much more information. Is a 4°C difference from the maximum value acceptable? A look at the display of Figure 11.2 gives the answer. It is not necessary to check a handbook to find out whether the displayed value is within the allowed range.

As necessary as visibility is to understand a concept the first time, **consistency** helps to transfer existing knowledge to new contexts. Consistency means that a command will always have the same or an equivalent meaning, independent of a specific situation. A command used to load a file should not delete the file in a different context of the same program.

An example of visibility and consistency is given in the standard symbols defined by the standardization organizations ISO and IEC (see Section 12.1.4) for use on electrical appliances (Figure 11.3). These symbols are simple graphical representations of common objects and help make the operation of the equipment as intuitive and visible as possible. Consistency is achieved if the symbols are used extensively on all devices, and especially

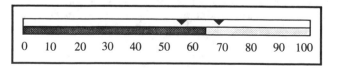

Figure 11.2   Example of intuitive display output.

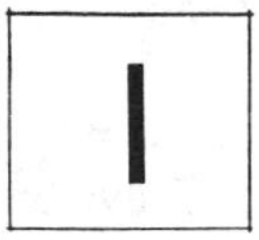

 On Control by which the appliance is connected to the energy source

 Off Control by which the appliance is disconnected from the energy source

 On/Off Control by which the appliance is alternatively connected to or disconnected from the energy source

 Stand-by Control for activation and deactivation of stand-by mode

 Variability Control which when moved sideways or turned changes a quantity

 Minimum Position of a control which gives the lowest value of an adjustable quantity

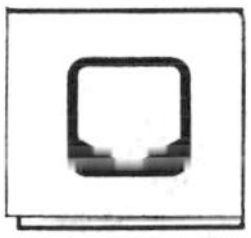

 Maximum Position of a control which gives the highest value of an adjustable quantity

 Temperature Indicates a thermometer or a control to adjust the temperature

 Illumination Control for putting on or off or changing the illumination

Time Control for starting or stopping at a given time, or for setting the correct time on a clock

Figure 11.3 Examples of standard control symbols found on different appliances (IEC 417 / ISO 3461 / ISO 7000-DAD / 1).

that no device carries different symbols for the same function if it is not justified by some special reason.

Another important factor of user-oriented design is **feedback**. As soon as a command has been given, there must be some reaction to indicate the new action. This is evident in the case of the steering wheel and the car, which provides immediate directional feedback. Other examples of feedback are the 'click' of a keyboard when a key is pressed and the tones in a touch-tone dialling telephone. The feedback is intuitive: nobody recognizes the dialled number from the tones, but we get enough information to tell whether all the numbers were dialled and whether a slip of the finger leads a number to be dialled twice.

Feedback is important also in another respect. According to Murphy's law, 'if an error is possible someone will make it': in looking for trouble, many people are quite ingenious. Feedback is not only there as confirmation, but should let the user understand if an action has to be halted or reversed. No single command should lead to a no-return situation.

Many designers have a bad habit of clogging their products with all possible features. It sometimes seems that if some company sells a stereo set with 100 different controls, then the competitor's product must have at least 20 more. Discernment is called for. How many of the controls are really necessary? When the conceptual model is defined, the basic operations should be made easy and immediate. Additional functions should be kept separate.

Three ergonomic aspects influence the acquisition of information by the user. They are perception facility, coding and structuring.

Perception facility is of a physical nature and depends on the hardware equipment: for a terminal screen important factors are brightness, absence of reflection, colour contrast, symbol size, etc. We do not want to work with devices that require some effort just to get to the information.

Coding is a way of transmitting information with help of symbols and clues. Message coding is also relatively easy to do with visual presentation and helps to transfer quite a large amount of information. For instance, the same symbol shown in different colours can convey different messages.

Organizing and **structuring** the presented information helps relieve the user from the voluntary effort of building a mental structure. Attention can be concentrated on the problem and its solutions. We will return to this aspect in relation to data presentation on displays.

The goal of a good man—machine interface is to draw attention to important facts and allow prompt and correct reaction on the basis of the information given. As a consequence of limited human ability to process large quantities of information at the same time, the presentation of data should draw attention to important items, so that they are immediately understood and acted upon.

For this reason, if an operator has to react quickly to new information, for instance, to select new commands, the presented information should

(a) be logically organized and
(b) not exceed about five distinct items at the same abstraction level

Five items does not mean 5 bits of information. The chunks can be highly abstract and carry a large quantity of information, but must still be homogeneous. Some examples and practical considerations in this respect will be discussed in Section 11.4.

Whereas the designer has the machine in mind and the user the task to be accomplished by the machine, the man—machine interface is of concern to both. A good interface is transparent and is noticed as little as possible. The best man—machine interface allows the user to concentrate on the process under control without being distracted by the interface and the control system themselves.

11.3 HARDWARE FOR MAN—MACHINE COMMUNICATION

There are several types of devices for man—machine communication. Some of them are in wide use, some are used more seldomly and found only in special applications. The response of the users to these devices has been quite selective, mostly favouring practical solutions rather than fancy findings. The most important I/O devices for man—machine communication in industrial applications are:

1. Devices for direct data I/O:
   - screen terminal
   - keyboard
   - special function keys
   - printing terminal
   - printer
   - control panel
2. Pointing devices with direct control of the surface of a screen terminal:
   - lightpen
   - touchscreen
3. Pointing devices with indirect control of markers on a screen terminal:
   - mouse
   - trackball
   - joystick
4. Other man—machine interface devices:
   - speech recognition systems
   - speech generators
   - optical, acoustic alarms

Screen terminals with keyboards are the most popular data exchange devices for practical control applications. The only major drawback of terminals is that their hardware can be sensitive to factory environments with high levels of electromagnetic disturbances, vibrations, moisture and dust present. Special 'rugged' terminals to be used in factory environments are available (Figure 11.4). Finger agility is necessary to use a keyboard, and there may be problems in environments where it is necessary to wear gloves.

Although it may sound antiquated, a cheap and practical I/O device is the **printing**

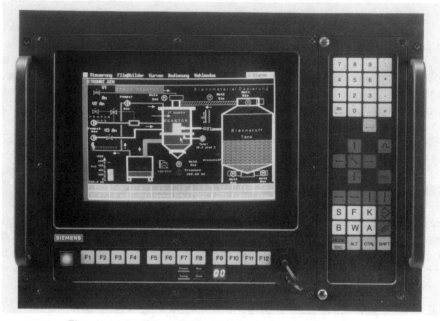

Figure 11.4   Rugged terminal for industrial applications.
(Courtesy of Siemens, Germany.)

terminal or **teletype**. It is used when information is exchanged at a low rate (1–2 events/minute) and each information item is self-contained (i.e. not related to other information), so that the user does not have to wait too long for completion of a printout in order to get a complete picture of the situation. The printing terminal has the advantage that the display medium (paper) can also be archived as an operational record with no need for further processing.

Control panels date from the time when the process interface devices under control were wired one by one to their commands located in central rooms. Control panels have indicators (lamps, needle instruments, etc.) for data output and switches or keypads for data input. Control panels can be used with a limited amount of data I/O and when the commands are clearly identified, with only a few possible alternatives.

Direct pointing devices (**lightpen** and **touchscreen**) have not found great interest among users. Continuously raising the arm from the worktable to the screen and moving it back is an effort and the precision of the movement is quite low. Pointing to a screen with a finger or with a lightpen takes more time than using an indirect control device such as a mouse or a trackball. In the first case there is also a 'fingerprint effect'. Direct pointing devices are of interest only when the use of a keyboard is not possible and the man–machine dialogue can be organized in the form of menus with a small number of basic selections. But even here, a small keypad could easily replace the touchscreen for input.

With indirect pointing devices such as the **mouse, trackball** and **joystick**, the position of a marker or pointer on the terminal screen is controlled directly by hand manipulation of the device. The precision of the movement is much higher than with touchscreens and lightpens. If one of several items has to be selected, the current choice is highlighted, thus ensuring correct operation. Confirmation of the selected option is given by pressing a button.

Much effort has been dedicated by the computer industry to the development of **speech recognition systems**. For decades a solution seemed to be very close, but it had to be periodically admitted that it was not the case. The systems available today can only recognize single words matching prerecorded patterns. The same word pronounced by different people (or by the same person with a cold) is usually not recognized. Until a machine can reliably decode normal speech with different intonations and accents, the use of speech recognition interfaces will remain limited.

Speech generation is technically easier than speech recognition and different systems are available on the market. Speech generation systems have a drawback in that messages from them can come at random, when a user is not expecting them and does not pay attention. It is also felt to be annoying when a conversation between people is interrupted by a machine delivering a synthesized message. The case of automated telephone answering systems is different, because the caller knows that a voice message is about to be delivered and is ready to pay the necessary attention.

Special alarm situations can be announced with help of **visual** and **acoustic devices** under computer control. They are used to draw immediate attention when the human operator is dedicated to some other task. In the case of acoustic devices, some parameters (tone level, frequency, loudness) should be adjustable by the operator. It is imperative that such devices have a quick and practical reset command.

The advantage of visual versus acoustic presentation is the large amount of information that can be transferred, but this requires the active participation of the human subject. An acoustic message does not require continuous attention by the addressed person, can come at any time and be perceived independently of what the person was doing at the moment. The amount of information carried is, however, very limited, down to just 1 bit (presence or absence of a sound). Acoustic signals carrying more information, for instance, a tone at different frequencies or a voice message, require attention in order to be interpreted correctly. If they occur too often, they might easily irritate.

11.4 INTERFACE DESIGN PRINCIPLES

Computer terminals are the most common devices for man—machine communication. There is much to gain from a good presentation of information on terminals and the definition of the operator's commands.

We will deal here with a man—machine interface based on a terminal with some basic features for highlighting data, a keyboard with control keys and a pointing device, e.g. a mouse. This kind of hardware is readily available, cheap and more than adequate to build a good user interface. Emphasis has to be placed on the correct interaction of system components and on the correct coding of messages and commands. Poor performance because of bad design should not be blamed on the I/O hardware.

11.4.1 General Principles

The typical computer user tends to treat everything that appears on a screen as important. Therefore screen displays should be simple and not contain useless information, otherwise there is a risk that unimportant items draw unwarranted attention. The information content should also focus on the user; the expert is not interested in 'trivial' information that may be vital to a novice. For instance, many systems allow the selection of different levels of help texts, length of menu and explanations for questions.

The key to functionality in the presentation of large amounts of complex data is correct selection and structuring. The original data is therefore divided into smaller groups, each to be presented on a screen page. Ideally, the system user will have to jump as few pages as possible in order to get the relevant information.

Each screen should show only one **main concept** in the easiest possible way. (It never pays to make simple things difficult!) Emphasis should be given to the most important information concerning a specific object. In a plant overview the functionality of the plant and of its subunits will be shown as information chunks. All parameters related to the same machine can be shown on a screen. The screen dedicated to a work cell will present only few basic data for each machine, such as whether it is operating correctly and the current production rate. At a higher abstraction level, the screen layout for a production line will present basic data for all the cells, whose detailed states do not have to be displayed. Although the main concept (the plant or production line) is complex, the functional idea on each screen can remain simple: *does the machine or plant operate correctly or not?*

At higher abstraction levels, it may be preferable to show not a state but the effect of a function. For instance, if a switch is used to connect two alternative devices (or a production line branches into two cells), the switch position itself can be shown, as well as which device (or cell) is connected, identifying it with a different feature such as an empty or framed symbol (Figure 11.5). The symbolic representation does not require explicit interpretation of the picture.

In general, a good layout has the following characteristics:

- It is adequate for the purpose; it does not present *more* or *less* information than necessary.
- It is, as far as possible, self-explanatory.
- It is consistent at more levels. The same coding (in symbols and colours) carries the same meaning on different screens and the user knows what to expect in different situations.

It is very important for the user to get an immediate 'feeling' that a command has

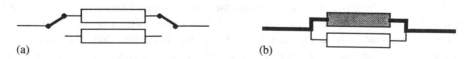

(a)   (b)

Figure 11.5   Symbolic vs functional graphical representation. (a) Symbolic representation; (b) functional representation.

been received and accepted, even if processing does not start immediately. The feeling starts right from the keyboard 'click' when a key is pressed and continues with the actual dialogue. If, because of long response times, nothing seems to happen after having pressed the ⟨RETURN⟩ or ⟨ENTER⟩ key, one may wonder whether the system operates correctly.

If a system cannot give immediate response to a command, as verification that the command was accepted and is currently being processed, some kind of partial response should be displayed. This could be, for example, a message of acknowledgment or a different highlight of the input data on the screen. The effect of a command should be immediately evident and there should always be the possibility of reversing it. For processes with longer time constants, the first reaction could be a message from the process computer: *New temperature setpoint value is 66&C. Estimated time to reach setpoint is **18** minutes.*

It is advisable for help to be available on line. Help should always be called from the same key, which should be distinct and clearly marked. Modern systems offer context-related help, that is, they recognize the current situation (the data or program which is currently active) and offer help related to that situation.

The disposition of one or more screens on a large panel should be integrated with additional keys and controls so that the *most frequently used* and *most important* controls are the easiest and first for the user to reach.

11.4.2 Parameter Description Syntax

Information to and from computers is often in one of a limited set of states. In most cases, the plant or process control engineer is responsible for the selection of appropriate texts to present this information. These texts are presented on the terminal screens and in the computer printouts and may form the basis for actions which have to be undertaken. This kind of information can be divided in a fixed (static) text (**lead text**) to indicate the type of selection together with a variable for the actual state (**dynamic variable**).

The lead text should create a small 'tension' to be released only in conjunction with the dynamic information. The lead text alone should not give complete information in a grammatical/syntactical sense, otherwise there is a risk that the lead text may be misunderstood to be itself the actual state information. Only the combination with dynamic information should make complete sense. Thus:

| instead of: | **device A11 powered:** | yes/no |
| use: | **device A11 power:** | on/off |

In other words, the dynamic information should not negate the static part, but complete it.

The lead text should not be too generic and should contain hints on what the dynamic part is going to be. Compare the following examples:

device A12 status:	on/off?
	ok/alarm?
	active/standby?

write instead:

device A12 power:	**on/off**
device A12 operation:	**ok/alarm**
device A12 connection:	**active/standby**

A clearer distinction between static and dynamic information is obtained on-screen when the lead texts are written with normal intensity and the dynamic variables in high intensity. If the dynamic variables are also used as input, the variable currently selected may, for example, be shown in reverse video (that is, where foreground and background colours are exchanged). Dynamic variables showing parameters in an alarm state can be highlighted using a different colour or reverse video.

Avoid the use of words with negative connotation ('ALARM', 'WARNING') in static texts unless they are clearly unrelated to the actual state of the controlled system. The texts should motivate and not unnecessarily alarm or irritate the user.

11.4.3 Screen Layouts

Some design principles related to screen layouts were primarily investigated in relation to other purposes, mainly mass communication techniques for advertising, yet they give useful hints for presenting information on a computer screen. The indications are formulated in such a way as to be easily incorporated in a software requirement document. The most important principles are listed below.

1. The natural movement of the eye is from left to right. A drawing developed along the horizontal axis is easier to perceive than one drawn vertically. Thus the evolution of a process should also be represented from left to right. In some cases a vertical drawing can be necessary, for instance, to highlight a particular form or topography.

2. The screen layouts must be consistent in their appearance. A layout can be considered to be divided into four sections for **work** (application related), **control** and **message** sections for the computer system and a general, static **information** area with the screen title (Figure 11.6). On a standard screen with 24 horizontal lines, about 16−20 lines are reserved for the work area and 1−3 lines each for the other sections. The areas should be separated with lines and boxes or highlighted in some other way. The size of each area should be the same in all layouts.

3. Graphic ways to indicate a structure are proximity, symmetry, similarity and grouping. In all cases it becomes much easier for the observer to recognize the underlying structure. **Grouping** is a very powerful method to structure otherwise disordered items, because it is natural to consider objects which are close to each other to be in mutual relation. Figures 11.7 and 11.8 show the same information as it would appear on a computer screen, in the first case without and in the second with structure. The structure may be additionally underlined by using upper and lower case letters.

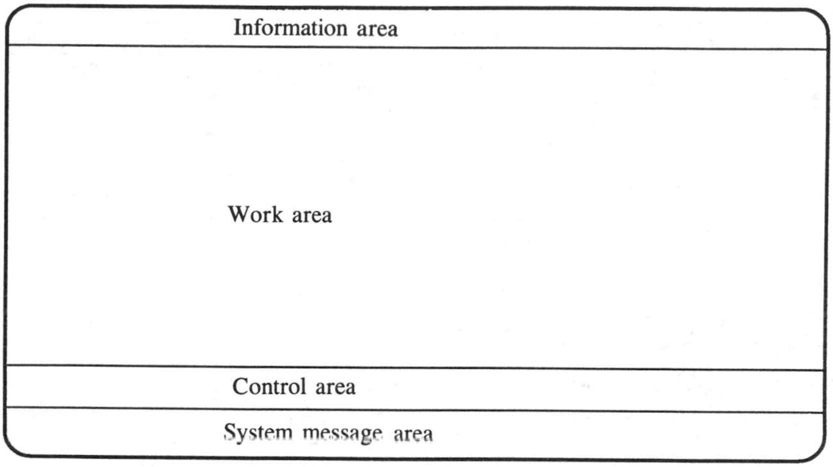

```
┌──────────────────────────────────────────────────────────┐
│                      Information area                       │
├──────────────────────────────────────────────────────────┤
│                                                            │
│                                                            │
│                                                            │
│                       Work area                            │
│                                                            │
│                                                            │
│                                                            │
├──────────────────────────────────────────────────────────┤
│                      Control area                          │
├──────────────────────────────────────────────────────────┤
│                  System message area                       │
└──────────────────────────────────────────────────────────┘
```

Figure 11.6   Functional divisions on a screen display.

4. Colours are a powerful way to code information, but must be used sparingly; four to five colours can be understood with no major effort and there are several indications that say seven different colours is an absolute maximum.

5. Colours must be used consistently: one colour — one meaning. Variations in the meaning of a colour depending on its position might lead to false analogies and wrong assumptions.

6. Colours can be used to display functional states. Choose natural colours to indicate different states. **Green** is generally perceived as an indication of security, permission or correctness (it may for instance indicate that a unit is in proper working order). **Red** is related to states of alarm, danger and prohibition. **Yellow** is understood as a warning and can indicate the presence of some minor problem.

```
WATER TREATMENT PLANT / CHEMICAL PRECIPITATION SECTION Page 24 14.18.04

PUMP 105 PROCESS WATER STATE=ON ALARM=NO OVERHEAT=NO
PUMP 118 WASHWATER STATE=ON ALARM=NO OVERHEAT=NO
PUMP 127 REACTION VESSEL OUTPUT STATE=ON ALARM=YES OVERHEAT=NO
PUMP 132 SLUDGE SILO FEED STATE=ON ALARM=NO OVERHEAT=NO
PUMP 138 SLUDGE SILO OUTPUT STATE=ON ALARM=NO OVERHEAT=YES
PUMP 139 SLUDGE FINAL OUTPUT STATE=OFF ALARM=NO OVERHEAT=NO
PUMP 143 VACUUM FILTERING STATE=ON ALARM=NO OVERHEAT=NO
PUMP 154 LIQUID WASTE STATE=ON ALARM=NO OVERHEAT=NO
PUMP 166 LIQUID FILTRATION STATE=ON ALARM=NO OVERHEAT=NO
PUMP 221 ALKALI INLET STATE=ON ALARM=NO OVERHEAT=NO
PUMP 226 NA-SULPHIDE INLET STATE=ON ALARM=NO OVERHEAT=NO
PUMP 232 POLYMER PROC.A INLET STATE=ON ALARM=NO OVERHEAT=NO
PUMP 237 POLYMER PROC.B INLET STATE=OFF ALARM=NO OVERHEAT=NO
PUMP 242 POLYMER PROC.C INLET STATE=ON ALARM=NO OVERHEAT=NO

REACTION VESSEL OUTPUT /127/ (m3/h) = 53
SLUDGE SILO FEED /132/ (m3/h) = 92
SLUDGE SILO OUTPUT /138/ (m3/h) = 74
NA-SULPHIDE INLET FLOW /226/ (m3/h) = 68
```

Figure 11.7   Example of a poorly structured screen display.

```
┌─────────────────────────────────────────────────────────────────────────┐
│ WATER TREATMENT PLANT Chemical Precipitation Section 14.18.04 │Page 45│   │
│ ┌───────────────────────────────────────────────────────────────────┐   │
│ │Main Reaction                       Oper.  Funct. Overheat   Flow  │   │
│                                                                           │
│  PUMP 105 PROCESS WATER              ON     OK     OK                     │
│  PUMP 118 WASHWATER                  ON     OK     OK                     │
│  PUMP 127 REACTION VESSEL OUTPUT     ON     RASTER │ OK        53 m3/h    │
│  PUMP 132 SLUDGE SILO FEED           ON     OK     OK         92 m3/h     │
│ ┌───────────────────────────────────────────────────────────────────┐   │
│ │Final Treatment                     Oper.  Funct. Overheat   Flow  │   │
│                                                                           │
│  PUMP 138 SLUDGE SILO OUTPUT         ON     OK     RASTER │   74 m3/h     │
│  PUMP 139 SLUDGE FINAL OUTPUT        OFF    OK     OK                     │
│  PUMP 143 VACUUM FILTERING           ON     OK     OK                     │
│  PUMP 154 LIQUID WASTE               ON     OK     OK                     │
│  PUMP 166 LIQUID FILTRATION          ON     OK     OK                     │
│ ┌───────────────────────────────────────────────────────────────────┐   │
│ │Reagents                            Oper.  Funct. Overheat   Flow  │   │
│                                                                           │
│  PUMP 221 ALKALI INLET               ON     OK     OK                     │
│  PUMP 226 NA-SULPHIDE INLET          ON     OK     OK         68 m3/h     │
│  PUMP 232 POLYMER PROC. A INLET      ON     OK     OK                     │
│  PUMP 237 POLYMER PROC. B INLET      OFF    OK     OK                     │
│  PUMP 242 POLYMER PROC. C INLET      ON     OK     OK                     │
│                                                                           │
│ <PgUp>=Previous  <PgDn>=Next  <Esc>=Menu  <Ctrl+P>=Data Printout          │
└───────────────────────────────────────────────────────────────────────────┘
```

Figure 11.8   The same information of Figure 11.7 in a structured display.

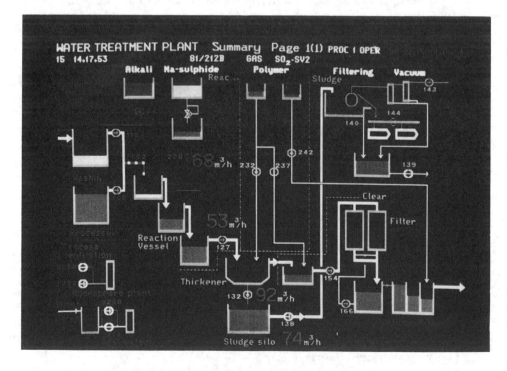

Figure 11.9   Example of graphical display output.
(Courtesy of ABB Automation, Sweden.)

7. Colour combinations should be pleasant and not fatiguing; all symbols should have a certain contrast with the background. Ease of perception varies much for different colour combinations. Yellow and black have a strong mutual contrast and are perceived easily, so are the combinations red/white, red/yellow, black/white and green/black. If the foreground and background colours on a screen display are controlled by separate program variables, check that the 'pink on white' and 'blue on black' colour combinations never take place.

8. It is important not to rely only on colours as a means of showing important information. A large number of people are blind to some colours and are therefore incapable of recognizing them. Environmental factors such as illumination and shadows may make the perception of some colours on a terminal screen difficult. The information to be shown should present some kind of **redundancy**, for instance, with help of labels, texts or other graphical symbols, to ensure that the conveyed meaning is understood.

9. The most natural representation for an object is its drawing. Compare the display of Figure 11.9 with those of Figures 11.7 and 11.8. The addition of colours to the symbols would convey more information than could fit on a single screen layout if everything had to be described in plain text.

10. An annoying method, subject to criticism, to call for attention is blinking. A text should never blink because it makes reading more difficult. It is sufficient for a small symbol to blink near it.

11. The display layouts should not be boring but interest and motivate the users of the control system. Motivation is one of the main keys to success in operations.

11.4.4 Commands

Communication from man to machine is as important as from machine to man. Communication from man to machine takes place by manipulating a device, such as a mouse or a joystick, by pushing buttons on a panel or by typing in command sequences on a keyboard. Some considerations regarding the design of man to machine communication follow, with special regard to controls typed in via a keyboard.

1. A command defines a reference value for a state; the actual value will later become equal to the reference value only if the control system, actuators, sensors and physical process all work correctly. It takes some time before a command reaches the actuators, is executed and the verification reported back to the user. **Actual** and **reference** values must be presented in a way clearly identifying them so that they cannot be confused with each other.

2. Actions following commands should be as far as possible **context-free** that is, they should lead to the same results, irrespective of past activities and system history.

3. When several input alternatives are possible, they should be clearly indicated. Input commands and data should be checked immediately for correctness by the control system and the result reported to the user.

4. One of the command alternatives — the current, previous, most common or safest one — should be shown as default selection for a command. The displayed value is then accepted by the system as valid unless explicitly changed by the user.

5. String commands to be typed explicitly on a keyboard should be as short as possible, without losing their meaning. A good method is to use the first letters of the command name, provided that different abbreviations are not confused with each other. In the VMS operating system for VAX computers this method is used for all commands and parameters, where the first four letters are sufficient to uniquely identify any command (these may also be typed in their entire length).

6. In fields where an alphanumeric input is requested, only a few combinations usually make sense. 'FGS' and 'OID' are letter combinations as 'ON' and 'OFF' are, but they will not be understood by a binary actuator. Possible alternatives for avoiding nonsense input data are:

    (a) displaying the correct values as part of the background information;
    (b) providing a window menu selection of the possible values;
    (c) displaying a message if the input is not understood by the system.

    Alternative (a) is not feasible where the number of possible commands is large; it leads easily to cluttering the screen with too much information. Alternative (c) may cause delays, depending on how often mistakes are made. Solution (b) may be optimal choice. A new value can be selected either by typing it in explicitly, or at least some of its characters, or by pointing at it with help of the arrow keys or the mouse. The selection is then confirmed by an ⟨ENTER⟩ command or a click on the mouse.

7. Typing a command from the keyboard requires some thinking and can lead to errors. It is good to ask for confirmation before execution of sensitive commands, e.g. with a question of the kind '*Do you really want to erase the directory [YES/NO]*?'

8. Highly sensitive and potentially dangerous commands should be validated with a password. Complicated protection schemes, such as making it explicitly difficult to give dangerous commands, should be avoided. A good control system should be at the same time *safe* and *easy* to use.

9. It is important to be able to stop a computer-controlled machine immediately in case of an emergency. In such a situation, nobody would have the patience to type in an ordered command sequence. A clearly marked emergency ⟨OFF⟩ button should be installed within easy reach for the operator. It is common practice to paint the button red on a yellow background. 'OFF' buttons are usually large enough to be operated with gloves on and without the need for careful aiming.

11.4.5 Menus

The principles of clarity and consistency required in screen and command design must of course also be followed in the design of menus. In particular, the following considerations can be made.

1. The menu structure should become clear to the user quickly. Each menu should be identified by a title/headline, possibly using the same text indicated in the menu selection item one level above.

2. A ⟨**BREAK**⟩ command should always be present, possibly with its own dedicated key. It should be possible at any moment to interrupt the current work and go back to the next higher menu level or even to jump to the highest level, or **root**, menu. Avoid situations where one has to go through a series of screens, or even reply to additional questions, just to leave the current menu.

3. The items in a menu should be at the same abstraction level. Do not mix together functions such as 'delete character', 'print file' and 'boot computer system' in the same menu.

4. Keep the number of choices limited, remember the rule of 7 ± 2. With too many items on the screen, the first ones may have already been forgotten by the time one is through scanning the list.

5. Do not put unrelated questions on the same screen. Remember, one screen — one concept. It is easier to go through three different menus than to think over three series of unrelated questions presented on the same screen. Again, keep just one basic idea on each screen.

6. Similar functions on different menus should be connected to the same keys; do not be afraid of leaving 'empty' spaces in the numeration of functions. if 'exit' is related to key ⟨6⟩ on one menu, also indicate it with ⟨6⟩ on a menu page with a total of two functions. Even better, the keys ⟨0⟩, ⟨**ESC**⟩ape or equivalent should always be reserved for 'exit'.

11.5 SUMMARY

So many systems, so many users. There is no 'standard' user profile. Some people are knowledgable about computers, curious, show a positive attitude and always want to learn more. Others are afraid of computers, or just unmotivated, and do not care much. Normally, and especially when a control system is built to retrofit an existing plant, users are technicians with a good knowledge of the process under control. They mistrust computer programmers who ignore the details of the industrial process; and at every occasion they air comments about 'their old way of doing things'.

The computer system is accepted by the user when it is seen as a tool not to disrupt but to enhance the plant control operations. Correct design of the man—machine interface plays a fundamental role in this acceptance. The ergonomic aspects are particularly important with untrained users. Here a trade-off must be made between assuming that the users do not know anything and that they will not learn, and requiring that the users learn something about the computer control system. The second way requires more attention and planning but also has the better pay-off. Instruction is always a good investment: the trained user will, in general, be more motivated and less likely to make mistakes.

After so many suggestions, there is one last one to end with. Do not exaggerate computer friendliness. Trying to be 'too nice' might have counter-productive effects. In time, beginners will become experienced users anyway.

FURTHER READING

Some knowledge of psychology is useful in order to design user-friendly applications. An introduction to psychology with comprehensive sections dedicated to perception, learning and the dual memory theory is Atkinson *et al.* (1990). This is recommended for the reader interested in background information. The dual-memory model is also described in Atkinson and Shiffrin (1971).

An excellent book to look at things from the point of view of the user is Norman (1988). It contains several examples about how **not** to design devices and tools: a captivating lecture about common sense in engineering.

A serious introduction to the implications and the possibilities of man—machine communication is Norman and Draper (1986). The subject of man—computer communication is extensively treated in Shneiderman (1987) which gives particular consideration to the design of computer screen layouts and the definition of interactive command languages. This book deals with different aspects of software ergonomics and user interfaces with particular regard to information search in databases. An easy, although somewhat outdated introduction to the subject of man—machine communication is Martin (1973). The Apple interface, with indications on how to develop practical applications, is described in Apple (1987). The development of the most important ideas behind the Apple user interface is described in Laurel (1990). This book also deals with current research and future developments.

Sanders and McCormick (1987) is a comprehensive reference textbook of the several aspects of applied ergonomics. Computer-related aspects play a relatively minor role here, the book is recommended for the reader interested in the broader view.

12

System Integration

Aim: To reconsider all the material presented in the book under the aspect of mutual interaction and integration in order to understand and utilize complex control systems

It is now time to put the ideas and concepts we have examined so far into a general framework. In complex systems everything has to fit together. It is enough for one part or component not to fit for the global performance of a system to be jeopardized. Mathematical analysis and modelling helps to identify physical limits, but no ready recipes can be given on how to structure a system. System design is more an art and a craft than a science, and the best school is undoubtedly to learn from experience the necessary 'do's' and 'don'ts'.

Section 12.1 deals with the integration of control systems. The technical description is completed with some background about standards and their importance. Sections 12.2 and 12.3 treat monitoring and control systems in general. Although the description is oriented to large industrial systems, many ideas can be used for different applications. Databases for monitoring and control applications are also treated in Section 12.3.

Non-technical aspects are at least as important as technical ones for the success of a monitoring and control project and therefore they are briefly considered in Section 12.4. Section 12.5 is a very brief introduction to the concept of computer integrated manufacturing (CIM). At this point, the reader should have a realistic view of the problems and the possibilities in the automation process. We hope that the reader sees this chapter more as an opening to new experiences than just as the final part of this book.

12.1 INTEGRATION OF COMPUTER CONTROL SYSTEMS

12.1.1 Integration Levels

A real-time control system may take different aspects depending on the complexity of the problem to be solved and the desired goals. Although most of the involved issues are similar, the technology and the tools available lead to many different implementations of automation systems. There is no generally applicable solution, only solutions which are valid in relation to specific problems.

It is not easy to differentiate between integration levels of control systems, because, as might be expected, the borders cannot be drawn all too precisely. However, there are major fields which can be identified with their technology, their industrial base and their marketing share. A brief overview is given here without pretence of completeness.

Integrated circuits (IC) and **hybrid circuits** contain the data processing logic and the control programs (**firmware**). ICs can be produced on customer specifications and are cost effective at production levels of some thousands upwards. Typical applications for ICs are in mass-produced units where there is no need to change the installed program for the lifetime of the product and where the control logic must be packed into a very small space, as, for instance, in microcontrollers for microwave ovens and engine ignition control.

An alternative to having the chips custom-built is to use general purpose **CPU boards** with **read-only memory (ROM)** to store the programs. The implementation of logical circuits in programmable logic devices were shown in Section 8.1.5. Such solutions are more cost effective for small scale production and where space does not set too many limits. Because the technology is not expensive or too complicated, this is the preferred solution by many smaller companies for their specialized production.

Controller boards are larger units for more complex applications, especially where the program or application data have to be changed during the life of the product. A typical application example is the logic for a **numerically controlled (NC)** machine (Figure 12.1). These compact, board-based units are typically built with the possibility for external communication via fieldbuses or LANs and can therefore be integrated in larger processes.

Controller boards are usually programmed directly on the target system (i.e. the system where the final application is run), but in some cases the support of a different machine with more resources may be needed, e.g. to run a large compiler. The ready programs are then downloaded and run on the target machine.

Bus systems, discussed in Chapter 6, are at the same integration level as controller boards. Several peripheral boards are installed on the bus and can be changed at any time with comparatively little effort.

Programmable logical controllers (PLC) are self-contained units for those automation applications where the input and output data are digital signals. For uncomplicated automation applications without too many parameters, PLCs represent cost-effective solutions. PLCs can also be connected to computers at higher level for integrated process control. PLCs were treated in detail in Chapter 8.

Local control systems such as PLCs, bus systems or other local control units can be connected together and to a central computer where the information from the whole plant is collected and recorded, and from where commands can be sent to the detached units. Integrated systems of this kind are known as **monitoring and control (M&C) systems**, where process control takes place at different hierarchical levels. M&C systems are described in detail in Section 12.2.

The implementation of automation systems does not require all the software to be

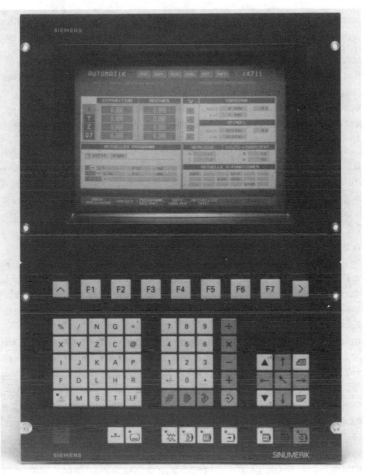

Figure 12.1 Control unit for a numerically controlled (NC) machine, SINUMERIK 805. (Courtesy of Siemens, Germany.)

written anew every time. For automation and supervisory computers software packages are available that are not programmed procedurally but with a description of the application ('fill the blanks' programming, see Sections 8.4.2, 9.6.8 and 12.3.7).

12.1.2 System Rack Integration

A system is composed of several units that operate together in order to fulfill a goal not reachable by the units working alone. When different units are connected together many factors have to fit, from physical dimensions to the type of information that is exchanged. The interconnectivity of different parts is ensured if common interfaces are used. Different types of interfaces have evolved in the course of time and, as already pointed out, are a combination of tradition and planning. The most important compatibility factors are:

- mechanical (rack dimensions)
- electrical (power supply, cabling)
- thermal (power dissipation, cooling)
- functional (communication, monitoring and control)
- operational (which functions to insert)

At the mechanical level, the most common industrial standard today is the 19″ rack (= 482.6 mm), where the figure refers to the rack width (Figure 12.2). Height is measured in height units (U), one U is 44.45 mm or 1.75 inches. In practice only multiples of three are used, e.g. 3U is 133.35 mm or 5.25 inch. Note that the term 'single height' means 3U, not one U. You can compare these dimensions with those for bus card formats (see Section 6.2.2). The Eurocard boards, designed to fit in a 19″ rack, have a minimum height of 100 mm and then increase in steps of 133.35 mm, or 3U.

The 19″ standard is internationally normed as ISO/IEC 297, it is also known as the German standard DIN 41494 and American standard ANSI/IEEE 1101. The 19″ standard is widely accepted and most producers of computing equipment and peripherals units follow it. For instance, several bus systems, such as VMEbus and Multibus II, are available in 19″ rack format.

The cabling between the units is drawn in the backside of the rack; the front panels are used for operator communication and should stay free of interfering cables. A power supply or power distribution unit is installed in the rack, usually on the base because of the weight of the transformers; it must seldom be reached for operation anyway. Sufficient ventilation to carry away the generated heat must be provided in the back of the racks and with free slots left between equipment units where necessary.

The units most commonly serviced by the operator will be installed at medium height, for easy reach. An emergency button to switch off the whole unit will also be located where it can be reached easily and marked in red on yellow background (Figure 12.3).

The single units can be purchased as well as also built by the user using standard electronic components all the way down to the chip level. Standard equipment such as measuring instruments, process computers and modems are available from several sources. These devices usually have an interface that allows remote operation, or can be upgraded with it. Yet many industrial and laboratory units may be based on new ideas so no dedicated instrument is available on the market; in such cases some special unit may be developed independently and combined with other, more readily available equipment.

12.1.3 Interface Design

An essential role in system interconnection is played by the interfaces between the different components. In a well-designed system all components act like black boxes, i.e. they can be substituted with different components that fulfill the same specifications and the system as a whole will not behave differently from the previous one.

Figure 12.2 A 19″ rack for industrial applications. (Courtesy of Schroff, Germany.)

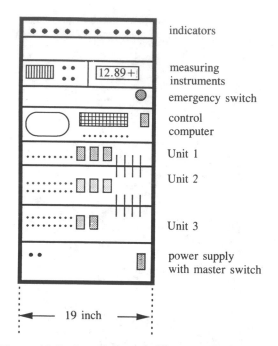

indicators

measuring instruments

emergency switch

control computer

Unit 1

Unit 2

Unit 3

power supply with master switch

19 inch

Figure 12.3   Installation of different units in a rack.

The specification of an interface must take into consideration all aspects that have to fit together in order for the function to be fulfilled. We have already got a taste of how involved and complicated specifications can be with OSI and MAP (see Sections 10.2 and 10.6). Normal project specifications need not be that complicated, although care must be taken to ensure that no important factor has been left out. Depending on the specific application, the description of an interface will have to include everything from mechanical and electrical specifications to the content and meaning of the exchanged data.

When different devices have to be connected together, there are in most cases readily available specifications that can be followed (e.g. the communications standards of Chapter 10). In the case that new specifications have to be defined, it is better to check whether there is already an established standard covering that function or one that can be taken as reference. There is no point in defining a sensor loop working between 5 and 28 mA, or a new way to code alphanumeric characters in a byte, when the 4–20 mA loop and ASCII or EBCDIC can do the job as well. In most cases the available components will force a specific choice.

12.1.4 Standards and Standardization Organizations

The need to standardize the interfaces of different components at several levels has long been recognized by users and by the industry. Different professional and government organizations work to define standards in different sectors for use as guidelines in the construction of components and services. Basically, standards are needed to ensure that different components can operate together and that the required quality is achieved. Current standards still have a drawback in the sense that they create compatibility islands without solving the more general problem of communication and interconnection between different systems.

Standards are not only the responsibility of special organizations. In many respects the 'anarchy' of industry and of the marketplace has led to better results, namely, fewer but more widely accepted standards, than the ordered and lengthy work of international bureaucrats. Some of the many possible constructions for hardware components have been so widely accepted in industrial environments as to become *de facto* standards, first unofficially and later sanctioned by official organizations. The most successful solutions have been those offering room to operate at reasonable cost together with good mechanical and electrical characteristics.

Standards are divided into *de jure* (legal, recognized) and *de facto* (actual or 'industrial'). Legal standards are defined by recognized government or industrial organizations. Actual standards 'happen' because almost everybody does the same or uses the same product without anyone officially taking the decision. This has been the case with the 'industry standard' computer (which means the IBM PC and its compatibles), the Hayes modem protocol, the Epson printer control language and the 19″ rack. The nice thing with *de facto* standards is that everybody agrees on them without being told to do so.

The case is different with standards meant to be such. As there is always disagreement on large international panels as to what to do, in many cases several incompatible standards are defined for the same issue, just take the data packet formats described in Section 10.4. Yet such a situation is better than nothing. It must be noted that the need for convergence is recognized, and the different standardization organizations do cooperate. The most important organizations responsible for the definition of standards recognized and used by the industry are briefly listed below.

ISO (International Organization for Standardization) is the association of national standardization organizations like ANSI (USA), DIN (Germany), AFNOR (France), BSI (Great Britain), UNI (Italy), SS (Sweden) and many others. These organizations define standards in all possible areas. Many of these standards are related to industrial and computer issues. Once they are decided, everybody in the participating countries should, in principle, conform to them. The ISO organization that specifically deals with standards for electrotechnics and electronics is the **IEC** (International Electrotechnical Commission).

Standards from some of the ISO participating organizations are so highly regarded that they are directly 'imported' into other countries; this is often the case with the German DIN standards. The Commission of the European Communities is encouraging the use of common standards in the perspective of political and economic integration, so that 'EC standards' are coming.

The **CCITT** recommendations are fundamental to telecommunications (the acronym means Comité Consultatif International de Télégraphie et de Téléphonie). In practice all member countries of the United Nations are represented in CCITT through their PTT administration. In most countries, telecommunication issues are the monopoly of the local PTT. A notable exception is the United States, where there is no such state monopoly but several private companies share the communications market. The United States is represented at CCITT via the State Department.

Note that the CCITT does not issue standards, but recommendations. In principle, each PTT administration can do as it wishes at home, but for international telephone traffic to function, compatibility is necessary. Even here there is a slow trend to common rules: we have seen such an example with the digital network ISDN, that is to be implemented throughout the world according to similar guidelines.

ISO standards cover everything possible and therefore they are the main reference for industry, together with the applicable national standards. CCITT recommendations are of interest for long distance data communication. When the competence areas overlap, as for instance in the definition of communication interfaces, a common standard has to be published twice, by both ISO and CCITT.

EIA is the Electrical Industries Association, an American trade association responsible for the development of standards at hardware level. Some of its standards, like the communication port interfaces RS, have also found wide acceptance outside the United States.

The **IEEE** (Institute of Electrical and Electronic Engineers) is a professional society based in the United States with local national chapters throughout the world. The IEEE issues its own set of standards related to the aspect and quality of hardware as well as to performance criteria and work methods. IEEE standards also enjoy worldwide recognition. In the United States, IEEE standards are issued jointly with ANSI (American National Standards Institute).

The **ISA** (Instrument Society of America) is another standard setting society in its specific field. They publish standards about instrumentation and control, not only on hardware but also on related operational practices.

12.1.5 Which Integration Level to Choose?

The ways to automation are many and it is not possible to define a few single, general solutions. Many problems can be solved in different ways and with different equipment. What is important is to recognize different levels of system integration and match the right hardware and software solution to the actual problem.

A car injection control system will have to be rugged, small and cheap. Such a system will be mass produced in thousands of units and it may be cost effective to implement it with custom-built integrated circuits. On the opposite side we may have a big chemical plant with thousands of sensors to monitor and actuators to control. It is unlikely that a company running the plant will invest huge sums and tens of man-years of equivalent work to build a 100% optimal monitoring and control system. In such a plant a standard M&C package can be installed (see Section 12.2), where programming is done by defining the interfaces toward the process (data collection and control parameters) and toward the operators (screen outputs, command syntax). Special control cases will be solved with dedicated hardware and software.

Many companies offer components for data acquisition and control systems. The catalogues from these companies are thick books that make interesting reading. Equipment performing the same function can take very different forms. To measure a voltage and deliver the value in digital form, everything is available from chip-built A/D converters to fancy digital voltmeters with remote data collection and programming.

External constraints and economics will indicate the final solution. Different costs reflect advantages and disadvantages that have to be weighed against each other. Higher flexibility and the potential for future expansion cost more money, but this money could turn out to be less in the longer perspective when an expansion of the system actually takes place. The solutions for automation are not only technical but largely economic and strategic. With the distraction of so many conflicting factors, it is best to keep in mind that set goals have to be reached and it is necessary to look at things in a longer temporal perspective. Today's problems should be seen as part of tomorrow's problems (and problems, of course, should be seen as 'challenges').

12.2 THE STRUCTURE OF COMPUTER CONTROL SYSTEMS

Structuring is probably the single most important issue in process control problems. This influences both the hardware configurations and the software modularization of the system.

12.2.1 Lessons Learned from History

From the short history of three decades of computer process control some important lessons can be learnt. The two main driving forces behind computer control have traditionally been

- technology and
- market

Many of the pioneering projects were driven by the new available technology. As remarked in Section 1.2, many processes were too complex for the computer control tasks, and the process control problems were not adequately structured in terms of hardware and software. Take, for example, the investments for software development of the Apollo space programme. The small computer memory (64 kbytes) had to be utilized to the last bit and therefore the programming task had to be enormous; in the end it turned out to be almost 1000 man-years. Today even the cheapest computers have at least ten times as much memory (and every kid knows how to fill it in almost no time).

Two successful areas of process computer control in the early 1960s were chemical process control and power generation and transmission. In the control of chemical plants the computer simply replaced analog PID controllers. The control strategy was already well understood and the computer basically performed the same role as the previous controllers. In the power industry system structuring also had a long tradition. The engineers had a good feel for how to use computers in the power systems and could formulate adequate performance criteria and specifications, releasing the computer manufacturers from this kind of responsibility.

The typical process engineer will see the plant in terms of partial systems and unit processes. The computer engineer, who often is unfamiliar with the specific plant application, would rather think in terms of computer structures which are usually organized hierarchically. **It is crucial, that the process structure is reflected into a proper computer structure and that computers should not just force their way into the plant.**

In the chemical and power industries standards have long been established. The operator's needs were quite obvious: to transfer information from the control panel to the computer terminal. Plant operators wanted to see the same information as in the old PID controllers, the same curves as those from the old plotters and the same process schemes as in the control room. The computer systems were programmed with 'fill-in-the-blank' languages that resembled previous ways of describing the regulators.

For logical circuits and sequencing networks it was obvious to replace the old relays with computers; the background for this is given in Chapter 8. The old ladder diagrams could be translated to similar symbols on the computer screen, but now implemented with software. The first PLCs were simply replacements of relays. Now the increasing demand for structuring has led to function charts such as Grafcet (see Section 8.4). In fact the PLC market almost exploded at the end of the 1970s and since then has continued to grow rapidly.

It became clear quite early on that feedback and sequential control had to be integrated. This was actually done in the first systems, although not in a structured way. The industrial control systems of today are integrated in a more structured way, with building blocks for both feedback controllers and logical circuits (see also Section 9.6.8).

For some years a lot of attention has been paid to computer integrated manufacturing (CIM), yet few real CIM applications have so far seen the light of day (Section 12.5). Once again, an important reason for this is structuring. Unlike the chemical process industry, the manufacturing industry has no traditional standard to formulate the production structure. In addition, the manufacturing industry is extremely diversified, so that it is so much more difficult to formulate proper performance criteria for integrated control. Today, only limited solutions are available, as for instance for the control of NC machines, of robots and of production cells. The overall manufacturing plant control problem, however, is in general much more unstructured.

The lesson we have learnt so far is that it is relatively easy to get adequate hardware components or software modules for computer control tasks. The real problems have to do with the overall objectives, the difficulty in getting a unified view of the system and structuring the control hardware and software in a relevant — goal-oriented — way. This kind of overall view has, in fact, been our own driving force for writing this book.

12.2.2 The Hierarchical Structure of the Process

In most processes several hierarchical levels can be identified. The levels correspond more or less to different decisions to be taken to control a process. In general, all entities located at the same level have intensive mutual data exchange; the data exchange between layers is usually reduced and not time critical.

As an example, let us consider again the chemical tank described in Section 2.3. The lowest level in the heat regulation loop decides whether or not additional heat is necessary in order to keep the tank temperature constant. The decision is made by a regulator based on the actual and the reference temperatures. The temperature at which the process should operate is decided at a higher level. At a still higher level the chemical process to be run is selected, etc. Obviously, it does not make any sense to have a very stable and optimal temperature for Process A when Process B is run. The hierarchical model is decentralized: while the decisions influence each other, each level is independent of the choice of how to carry them out.

A similar hierarchical model is that of a company. As in the case of the chemical tank, the low level regulation is only part of the picture. Additional levels have to be considered in order to get the whole view. This structuring is by far not only an academic issue: the design of the exchange of data of different nature between real-time and administrative systems is a technical challenge.

A single hierarchical model for a manufacturing company or process industry is shown in Figure 12.4. The number of decision and operational levels there are, how much these levels depend on each other and how much autonomy is left to each single entity varies from one case to another.

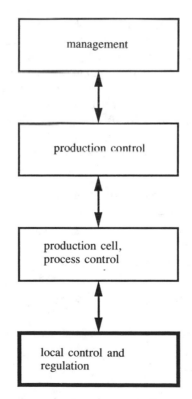

Figure 12.4 The hierarchical structure of a complex process (e.g. factory or plant).

Situated at the lowest level in the plant hierarchy are the machines and units that are directly in contact with the process. Production cells are one level above the first. In the case of a production cell where different machines are connected together to carry out a certain procedure, there is a high horizontal exchange of data (i.e. within the same level), to coordinate the operations of different machines, but less vertical exchange (see Section 8.5). Basically, only the materials and parts input and output from the cell are of interest for the higher levels. At the production control level, the activity of several cells is coordinated to reach a uniform material flow (the output of a cell is the input of another). Finally, at the management level the global decisions that affect the whole plant are taken.

Although necessarily very general, the hierarchical model offers a useful frame for analysing and structuring control systems. The model is not necessarily restricted to a manufacturing plant; equivalent levels are found in all complex control systems. For example, in an aircraft autopilot the regulation loops for the aerodynamic surfaces and for the engines are at the lowest hierarchical level; where to fly is a management decision left to the pilot.

12.2.3 Data Collection and Information Flow in Process Control

The requirements for data exchange within each hierarchical level vary considerably. We have already pointed out several times that real-time systems must process data at a speed higher than that at which changes can occur in the process under control. Something similar holds for the process at large. Here, naturally, the reaction times are different and become longer at higher levels where, on the other hand, there is more data to process. In Table 12.1 the typical data quantities, the response times and the frequencies are compared for new actions at different hierarchical levels of a plant. The quantities indicated in Table 12.1 have to be understood as orders of magnitude. The borders between the levels can be drawn differently and the description could be related to entities other than manufacturing plants.

Table 12.1   Typical requirements for process data

	Data quantity	Response time	Frequency
Management	Mbyte	hours/ minutes	day
Production control	kbyte	seconds	hours/ minutes
Process control	byte	100 ms	seconds
Local control and regulation	bit	ms	ms

12.2.3.1 Data collection

A major feature of control system architecture is the number of installed processors. There are control systems with only one central processor and systems with several CPUs. In distributed control, different processors are dedicated to the control of only one part of the physical process; the central unit coordinates the general operations (Figure 12.5).

Local processors or process interface modules are in direct contact with the physical process and collect the process data via sensors and A/D converters. The local processors also control the process via actuators. The function of the local processors may also be exerted by PLCs, local controllers or other types of units connected to the process, e.g. NC machines.

Data collected by the local processors may be conveyed to **front-end processors (FEPs)** which relay it to the central control unit. The FEPs may act as pure communication units and interface between different network types but also have local control functions.

The FEPs are connected to the local processors with direct lines, via LANs or fieldbuses. At this level, time responses may be critical and the choice of whether the network has to be deterministic or more random is important. There is usually no similar requirement for communication between the FEPs and the central control unit, so that comparatively simple and cheap networks such as Ethernet or Token Bus are often selected to act as 'backbone'.

There are basically three ways to transmit the process data from FEPs to the higher levels. With **polling**, a central unit periodically asks the sensors or other peripheral units about their state and waits for their answer. In this way it is guaranteed that the data is periodically updated, but the procedure is feasible only when the number of units to poll is not too high.

A second method for data collection is used in **telemetry**. All data is reported

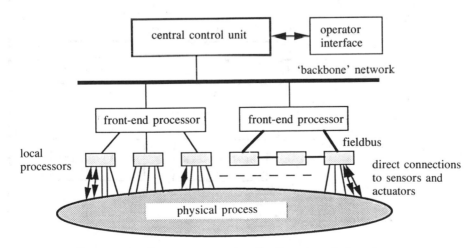

Figure 12.5 The structure of a distributed control system.

continuously in a predetermined form. After a cycle is completed, a new one starts. The address of a parameter is its position in the incoming data flow.

A third method is to report only the variables which change their value with respect to the preceding cycle. Digital variables are reported at each change, while for analog variables a transition band is defined. Only when the value of a variable has changed by some percentage in comparison to the last reported value, is the new information sent to the central unit.

The main advantage of the distributed data collection method is that the central unit and the communication channels are not overloaded with sampling and transmission of static data. On the other hand, if many monitored values change at the same time, the communication channels of the distributed systems might become overloaded.

12.2.3.2 Information flow

An important aspect in distributed data collection is dimensioning of the channel and of the data processing units. There must be sufficient capacity to process the required quantity of data and safety margins must also be provided. A simple rule of thumb is to oversize the minimal required capacity by a factor of 3 to 10 times.

Let us consider the example of Figure 12.6. There are local processors, FEPs and a backbone network connected to a central processor. Let us assume that 20 analog input channels are connected to every local processor and that their values are sampled 100

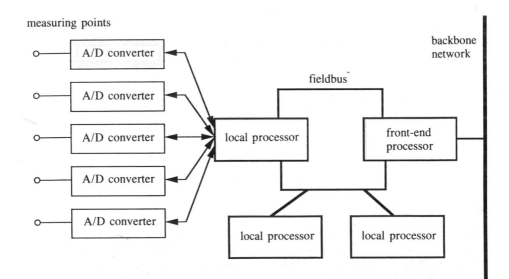

Figure 12.6   Information flow in hierarchical monitoring.

times per second via 12-bit A/D converters. The total quantity of collected data is 20*100*12 = 24 000 bit s^{-1}. The local processor must be able to process this amount of data and to organize it in a form that can be transmitted further along the fieldbus. Considering the safety margin and the processing overhead, the fieldbus should operate at a speed of at least 100 kbit s^{-1}. If a total of 15 units is connected to the FEP and produces an equivalent amount of data, the total quantity that has to be carried is 360 000 bit s^{-1}. A backbone network of capacity 5−10 Mbit s^{-1} should be considered.

The amount of data to carry to the higher levels may be reduced if local processors select the input data. The local processors could, for instance, relay only one out of ten values for each measuring point, which would obviously reduce the total volume of data traffic by a factor of ten.

More attention is needed if sensor data is reported only when it changes. There might be no need to communicate anything for a long time, and then a sudden change in the process could lead to so much data production that it effectively clogs the network. If the process data cannot be buffered so that some state information is lost, the central system may end up operating on wrong information.

The problem is solved in practice by combining a periodical and an event-controlled update method. Process values are reported whenever they change; in addition a general update takes place at long intervals, e.g. every few minutes. In this way it is ensured that the centrally stored data is *reasonably* consistent. In general, the choice of a data collection philosophy requires careful analysis of normal operations as well as of the special operating cases.

The trade off in distributed control systems is communication versus local intelligence. The present trend is to have local intelligence wherever possible, with the provision that the central control unit may always override local control decisions. Such a solution is both economic and safe. Breakdowns of the central control unit or of the communication links will not lead to a general system halt. Accurate control also requires as little delay as possible in the control loop, and this is less feasible if all messages from the local processors have to be sent to the central control unit for processing and then sent back. Finally, in distributed systems the processors can take care of more data than one single central unit, although powerful, can cope with.

12.3 THE FUNCTIONS OF A COMPUTER CONTROL SYSTEM

A computer system for process control can carry out many different functions but seldom all of them are present, normally only a few functions are implemented. The functions can be divided into three major groups (Figure 12.7):

- collection and interpretation of data from the physical process (monitoring)
- control of some parameter of the physical process
- connection of the process input and output data: feedback, automatic control

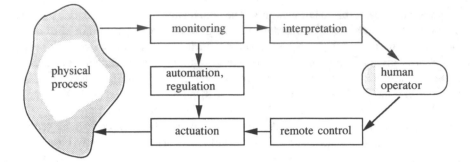

Figure 12.7   The main functions of a control system.

12.3.1 Monitoring

Monitoring, or data collection, is the basic function common to all control systems. Monitoring is the collection of the physical process variables, their storage as well as their presentation in a suitable form for the human operator. Monitoring is fundamental for all other data processing.

Monitoring can be limited to displaying raw or processed data on the terminal screen or printing them out on paper, but can also include more sophisticated analysis and presentation functions. For example, variables that cannot be read directly have to be calculated from the available measurements. Other basic monitoring operations include verifying that monitored or computed process values do not exceed preset limits as well as diagnostic operations.

When the plant operator bases the control decisions on the computer monitoring, the control is known as **supervisory control**. This control mode was quite common in the first applications of process computers. Supervisory control is still useful, particularly in relatively slow processes of great complexity and where human interaction is needed. Examples are biological processes, where part of the observations cannot be carried out automatically.

As soon as new data is collected, its value is examined with respect to alarm limits. In more advanced testing, several observations may be combined in more or less complex patterns in order to check whether the process is in a 'normal' state or whether it has exceeded some allowed limits. In the most advanced solutions, observations made by the operators are combined with the on-line information from the sensors (**expert** or **knowledge-based systems**).

12.3.2 Actuation

Actuation, or **command**, is the opposite function to monitoring. Commands from the computer reach the actuators in order to influence the physical process (see Sections 4.5 and 4.6). It may be that some process parameters cannot be influenced directly and that other parameters have to be controlled instead (compare with controllability, Section 3.5.1). The control structure is of major importance, i.e. which sensors to connect to which actuators.

12.3.3 Feedback Control

A system that acts autonomously and without the direct intervention of a human operator is **automatic**. Automatic control may be structured as simple controllers, one for each input/output pair, or more complex controllers with many inputs and outputs (see Chapter 9).

There are two different techniques to implement feedback control in computer systems. In the traditional **direct digital control (DDC)** the central computer calculates the control action and the signals for the actuators. The monitoring data has to be transmitted all the way from the sensors to the central control unit and the control signal back to the actuators.

In **distributed direct digital control (DDDC)** the computer system has a distributed architecture and the digital process regulators are implemented in the satellite processors. The central control unit defines the reference values and the satellite processors compute the appropriate control values for actuators based on the local monitoring data.

An early form of computer control was the so-called **setpoint control**. The computer only calculates the setpoint to be fed into a conventional analog controller. In such a case, the computer is used only for carrying out calculations and not for measurement or actuation.

12.3.4 A Process Database for Monitoring and Control

The software of the central control unit can be quite complex (Figure 12.8). The following modules may be identified:

- database for all the physical process I/O data
- database for all derived parameters
- data input modules and database interface
- data output modules (command transmission to the actuators)
- data presentation modules
- command input
- event recognition
- time reaction recognition
- automatic process control

12.3.4.1 The process database

A medium or large control system has hundreds or thousands of interface points to the physical process. It is practically impossible to manage the amount of related information with program modules written specifically for each point. What is needed is a systematic

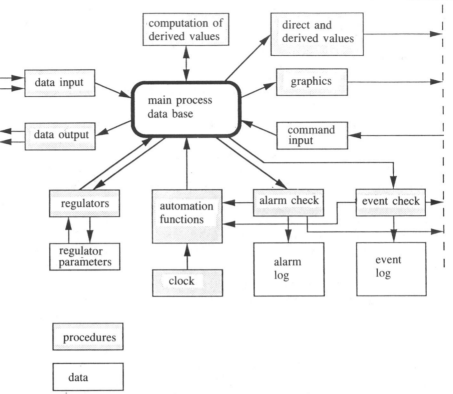

Figure 12.8   Software structure of a digital M&C system.

processing method for all the input data.

Help in this sense comes from considering the nature of the process-related information. Very often this comprises measured values or binary information of the kind **ON/OFF** or **OK/ALARM**. Due to the regularity of these representations, the input data can be processed via a general acquisition and interpretation program that acts on the basis of specific parameters for each point. The parameters that describe all single points are stored in the **process database**, which is the central component of the control system software.

Sophisticated databases may contain up to some tens of description parameters for every I/O point. Some of the description parameters are particularly important and are found in all database implementation. (Note that different terminologies are used in different systems although the concepts remain the same.)

- Code
- Name/description
- Type

- Address or physical reference: channel/message
- Event class
- Alarm class
- Sampling time
- Raw value
- Converted value
- Alarm state

For **analog** points the following additions are necessary (See Section 5.6.6):

- Scaling coefficients
- Measuring units
- Minimum and maximum limit values

and for **digital** points there are also:

- The states, in clear text
- Alarm states

An example of a record for an analog input follows:

```
CODE:              T439
DESCRIPTION:       IN TEMP PRIMARY CIRCUIT
TYPE:              AI
ADDRESS.           # 7.12.2
EVENT CLASS:       0
ALARM CLASS.       3
SAMPLE (SEC):      30
RAW VALUE:         3228
CONV. VALUE:       78.8
ALARM STATE;       1
A. COEFF.:         0
B. COEFF.:         0.0244
UNITS:             °C
MIN. VALUE:        50
MAX. VALUE:        75
```

The **code** is used to univocally identify the point in space; in the database, it acts as a key for the related record. The **name** (or **description**) is a mnemonic with a clear text reference to define the point and is used in printouts and screen presentations. The **type** shows whether the point is input or output and the kind of data: analog, digital, counter, etc. In this example, **AI** indicates that the point is an analog input.

The point must be associated with a certain input channel and a place in the incoming messages. The reference is given by the **address** (or **physical reference**). The address of the example could therefore mean unit 7, module 12, point 2.

The **event class** shows whether the point is connected to automated functions to be started when its value changes. The **alarm class** indicates whether some state is to be understood as an alarm. It may also show the relative importance of the alarm.

Some of the parameters are the same as the record for input data processing described in Section 5.6.6. This is the case for **sampling time, scaling factors** and the **limit** and **alarm values**. These parameters are needed for first processing of the signal at a level as close as possible to the sensors and the process.

In the case of the analog input point above, the scaling coefficients for the linear conversion from raw data to temperature are: $a = 0$ and $b = 100/4095 = 0.0244$. The conversion is assumed to be 12-bit A/D, with 0 corresponding to 0°C and 4095 to 100°C. The raw input value 3228 is converted in $(3228*0.0244) = 78.8$°C. This is higher than the maximum allowed value 75°C, so that an alarm state is reported.

For digital input (DI) some of the record parameters differ:

CODE:	K010
DESCRIPTION:	PRIMARY LOOP PUMP
TYPE:	DI
ADDRESS:	# 7.45.01
EVENT CLASS:	0
ALARM CLASS:	0
SAMPLE (SEC):	0
RAW VALUE:	1
CONV. VALUE:	ON
ALARM STATE:	0
STATE = 0:	OFF
STATE = 1:	ON
ALARM STATE:	-

For a point of digital type the states **0** and **1** are associated with descriptions in clear text, for example, as **0 = OFF** and **1 = ON**, or **0 = OK, 1 = ALARM**. Depending on how the system is built, the data from peripheral units can be more or less homogeneous. An analog measurement from one unit could reach the central system unit in raw format, and another already be converted and coded in ASCII. The database gives uniformity to the data it contains. The program modules that access the database do not need to refer to the physical details of the sensors.

Abstraction and separation of the measured values from the physical data is useful when some parameter has to be updated. It is not necessary to modify a program or halt the control system, but it is sufficient to redefine the conversion parameters stored in the database.

12.3.4.2 Derived variables

In system control, the ideal situation is when all state variables are directly measurable with appropriate sensors. This does not happen often in practice, and one has to find important values based on others. There are no sensors for 'energy' or 'efficiency', although their values have a physical meaning and are of practical importance.

The estimation of non-measurable state variables was explained in Section 3.5.2. The control system software has to allow the computation of new variables from those which can be collected directly. Whenever new data is available for some of the input variables, even the derived variable is computed anew. Derived variables may refer to other derived variables.

An important type of derived variables are **operational statistics** for the different hierarchical monitoring levels described earlier in this section. What is important practically is the knowledge of total quantities, such as production figures for a whole day, used materials or consumed energy. There is, in a sense, an integration of the collected variables, the higher one goes in the hierarchical structure. Other important global variables, e.g. the total energy consumed per product unit or per machine, can be derived easily from the integrated values. These values offer a better insight into the operations than the raw data.

12.3.5 Access to the Process Database, Queries and Protocols

The data contained in a database can be accessed by three fundamental operations which can be combined together: selection, projection and sorting.

Selection indicates that only data matching specific criteria is chosen, e.g. all points with code beginning with the 'A'. A comfortable feature of selection is when different criteria are combined together, e.g. for the selection of all digital inputs starting with 'A' and located in a determined area.

Projection is the choice of part of the stored information for printout. For instance, from the general database record described above, only code, description, converted value and units could be chosen (projected) for tabulation.

Sorting means ordering of the selected data following some principle; alphabetical ordering or ordering in increasing/decreasing values are the most common methods. Sorting is done with reference to one or more fields selected in the projection operation.

The three basic database operations allow a great number of possibilities when they are combined. There is too much data contained in a database to make sense and be understood, but once it is approached with the right tool, all kinds of information can be extracted. The database access operations provide this tool.

An operation on the database to extract information is called **query**. An example of a query follows:

```
FOR ALL POINTS WITH CODE = "A*", "B*"
                AND TYPE = AI
                PRINT CODE, DESCRIPTION, VALUE, UNITS
                ORDERED BY CODE
END.
```

The reader will recognize that 'WITH CODE = "A*", "B*" AND TYPE = AI' is the selection operation, 'PRINT CODE, DESCRIPTION, VALUE, UNITS' is the projection and 'ORDERED BY CODE' is the sorting. For newcomers to databases it should be remembered that sorting may take a long time, depending on the amount of records to sort, the quality of the software and the machine capacity. Do not start a large database query towards the end of the work day and wait for the result if you have an important date in the evening.

It helps to follow some logical principle in the code structure for the process points. The most important types of selection are geographic (all points at a specific location), for system/subsystem and for kind of sensor. In the definition of the code structure a precise meaning can be given to all character positions.

Although there are many query languages, the basic operations are more or less the same and the syntaxes do not differ too much. In order to effectively use a database access program it is necessary to select in advance a subset for the data of interest. It is useless to list the state of all the points of a system if it is known in advance where the information of interest is located.

Usually only a few combinations of data printouts from the database are of interest, so that a small number of standard queries can be identified. These queries are known as **protocols** (no relation to the protocols defined as a set of rules and procedures for data communication in Chapter 10). Protocols are basically queries where the projection and sorting operations (what information to print out and in what order) are predefined, and only the selection parameters need be given when the function is started (Figure 12.9). Note that the fields to printout and the sorting order are not explicitly indicated in the query.

One of the most important functions of a control system is to quickly recognize unallowed states and to alert the human operators. Every variation of the state of points classified as alarm should also be recorded with an indication of the time in a special file, the **alarm log**.

12.3.5.1 Maintenance protocols

In production and process operations, there are many maintenance actions which have to be carried out by personnel. Examples of such actions are changing machine tools, calibrating sensors and controlling the level of fuel and lubricants. Maintenance actions can even be quite complicated, up to the point of disassembling a device and building

```
STATE PROTOCOL == BEGIN AT 28-AUG-91 10:30:05
SELECTION = K*, T*

DI   K010   PRIMARY CIRCUIT PUMP = OPERATING
DI   K012   PRIMARY CIRCUIT PUMP = NORMAL
DI   K014   SECONDARY CIRCUIT PUMP = OPERATING
DI   K016   SECONDARY CIRCUIT PUMP = NORMAL
DI   K023   SAFETY SWITCH 1  = CLOSED
DI   K024   SAFETY SWITCH 2  = CLOSED
DI   K025   SAFETY SWITCH 3  = CLOSED
DI   K098   FIRE  SENSOR = NORMAL
DI   K099   PLANT VENTILATION  = NORMAL
AI   T439   PRIM. CIRC. IN TEMP. = 78.8 (75) C      ***
AI   T442   PRIM.Y CIRC. OUT TEMP.  = 59.4 (60) C
AI   T444   SECONDARY CIRC. IN TEMP.  = 38.8 (45) C
AI   T445   SECONDARY CIRC. OUT TEMP.  = 54.0 (60) C

STATE PROTOCOL END AT 28-AUG-91 10:30:12
```

Figure 12.9   Example of process state protocol.

```
MAINTENANCE PLAN FOR: 25-MAR-91 MONDAY  == BEGIN AT 22-MAR-91 09:05:12

K022 DRILL VERIFY TOOLS (200 HRS). ACCTD TIME = 228.4 HOURS
MA12 DIESEL GENERATOR VERIFY OIL LEVEL (50 HRS). ACCTD TIME = 47.2 HOURS
LA05 LATHE MAIN REVISION (1500 HRS). ACCTD TIME = 1502.0 HOURS
CO37 COMPRESSOR UNIT 01 CHANGE AIR FILTER (1 MTH). LATEST = 22-FEB-91
CO38 COMPRESSOR UNIT 02 CHANGE AIR FILTER (1 MTH). LATEST = 22-FEB-91
P101 WATER PUMP MAIN INLET, VERIFY (100 HRS). ACCTD TIME = 98.2 HOURS
P102 WATER PUMP MAIN INLET, VERIFY (100 HRS). ACCTD TIME = 102.7 HOURS

MAINTENANCE PLAN END AT 22-MAR-91  09:05:28
```

Figure 12.10   Example of maintenance plan.

it again after having performed the required checks. This type of maintenance is called **preventive maintenance**; it aims to keep the equipment at an optimum operational stand (actions to repair broken devices are known as **corrective maintenance**).

Depending on the type of device, preventive maintenance operations are carried out either on a fixed schedule (e.g. once every 30 days) or after a certain accumulated work time (a tool may have to be changed after 200 hours of operation). In large plants maintenance is required for hundreds or even thousands of devices. To keep tab of the right schedule manually is a complicated and not so satisfying a job.

It is straightforward to prepare schedules for planned maintenance work for a day or some other selected time period from data stored in the process database. Extrapolation of what maintenance action is required when the workload is known is also trivial. Obviously, corrective maintenance cannot be planned in the same way. An example of a preventive maintenance schedule is shown in Figure 12.10.

12.3.5.2 Data analysis and trends

An important task in the production and process industry is to keep tabs on the production rates and form statistics. The data contained in the database can be the first input to the statistical routines. The basic operation is to integrate the data, that is, to transform single inputs in accumulated values for a given period of time. The total accumulated values can then be printed out as statistical tables and other relations, such as efficiencies or quality values, extracted from them (Figure 12.11).

12.3.6 Control Operations via the Process Database

In some control systems, automated reactions can be carried out via the central database. A special table indicates when variations in the value of input parameters are associated with output commands. This table operates in a way similar to PLCs, although the data it refers to is at a much higher level and can include derived variables.

A particular kind of event is when a certain time has been reached or when a period of time has elapsed. The action table then has the form time → reaction. In a plant the lights can be automatically turned on and the machines powered shortly before the beginning of the first work shift. In automation systems designed for use in production plants, it

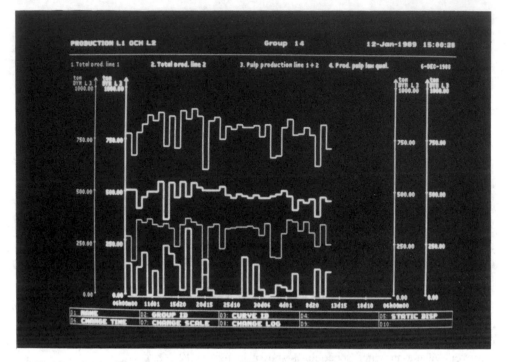

Figure 12.11   Data trend curves for different process variables on an operator
display. (Courtesy of ABB Automation, Sweden.)

must be possible to relate the schedule to workdays and shift periods. Holidays should
be recorded for separate treatment to avoid a plant warming up and starting all the automated
equipment at 6.30 a.m. on Monday, 25 December.

The most common types of digital regulators are implementations of PID regulators
where the integration and derivation operations are approximated with numerical methods.
In direct digital control (DDC) the regulators are built in the database. DDC packages
are implemented with a single program module acting on the basis of parameters for each
single control loop; the parameters are structured as records to be stored in the database.
Examples of such data structures and of the code for a digital PID regulator are given
in Chapter 9.

There is an important practical difference in automation function and process regulation
via the database or with PLC systems or local regulators. The latter are installed close
to the process inputs and outputs and react quickly to a change in the input data. The database
of a hierarchical control system has a slower reaction time because the information must
follow the communication chain up and down, and go through several processing steps
in the central processor. It is therefore worth programming automated reactions into the
central computer only when several parameters have to be compared with each other and
when the same operation cannot be performed in the local control units.

12.3.7 Advanced Languages for Process Control

For a long time, control systems were programmed almost exclusively in assembler. Two major reasons for this were the necessity to exploit all the resources of machines of limited capacity (at least in relation to the task) and the lack of adequate languages for real-time programming.

Current languages for control system programming can be divided into two main types: procedural and descriptive. **Procedural languages** — that includes ordinary programming language — require the definition of all the single instructions that the computer must execute. **Descriptive languages** require only the definition of data, relations and parameters to store in the different databases. The desired input/output relations are described in a table-like fashion and the program execution internals are left to the machine. An example of a descriptive language is the database described earlier in Section 12.3.4.1. The related programming is also commented on in Section 9.6.8.

Using and programming a control system via descriptive languages does not require detailed computer knowledge. Good knowledge of the process to be controlled is of greater importance, as well as an understanding of what the control system can and *cannot* do. The success of the system depends strongly on how well the model of the physical process and the I/O parameters have been defined. Descriptive programming also requires the correct tuning of the operating parameters in order to best match the control system's capabilities to the process under control. Giving different priorities to different tasks may also help in that respect. It does not make sense to try to run hundreds of regulating loops every second in a computer of limited capacity.

Descriptive languages are easier to use than procedural languages. As long as the user can add new definitions of functions they are also quite flexible. Examples of programming methods for process computers have been shown in Sections 8.4 and 9.6.8.

12.4 NON-TECHNICAL ASPECTS OF COMPLEX SYSTEMS

A control system is not only a computer program but includes processors, networks, terminals, sensors, actuators and many more components. In some cases the vendors of industrial automation systems require that everything is bought from them, from the software down to the nuts and bolts. Others take a more systematic, and customer-friendly, approach and deliver single modules on request. The current trend is to support standardization, that is, the interconnection and interchangeability of components.

The practicing automation engineer will dedicate a large portion of time to the construction of systems for solving special problems. As problems tend to become more complicated and specialized, the chance of finding a complete solution in a ready package is rather slim, despite affirmations to the contrary by most salespeople.

There are no 'optimal' solutions; the problems may, in fact, be approached more or less satisfactorily in many different ways. External factors — many of them indeed seem esoteric, from the climate to the availability of spare parts and to the education of maintenance personnel — are nevertheless relevant to the success of a complex solution. Everything should be considered in order to avoid malfunctions and costly surprises at a later stage.

Among the non-technical aspects that directly determine the success of an automation application is the overall organization, that becomes more and more important the larger the project is. Normally, a project is divided into planning, implementation, documentation, testing and operation phases. Each phase is critical to the success of the project at large.

The **planning** phase includes a general analysis of the problem to be solved. At the end of planning, a **specification document** is produced. Another type of document is the **feasibility study**, an analysis of whether and how a problem can be solved, what the possible solutions are and how much they cost. A feasibility study does not need to be as detailed as a specification document. The feasibility study is intended for company management while the specifications are for external suppliers.

The final contracts should specify exactly what the system has to do: 'monitoring' should not be left as a generic word, it should instead indicate whether alarm states are to be processed and presented in a special way, how the graphic presentation will look, etc.

The **operational specifications** are partly the responsibility of the buyer of a control system and partly that of the delivering company. To avoid conflicts, it is better to define in advance the respective competence areas. It is also advisable that a person or a group of people representing the customer follows the execution of the project. The deliverer can offer generic knowledge about control systems; while the customer has to check that all needs determined from experience are satisfied.

A solution must fulfill several requirements at once, many of which are quite contradictory. The most important aspects are:

- Goal fulfillment
- Reliability
- Safety (The system will not act in a dangerous way under unforeseen conditions)
- Maintainability (Almost all systems need some kind of periodic human intervention for cleaning, checking, calibration, etc.; these operations should be facilitated)
- Economy (More in the sense of getting value for money than just spending as little money as possible)

It is imperative that all parts of a system are correctly documented in relation to the hardware and software implementation as well as to the functions to be performed. **Documentation** is the interface between the different people working on the project when they cannot communicate directly. The documentation should not be considered as an uncomfortable duty to fulfill after a system has been developed and built, but rather as a key element of its development and its success.

Testing a system is a very important step. It must be verified that the operations satisfy the general planning indications and the specification documents. Testing is, of course, necessary to identify errors for immediate correction.

The steps of planning, project, implementation and testing are ideally separated, but there are actually strong interactions and feedback among them. During the project phase some new facts may appear which might lead to changes in planning. Practical, but previously unknown, implementation constraints may also influence the course of the project. With new experience gained in the field it may become necessary to change again parts already considered as completed. Unfortunately, changes in the specifications become harder and costlier to implement the later they are done, but on the other side a complex system is constantly modified anyway. There is no final version as such, only different grades of functionality.

The designer and the test engineer have two different goals. The designer has to design a **robust** and **efficient** system; the test engineer has to work out a test to find as many errors as possible. Both engineers have to follow the original requirements document, although changes dictated by practicality and common sense may be included.

12.5 TOWARDS THE AUTOMATED FACTORY?

Computer Integrated Manufacturing (CIM) has become one of the key concepts in factory automation. What is meant is the integration of all production-related information in a plant. But there is still a long time left before a factory can be switched on just by pressing one button!

CIM is not a simple business. Automation is not an 'all or nothing' fact, but a gradual development. And this development is not limited to the factory building. As an example it is instructive to compare what happened in the last decade in the United States, in Europe and in Japan. In Japan, at the same time as automation was introduced the new products were designed to be easier to build. In Europe, progressive changes were made on existing plants. In the United States the approach was to automate everything at once. The result of the 'competition' is now clear for everybody. At Toyota, 97 000 workers produce almost as many cars per year as more than 750 000 workers at General Motors, yet in Japan there is no widespread unemployment. The moral of the story is that 'system thinking' is not limited to plant technology, but involves a great number of other factors which may be harder to define and describe but are no less important for that. A constant verification of the proposed solutions against the set goals is therefore imperative.

In fact, there is no real 'automated factory'. A microprocessor could well be installed in the coffee machine, but that does not mean that the drinking of coffee is much different than if the water had been boiled on a stove. What is essential is to define exactly measurable goals and design the automation plant accordingly. The goals should be clearly identified, for example:

- Mean device production time: now — 2 h; automated — 45 m
- Total production cost per unit: reduction by 10%
- Quality: reduction of wrong parts from 20% to 5%
- Dead time (a machine is not utilized because another has to terminate its operation): reduction by 50%

If the goals cannot be reached with the automation design, or if they can be reached with solutions other than automation, then don't automate!

With that in mind, it should be clear that CIM is not a computer program that can be bought more or less off the shelf or even especially tailored for a particular company. CIM means planning and organizing the activity in a plant so that automation technology can be gradually introduced, first in a cell and then, step by step, to extend to several areas. CIM is no quick solution; it must be seen as *evolution* rather than *revolution*.

12.6 SUMMARY

Computers are used in process control for data collection, analysis and archiving, but apart from process regulation there are comparatively few applications in automatic control. There is still much to do in this respect.

In a digital control system it is comparatively easy to try new control strategies, because a digital computer may completely change the way it operates when its software is modified without the need to redesign and rewire all the hardware. Digital control systems are thus not just a new way to implement established principles for control, but rather an entirely new technique that offers greater flexibility and new possibilities.

The implementation of control systems takes place at many levels. The final choice of a particular solution may be dictated not only by technical aspects, but also by factors as different as personnel availability, operating environment, etc. For large applications typical of both manufacturing and process industry, generalized monitoring and control packages programmed on the basis of parameter tables can be used. The advantage of ready packages for M&C and automation lies in simplifying their programming and operation. Often such systems also support the writing and updating of documentation.

Automation is a key technology to support quality and efficiency in industrial and processing applications and computers are the basic components for advanced automation. The future of automation is not restricted to manufacturing plants: automation will play a key role in the efficient use of natural resources and in the protection of the environment from the negative sides of too-fast and indiscriminate an industrial development. Automation is one of the few realistic answers to the problem of how to enjoy a high quality of life in a world worth living in.

FURTHER READING

Groover (1987) is about the use of computers in manufacturing. It is very informative but also lengthy, it would have been better had it been more concise.

Brooks (1975) is a witty and stimulating pamphlet about the real-life aspects (read: delays) of software project planning. It is a recommended lecture for software development managers, especially when they do not have direct programming experience.

Databases are described in Ullman (1988). This book offers a high-level description of different database types (such as relational, hierarchical, etc.) as special cases of a general concept. Applications of expert systems for process fault diagnosis are discussed in Hayes-Roth, Waterman and Lenat (1983).

Standards

Throughout this book we have referred quite liberally to standards. The original publications can be ordered from the respective organizations, and some of the addresses are given below for convenience. In addition, national standards organizations also act as points of sale for literature from other countries.

Copies of the ISO/IEC standards are available from:

ISO Central Secretariat, 1 rue Varembé, CH-1211 Geneva 20, Switzerland

or from

IEC Central Office, 3 rue Varembé, CH-1211 Geneva 20, Switzerland.

The standards are also available from any of their national members.
ANSI/EIA publications can be obtained from

Sales Department, American National Standards Institute, 1430 Broadway, New York, NY 10018, USA

They can also be ordered from

Standards Sales Office, Electrical Industries Association, 2001 I Street, NW, Washington, DC 20006, USA

IEEE standards may be purchased from

Institute of Electrical and Electronics Engineers, Inc., 345 East 47th Street, New York, NY 10017, USA

ISA standards are obtained from

ISA (Instrument Society of America), 67 Alexander Drive, PO Box 12277, Research Triangle Park, NC 27709, USA.

Bibliography

Alloca, J. and A. Stuart 1984. *Transducers — Theory and Applications* Reston Publishing Company, Reston, VA.

Apple Company 1987. *Human Interface Guidelines: The Apple Desktop Interface* Addison-Wesley, Reading, MA and Apple Computer, Inc., Cupertino, CA.

Asada, Haruhiko and Jean-Jacques E. Slotine 1986. *Robot Analysis and Control* John Wiley & Sons, New York.

Åström, Karl J. and Tore Hägglund 1988. *Automatic Tuning of PID Controllers* Instrument Society of America, Research Triangle Park, NC 27709.

Åström, Karl J. and Björn Wittenmark 1989. *Adaptive Control* Addison-Wesley, Reading, MA.

Åström, Karl J. and Björn Wittenmark 1990. *Computer Controlled Systems, Theory and Design* (2nd edition) Prentice Hall, Englewood Cliffs, NJ.

Atkinson, Richard C. and R.M. Shiffrin 1971. The control of short-term memory *Scientific American* vol. 224, pp. 82—90.

Atkinson, Rita L., Richard C. Atkinson, Edward E. Smith and Ernest R. Hilgard 1990. *Introduction to Psychology* (10th edition) Harcourt Brace Jovanovich, San Diego, CA.

Bailey, S.J. 1989. MacIntosh II: a candidate for plant floor workstation uses standard NuBus cards *Control Engineering* January, pp. 77—9.

Barney, G.C. 1985. *Intelligent Instrumentation* Prentice Hall, Englewood Cliffs, NJ.

Bellman, Richard and R. Kalaba (eds) 1964. *Mathematical Trends in Control Theory* Dover, New York.

Ben-Ari, M. 1982. *Principles of Concurrent Programming* Prentice Hall, Englewood Cliffs, NJ.

Bendat, Julius and Allan Piersol 1971. *Random Data, Analysis and Measurement Procedures* Wiley-Interscience.

Bendat, Julius and Allan Piersol 1980. *Engineering Application of Correlation and Spectral Analysis* Wiley-Interscience.

Black, Uyless D. 1989. *Data Networks: Concepts, Theory and Practice* Prentice Hall, Englewood Cliffs, NJ.

Bode, H.W. 1964. 'Feedback: the history of an idea', a lecture given at the Conference on Circuits and Systems, New York, 1960 and reprinted in Bellman and Kalaba (1964) *Mathematical Trends in Control Theory* Dover, New York.

Borrill, Paul L. 1989. High-speed 32-bit buses for forward-looking computers *IEEE Spectrum* July, pp. 34—7.

Brett Glass, L. 1989. Inside EISA *BYTE* November, pp. 417—25.

Brinch Hansen, Per. 1973. *Operating Systems Principles* Prentice Hall, Englewood Cliffs, NJ.

Brooks, Frederick P. 1975. *The Mythical Man-Month* Addison-Wesley, Reading, MA.

Buckley, Page S. 1964. *Techniques of Process Control* John Wiley & Sons, New York.

BYTE 1987. Programmable Hardware (seven articles on PAL systems), *BYTE* January, pp. 194–286.

Cannon, R.H. Jr. 1967. *Dynamics of Physical Systems* McGraw-Hill, New York.

Ciarcia, Steve 1987. Under the covers *BYTE* August, pp. 101–10.

Coffmann, E.G., M.J. Elphick and A. Shoshani 1971. System deadlocks *Computing Surveys* vol. 3, no. 2, pp. 67–78.

Coffmann, Edward G. and Peter J. Denning 1973. *Operating Systems Theory* Prentice Hall, Englewood Cliffs, NJ.

Cornejo, Ciro and Raymond Lee 1987. Comparing IBM's Micro Channel and Apple's NuBus, *BYTE* Extra Edition 'Inside the IBM PC', pp. 83–92.

Craig, John J. 1989. *Introduction to Robotics: Mechanics and Control* (2nd edition) Addison-Wesley, Reading, MA.

DEC 1982. *Introduction to Local Area Networks* Digital Equipment Corporation, Bedford, MA.

Derenzo, Stephen E. 1990. *Interfacing: A Laboratory Approach Using the Microcomputer for Instrumentation, Data Analysis and Control* Prentice Hall, Englewood Cliffs, NJ.

Deppert, W. and K. Stoll 1988. *Cutting Costs with Pneumatics* Vogel Buchverlag, Würzburg (Germany).

de Silva, Clarence W. 1989. *Control Sensors and Actuators* Prentice Hall, Englewood Cliffs, NJ.

Desrochers, Alan A. 1990. *Modeling and Control of Automated Manufacturing Systems* IEEE Computer Society Press, Washington, DC.

Dijkstra, Edsger W. 1968. Co-operating sequential processes, in F. Genuys (ed.) *Programming Languages* Academic Press, London and New York.

DIN 1990. PROFIBUS Standard, DIN 19245 Teil 1 und 2, Beuth Verlag, Berlin.

Doebelin, Ernest, O. 1983. *Measurement Systems, Application and Design* (3rd edition) McGraw-Hill, New York.

Elgerd, Olle I. 1982. *Electric Energy Systems Theory: An Introduction* McGraw-Hill, New York.

Farowich, Steven A. 1986. Communicating in the technical office *IEEE Spectrum* April, pp. 63–7.

Finger, Roger 1987. The interconnect space of Multibus II simplifies hardware/software use *Computer Technology Review* Fall.

Fitzgerald, Arthur E., Charles Kingsley Jr and Stephen D. Umans 1983. *Electric Machinery* (4th edition) McGraw-Hill, New York.

Fletcher, D.I. 1980. *An Engineering Approach to Digital Design* Prentice Hall, Englewood Cliffs, NJ.

Franklin, Gene F. and J. David Powell 1980. *Digital Control of Dynamic Systems* Addison-Wesley, Reading, MA.

Franklin, Gene F., J. David Powell and Abbas Emami-Naemi 1986. *Feedback Control of Dynamic Systems* Addison-Wesley, Reading, MA.

Fuller, A.T. 1976. The early development of control theory (Parts 1 and 2) *Transactions of ASME Journal of Dynamical Systems Measurement and Control* vol. 98, pp. 109–18, 224–35.

Ginzberg, Eli 1982. The mechanization of work *Scientific American* vol. 247 no. 3, pp. 38–47.

Glasford, Glenn G. 1986. *Analog Electronic Circuits* Prentice Hall, Englewood Cliffs, NJ.

Glass, Robert L. (ed.) 1983. *Real-Time Software* Prentice Hall, Englewood Cliffs, NJ.

Göddertz, Joachim 1988. PROFIBUS — Kommunikationsmedium der mittleren und unteren Feldgeräteebene *etz elektrotechnische Zeitschrift* Bd. 109, Heft 7/8.

Göddertz, Joachim 1990. *PROFIBUS* VER 27.759 GB, Klockner-Möller GmbH, Postfach 1880, W-5300 Bonn 1, Germany (in English).

GRAFCET — A Function Chart for Sequential Processes ADEPA, 17 Rue Perier, B.P. No. 54, 92123 Montrouge Cedex, France, 1979.

Groover, Mikell P. 1987. *Automation, Production Systems and Computer Integrated Manufacturing* Prentice Hall, Englewood Cliffs, NJ.

Hassel, K. and G. Tuvstedt 1978. Microcomputer problems in real-time *Elteknik med aktuell elektronik* No. 14 (in Swedish).

Hayes-Roth, Frederick, Donald A. Waterman and Douglas B. Lenat (eds) 1983. *Building Expert Systems* Addison-Wesley, Reading, MA.

Held, Gilbert 1989. *Data Communication Networking Devices* John Wiley & Sons, New York.

Henze, Mogens, C.P. Lesley Grady, Willi Gujer and G. v. R. Marais 1987. *Activated Sludge Model No. 1* IAWPRC Scientific and Technical Report, International Association for Water Pollution Research and Control, London.

Hofstadter, Douglas R. 1979. *Goedel, Escher, Bach: An Eternal Golden Braid* Basic Books Publishers, New York.

Hufault, John R. 1986. *Op Amp Network Design* John Wiley & Sons, New York.

Hyde, John 1988. Message passing on Multibus II *Control Engineering Magazine* March.

Irvine, Robert G. 1987. *Operational Amplifier Characteristics and Applications* Prentice Hall, Englewood Cliffs, NJ.

Jones, Brian K. 1986. *Electronics for Experimentation and Research* Prentice Hall, Englewood Cliffs, NJ.

Juds, S.M. 1988. *Photoelectric Sensors and Controls* Dekker, New York.

Kaminski, Michael A. Jr. 1986. Protocols for communicating in the factory *IEEE Spectrum* April, pp. 56–62.

Kay, Alan 1977. Microelectronics and the Personal Computer *Scientific American*, vol. 237 no. 3, pp. 230–44.

Kay, Alan 1984. Computer software *Scientific American* vol. 251 no. 3, pp. 41–7.

Kenjo, Takashi 1984. *Stepping Motors and their Microprocessor Controls* Clarendon Press, Oxford.

Klir, George J. and Tina A. Folger 1988. *Fuzzy Sets, Uncertainty and Information* Prentice Hall, Englewood Cliffs, NJ.

Kosko, B. 1990. *Neural Networks and Fuzzy Systems* Prentice Hall, Englewood Cliffs, NJ.

Kreutzer, Wolfgang 1986. *System Simulation: Programming Styles and Languages* Addison-Wesley, Reading, MA.

Kuo, Benjamin C. 1991. *Automatic Control Systems* (6th edition) Prentice Hall, Englewood Cliffs, NJ.

Laurel, Brenda (ed.), 1990. *The Art of Human–Computer Interface Design*, Addison-Wesley, Reading, MA.

Lawrence, Peter D. and Konrad Mauch 1987. *Real-Time Microcomputer System Design* McGraw-Hill, New York.

Lee, S.C. 1978. *Modern Switching Theory and Digital Design* Prentice Hall, Englewood Cliffs, NJ.

Leonhard, W. 1985. *Control of Electrical Drives* Springer Verlag, Berlin.

Levy, Henry M. and Richard H. Eckhouse 1980. *Computer Programming and Architecture: The VAX-11* Digital Press, Bedford, MA.

Ljung, Lennart 1987. *System Identification: Theory for the User* Prentice Hall, Englewood Cliffs, NJ.

Ljung, Lennart and Torsten Söderström 1983. *Theory and Practice of Recursive Identification* MIT Press, Cambridge, MA.

Luenberger, David G. 1979. *Introduction to Dynamic Systems: Theory, Models and Applications* John Wiley & Sons, New York.

Luyben, W.L. 1973. *Process Modelling, Simulation and Control for Chemical Engineers* McGraw-Hill, New York.

Manual 02.1987: Book 3 — Grafcet Language, Telemecanique Inc., 901 Baltimore Boulevard, Westminster, MD 21157, 1987.

Martin, James 1973. *Design of Man–Computer Dialogues* Prentice Hall, Englewood Cliffs, NJ.

Mohan, Ned, Tore M. Undeland and William P. Robbins 1989. *Power Electronics: Converters, Applications and Design* John Wiley & Sons, New York.

Morrison, R. 1967. *Grounding and Shielding Techniques in Instrumentation* John Wiley & Sons, New York.

Newsweek 1990. Can we trust our software?, January 29.

Norman, Donald A. 1988. *The Psychology of Everyday Things* Basic Books Publishers, New York.

Norman, Donald A. and Stephen W. Draper (eds.) 1986. *User Centered System Design: New Perspectives on Human–Computer Interaction*, Lawrence Erlbaum Associates, London and Hillsdale, NJ.

Norton, Harry N. 1989. *Handbook of Transducers* Prentice Hall, Englewood Cliffs, NJ..

Olsson, Gustaff 1985. Control strategies for the activated sludge process, in *Comprehensive Biotechnology, The Principles of Biotechnology* Charles L. Cooney and Arthur E. Humphrey (eds), Pergamon Press, New York, Chapter 65.

Ott, Henry 1976. *Noise Reduction Techniques in Electronic Systems* John Wiley & Sons, New York.

Pessen, David S. 1989. *Industrial Automation: Circuit Design and Components* John Wiley & Sons, New York.

Peterson, Wade D. 1989. *The VMEbus Handbook* VMEbus International Trade Association (VITA), Scottsdale, AZ.

Pirsig, Robert M. 1974. *Zen and the Art of Motorcycle Maintenance: an Inquiry into Values* Morrow, New York and Bantam, New York.

Sanders, Mark S. and Ernest J. McCormick 1987. *Human Factors in Engineering and Design* McGraw-Hill, New York.

Sargent, Murray III and Richard L. Shoemaker 1984. *The IBM Personal Computer from the Inside Out* Addison-Wesley, Reading, MA.

Seborg, Dale E., Thomas F. Edgar and Duncan A. Mellichamp 1989. *Process Dynamics and Control* John Wiley & Sons, New York.

Shannon, Claude E. and Warren Weaver 1963. *The Mathematical Theory of Communication* University of Illinois Press, Urbana, IL.

Sheingold, Daniel H. (ed.) 1986. *Analog–Digital Conversion Handbook* Prentice Hall, Englewood Cliffs, NJ.

Shiell, Jon 1987. The 32-bit Micro Channel *BYTE* Extra Edition 'Inside the IBM PC', pp. 59–64.

Shinskey, F. Greg 1967. *Process Control Systems* McGraw-Hill Inc., New York.

Shneiderman, Ben 1987. *Designing The User Interface* Addison-Wesley, Reading, MA.

Smith, Otto J.M. 1957. Close control of loops with deadtime *Chem. Eng. Progr.* vol. 53, pp. 217–19.

Söderström, Torsten and Petre Stoica 1989. *System Identification* Prentice Hall, Englewood Cliffs, NJ.

Spector, Alfred Z. 1984. Computer software for process control *Scientific American* vol. 251, pp. 127–38.

Spong, Mark W. and M. Vidyasagar 1989. *Robot Dynamics and Control* John Wiley & Sons, New York.

Stearns, Samuel D. and Ruth A. David 1988. *Signal Processing Algorithms* Prentice Hall, Englewood Cliffs, NJ.

Stephanopoulos, George 1984. *Chemical Process Control: An Introduction to Theory and Practice* Prentice Hall, Englewood Cliffs, NJ.

Stroustrup, Bjarne 1991. *The C++ Programming Language* (2nd edition) Addison-Wesley, Reading, MA.

Tanenbaum, Andrew S. 1987. *Operating Systems Design and Implementation* Prentice Hall, Englewood Cliffs, NJ.

Tanenbaum, Andrew S. 1989. *Computer Networks* Prentice Hall, Englewood Cliffs, NJ.

Tanenbaum, Andrew S. 1990. *Structured Computer Organization* (3rd edition) Prentice Hall, Englewood Cliffs, NJ.

Ullman, Jeffrey D. 1988. *Principles of Database and Knowledge-base Systems* Computer Science Press, Rockville, MD.

Voelcker, John 1986. Helping computers communicate *IEEE Spectrum* March, pp. 61–70.

Vögtlin, B. and Tschabold, P. 1990. Direct Measurement of Mass Flow Using the Coriolis Force, Flowtec AG, Kägenstr. 7, CH — 4153 Reinach BL 1, Switzerland.

Warnock, Ian G. 1988. *Programmable Controllers Operation and Application* Prentice Hall, Englewood Cliffs, NJ.

Warrior, Jay and Jim Cobb 1988. Structure and flexibility for fieldbus messaging, *Control Engineering* October, pp. 18–20.

White, George 1989. A bus tour *BYTE* September, pp. 296–302.

Wilson, J. and J.F.B. Hawkes 1983. *Optoelectronics: An Introduction* Prentice Hall.

Wood, G.G. 1988. International standards emerging for fieldbus *Control Engineering* October, pp. 22–5.

Yager, R.R. (ed.) 1987. *Fuzzy Sets and Applications: Selected Papers by L.A. Zadeh* John Wiley & Sons, New York.

Young, Stephen J. 1982. *Real-Time Languages: Design and Development* Ellis Horwood Ltd, Chichester and John Wiley & Sons, New York.

Index